Centre de Documentation

INRIA   Rocquencourt

**WILEY-INTERSCIENCE
SERIES IN DISCRETE MATHEMATICS AND OPTIMIZATION**

ADVISORY EDITORS

RONALD L. GRAHAM
AT & T Bell Laboratories, Murray Hill, New Jersey, U.S.A.

JAN KAREL LENSTRA
*Department of Mathematics and Computer Science,
Eindhoven University of Technology, Eindhoven, The Netherlands*

ROBERT E. TARJAN
*Princeton University, New Jersey, and
NEC Research Institute, Princeton, New Jersey, U.S.A.*

A complete list of titles in this series appears at the end of this volume.

Interior Point Algorithms
Theory and Analysis

YINYU YE

A Wiley-Interscience Publication
JOHN WILEY & SONS, INC.
New York • Chichester • Weinheim • Brisbane • Singapore • Toronto

This text is printed on acid-free paper.

Copyright © 1997 by John Wiley & Sons, Inc.

All rights reserved. Published simultaneously in Canada.

Reproduction or translation of any part of this work beyond that permitted by Section 107 or 108 of the 1976 United States Copyright Act without the permission of the copyright owner is unlawful. Requests for permission or further information should be addressed to the Permissions Department, John Wiley & Sons, Inc., 605 Third Avenue, New York, NY 10158-0012

Library of Congress Cataloging in Publication Data:

Ye, Yinyu.
    Interior point algorithms : theory and analysis / Yinyu Ye.
       p.   cm. — (Wiley-Interscience series in discrete mathematics and optimization)
    "A Wiley-Interscience publication."
    Includes index.
    ISBN 0-471-17420-3 (alk. paper)
    1. Programming (Mathematics)   2. Linear programming.   I. Title. II. Series.
QA402.5.Y4   1997
519.7'2—dc21   97-18492
                                                                                    CIP

Printed in the United States of America
10 9 8 7 6 5 4 3 2 1

To my family and parents

Contents

Preface		xiii
List of Figures		xv
1	**Introduction and Preliminaries**	**1**
	1.1 Introduction	1
	1.2 Mathematical Preliminaries	5
	1.2.1 Basic notations	5
	1.2.2 Convex sets	8
	1.2.3 Real functions	12
	1.2.4 Inequalities	14
	1.3 Decision and Optimization Problems	15
	1.3.1 System of linear equations	15
	1.3.2 System of nonlinear equations	16
	1.3.3 Linear least-squares problem	16
	1.3.4 System of linear inequalities	17
	1.3.5 Linear programming (LP)	18
	1.3.6 Quadratic programming (QP)	22
	1.3.7 Linear complementarity problem (LCP)	23
	1.3.8 Positive semi-definite programming (PSP)	24
	1.3.9 Nonlinear programming (NP)	27
	1.3.10 Nonlinear complementarity problem (NCP)	27
	1.4 Algorithms and Computation Models	28
	1.4.1 Worst-case complexity	29
	1.4.2 Condition-based complexity	31
	1.4.3 Average complexity	32
	1.4.4 Asymptotic complexity	33
	1.5 Basic Computational Procedures	34
	1.5.1 Gaussian elimination method	35
	1.5.2 Choleski decomposition method	35

	1.5.3	The Newton method	36
	1.5.4	Solving ball-constrained linear problem	37
	1.5.5	Solving ball-constrained quadratic problem	37
1.6	Notes		38
1.7	Exercises		39

2 Geometry of Convex Inequalities 43
2.1	Convex Bodies		44
	2.1.1	Center of gravity	44
	2.1.2	Ellipsoids	45
2.2	Analytic Center		48
	2.2.1	Analytic center	48
	2.2.2	Dual potential function	50
	2.2.3	Analytic central-section inequalities	53
2.3	Primal and Primal-Dual Potential Functions		59
	2.3.1	Primal potential function	59
	2.3.2	Primal-dual potential function	62
2.4	Potential Functions for LP, LCP, and PSP		63
	2.4.1	Primal potential function for LP	63
	2.4.2	Dual potential function for LP	66
	2.4.3	Primal-dual potential function for LP	66
	2.4.4	Potential function for LCP	67
	2.4.5	Potential function for PSP	68
2.5	Central Paths of LP, LCP, and PSP		70
	2.5.1	Central path for LP	70
	2.5.2	Central path for LCP	74
	2.5.3	Central path for PSP	74
2.6	Notes		75
2.7	Exercises		77

3 Computation of Analytic Center 81
3.1	Proximity to Analytic Center		81
3.2	Dual Algorithms		87
	3.2.1	Dual Newton procedure	87
	3.2.2	Dual potential algorithm	89
	3.2.3	Central-section algorithm	90
3.3	Primal Algorithms		94
	3.3.1	Primal Newton procedure	94
	3.3.2	Primal potential algorithm	95
	3.3.3	Affine scaling algorithm	100
3.4	Primal-Dual (Symmetric) Algorithms		101
	3.4.1	Primal-dual Newton procedure	102

		3.4.2 Primal-dual potential algorithm	103
	3.5	Notes	106
	3.6	Exercises	108

4 Linear Programming Algorithms — 109
- 4.1 Karmarkar's Algorithm — 109
- 4.2 Path-Following Algorithm — 117
- 4.3 Potential Reduction Algorithm — 120
- 4.4 Primal-Dual (Symmetric) Algorithm — 126
- 4.5 Adaptive Path-Following Algorithms — 128
  - 4.5.1 Predictor-corrector algorithm — 131
  - 4.5.2 Wide-neighborhood algorithm — 134
- 4.6 Affine Scaling Algorithm — 136
- 4.7 Extensions to QP and LCP — 141
- 4.8 Notes — 142
- 4.9 Exercises — 145

5 Worst-Case Analysis — 147
- 5.1 Arithmetic Operation — 148
- 5.2 Termination — 152
  - 5.2.1 Strict complementarity partition — 153
  - 5.2.2 Project an interior point onto the optimal face — 155
- 5.3 Initialization — 157
  - 5.3.1 A HSD linear program — 159
  - 5.3.2 Solving (HSD) — 164
  - 5.3.3 Further analysis — 167
- 5.4 Infeasible-Starting Algorithm — 169
- 5.5 Notes — 173
- 5.6 Exercises — 176

6 Average-Case Analysis — 179
- 6.1 One-Step Analysis — 181
  - 6.1.1 High-probability behavior — 182
  - 6.1.2 Proof of the theorem — 183
- 6.2 Random-Problem Analysis I — 187
  - 6.2.1 High-probability behavior — 189
  - 6.2.2 Random linear problems — 193
- 6.3 Random-Problem Analysis II — 196
  - 6.3.1 Termination scheme — 197
  - 6.3.2 Random model and analysis — 201
- 6.4 Notes — 206
- 6.5 Exercises — 208

7 Asymptotic Analysis — 209
7.1 Rate of Convergence — 209
7.1.1 Order of convergence — 210
7.1.2 Linear convergence — 211
7.1.3 Average order — 211
7.1.4 Error function — 212
7.2 Superlinear Convergence: LP — 213
7.2.1 Technical results — 213
7.2.2 Quadratic convergence — 216
7.3 Superlinear Convergence: Monotone LCP — 218
7.3.1 Predictor-corrector algorithm for LCP — 218
7.3.2 Technical results — 220
7.3.3 Quadratic convergence — 222
7.4 Quadratically Convergent Algorithms — 224
7.4.1 Variant 1 — 224
7.4.2 Variant 2 — 225
7.5 Notes — 227
7.6 Exercises — 229

8 Convex Optimization — 231
8.1 Analytic Centers of Nested Polytopes — 231
8.1.1 Recursive potential reduction algorithm — 232
8.1.2 Complexity analysis — 235
8.2 Convex (Non-Smooth) Feasibility — 236
8.2.1 Max-potential reduction — 238
8.2.2 Compute a new approximate center — 239
8.2.3 Convergence and complexity — 242
8.3 Positive Semi-Definite Programming — 246
8.3.1 Potential reduction algorithm — 248
8.3.2 Primal-dual algorithm — 254
8.4 Monotone Complementarity Problem — 256
8.4.1 A convex property — 257
8.4.2 A homogeneous MCP model — 260
8.4.3 The central path — 265
8.4.4 An interior-point algorithm — 268
8.5 Notes — 273
8.6 Exercises — 275

9 Nonconvex Optimization — 277
9.1 von Neumann Economic Growth Problem — 277
9.1.1 Max-potential of $\Gamma(\gamma)$ — 279
9.1.2 Approximate analytic centers of $\Gamma(\gamma)$ — 284

	9.1.3 Central-section algorithm	286
9.2	Linear Complementarity Problem	291
	9.2.1 Potential reduction algorithm	292
	9.2.2 A class of LCPs	295
	9.2.3 P-matrix LCP	298
9.3	Generalized Linear Complementarity Problem	303
	9.3.1 Potential reduction algorithm	304
	9.3.2 Complexity analysis	306
	9.3.3 Further discussions	309
9.4	Indefinite Quadratic Programming	310
	9.4.1 Potential reduction algorithm	312
	9.4.2 Generating an ϵ-KKT point	316
	9.4.3 Solving the ball-constrained QP problem	318
9.5	Approximating Quadratic Programming	325
	9.5.1 Positive semi-definite relaxation	325
	9.5.2 Approximation analysis	327
9.6	Notes	332
9.7	Exercises	335

10 Implementation Issues — **337**

10.1	Presolver	337
10.2	Linear System Solver	340
	10.2.1 Solving normal equation	340
	10.2.2 Solving augmented system	343
	10.2.3 Numerical phase	345
	10.2.4 Iterative method	349
10.3	High-Order Method	350
	10.3.1 High-order predictor-corrector method	350
	10.3.2 Analysis of a high-order method	352
10.4	Homogeneous and Self-Dual Method	355
10.5	Optimal-Basis Identifier	356
	10.5.1 A pivoting algorithm	356
	10.5.2 Theoretical and computational issues	358
10.6	Notes	359
10.7	Exercises	363

Bibliography — **365**

Index — **409**

Preface

On a sunny afternoon in 1984, one of my officemates told me that there would be a seminar given by N. Karmarkar, an AT&T scientist, on linear programming. At the time, my knowledge of linear programming was limited to one optimization course and one research project with Prof. David Luenberger in the Department of Engineering-Economic Systems, Stanford University. As a second-year Ph.D. student, I was familiar with economic terms like cost, supply, demand, and price, rather than mathematical terms like vertex, polyhedron, inequality, and duality.

That afternoon the Terman auditorium was packed. I could not find a seat and had to sit on the floor for about one and a half hours, where I saw Prof. George Dantzig and many other faculty members. I was not particular enthusiastic about the statement from the speaker that a new interior-point method would be 40 times faster than the simplex method, but I was amazed by the richness and applicability of linear programming as a whole. That was how and when I determined to devote my Ph.D. study to mathematical programming.

I immediately took a series of courses from Prof. Dantzig. In those courses I was fortunate to learn from many distinguished researchers. I also went to Cornell to work under the guidance of Prof. Michael Todd. Since then, my interest in linear programming has become stronger and my knowledge of interior-point algorithms has grown broader.

I decided to write a monograph about my understanding of interior-point algorithms and their complexities. I chose to highlight the underlying interior-point geometry, combinatorica, and potential theory for convex inequalities. I did not intend to cover the entire progress of linear programming and interior-point algorithms during the last decade in this write-up. For a complete survey, I refer the reader to several excellent articles or books by Goldfarb and Todd [151], Gonzaga [163], den Hertog [182], Nesterov and Nemirovskii [327], Roos, Terlaky and Vial [366], Terlaky [404], Todd [406, 412], Vanderbei [443], and Wright [457].

Information, supporting materials, and computer programs related to

this book may be found at the following address on the World-Wide Web:
http://dollar.biz.uiowa.edu/col/ye/book.html
Please report any question, comment and error to the address:
yinyu-ye@uiowa.edu

I am specially indebted to Steve Benson and Andrew Prince, who made a careful and thorough reading of the manuscript and provided me numerous comments and suggestions on organizing the materials presented in this book. Special thanks also go to Erling Andersen, for giving me detailed feedback from a course he taught in Odense University using an early version of the manuscript, and to Nadine Castellano, for her proofreading of the manuscript. I wish to thank my colleagues Kurt Anstreicher, Dingzhu Du, Ken Kortanek, Tom Luo, Panos Pardalos, Florian Potra, Mike Todd, Steve Vavasis, and Guoliang Xue, for their support and advice on the process of writing the manuscript. I am grateful to the University of Iowa, for the environment and resources which made this effort possible. For their contribution to this effort, I acknowledge the students and other participants in the course 6K287 at the Department of Management Science. I also wish to thank the Institute of Applied Mathematics of Chinese Academy and Huazhong University of Science and Technology, for their support and assistance during my many visits there. It has been a pleasure to work with Jessica Downey and her colleagues of John Wiley & Sons involved in the production of this book.

My research efforts have been supported by the following grants and awards at various institutes: National Science Foundation Grants DDM-8922636, DDM-9207347 and DMI-9522507; the summer grants of the Iowa College of Business of Administration, the Obermann Fellowship, and the Center for Advanced Studies Interdisciplinary Research Grant at the University of Iowa; the Cornell Center for Applied Mathematics and the Advanced Computing Research Institute, a unit of the Cornell Theory Center, which receives major funding from the NSF and IBM Corporation, with additional support from New York State and members of its Corporate Research Institute; NSF Coop. Agr. No. CCR-8809615 at the Center for Research in Parallel Computation of Rice University; NSF Grant OSR-9350540 at the University of Vermont; the award of K. C. WONG Education Foundation at Hong Kong; and Dutch Organization for Scientific Research (NWO).

Finally, I thank my wife Daisun Zhou, and daughter Fei, for their constant love and support, and my parents Xienshu Ye and Chunxuan Wang, for their inspiration.

YINYU YE

Iowa City, 1996

List of Figures

1.1 Coordinate-aligned and non-coordinate-aligned ellipsoids.   8
1.2 Polyhedral and nonpolyhedral cones.   9
1.3 Representations of a polyhedral cone.   10
1.4 A hyperplane and half-spaces.   11
1.5 Illustration of the separating hyperplane theorem; an exterior point y is separated by a hyperplane from a convex set C.   12

2.1 A hyperplane H cuts through the center of gravity of a convex body.   45
2.2 The max-volume ellipsoid inscribing a polytope and the concentric ellipsoid circumscribing the polytope $(R/r \leq m)$.   46
2.3 Illustration of the min-volume ellipsoid containing a half ellipsoid.   47
2.4 Illustration of the Karmarkar (simplex) polytope and its analytic center.   51
2.5 Coordinate-aligned (dual) ellipsoids centered at points s's on the intersection of an affine set A and the positive orthant; they are also contained by the positive orthant.   52
2.6 Translation of a hyperplane.   54
2.7 Addition of a new hyperplane.   56
2.8 Illustration of $r(\Omega)_1$: the distance from y^a to the tangent plane a_{n+1}.   59
2.9 Coordinate-aligned (primal) ellipsoids centered at points $\bar{s}$'s on an affine set A; they also contain the intersection of A and the positive orthant.   61
2.10 Intersections of a dual feasible region and the objective hyperplane; $b^T y \geq z$ on the left and $b^T y \geq b^T y^a$ on the right.   64
2.11 The central path of $y(z)$ in a dual feasible region.   71

LIST OF FIGURES

3.1 Illustration of the dual potential algorithm; it generates a sequence of contained coordinate-aligned ellipsoids whose volumes increase.   91

3.2 Illustration of the primal potential algorithm; it generates a sequence of containing coordinate-aligned ellipsoids whose volumes decrease.   100

3.3 Illustration of the primal-dual potential algorithm; it generates a sequence of containing and contained coordinate-aligned ellipsoids whose logarithmic volume-ratio reduces to $n \log n$.   106

4.1 Illustration of the predictor-corrector algorithm; the predictor step moves y^0 in a narrower neighborhood of the central path to y' on the boundary of a wider neighborhood and the corrector step then moves y' to y^1 back in the narrower neighborhood.   132

5.1 Illustration of the projection of y^k onto the (dual) optimal face.   156

6.1 Illustration of the projection of r onto a random subspace.   184

8.1 Illustration of the central-section algorithm for finding an interior-point in a convex set Γ; the sizes of a sequence of containing polytopes decreases as cuts added at the analytic centers.   237

9.1 Illustration of the level set $\Gamma(\gamma)$ on the simplex polytope; the size of $\Gamma(\gamma)$ decreases as γ increases.   284

9.2 Illustration of the product $\sigma(\frac{v_i^T u}{\|v_i\|}) \cdot \sigma(\frac{v_j^T u}{\|v_j\|})$ on the 2-dimensional unit circle. As the unit vector u is uniformly generated along the circle, the product is either 1 or -1.   329

10.1 Illustration of dense sub-factors in a Choleski factorization.   346

Chapter 1

Introduction and Preliminaries

1.1 Introduction

Complexity theory is the foundation of computer algorithms. The goal of the theory is to develop criteria for measuring the effectiveness of various algorithms and the difficulty of various problems. The term "complexity" refers to the amount of resources required by a computation. In this book, running time or number of arithmetic operations is the major resource of interest.

Linear programming, hereafter LP, plays a very important role in complexity analysis. In one sense it is a continuous optimization problem in minimizing a linear objective function over a convex polyhedron; but it is also a combinatorial problem involving selecting an extreme point among a finite set of possible vertices. Businesses, large and small, use linear programming models to optimize communication systems, to schedule transportation networks, to control inventories, to plan investments, and to maximize productivity.

Linear inequalities define a polyhedron, properties of which have been studied by mathematicians for centuries. Ancient Chinese and Greeks studied calculating volumes of simple polyhedra in three-dimensional space. Fourier's fundamental research connecting optimization and inequalities dates back to the early 1800s. At the end of 19th century, Farkas and Minkowski began basic work on algebraic aspects of linear inequalities. In 1910 De La Vallée Poussin developed an algebraic technique for minimizing the infinity-norm of $b - Ax$ that can be viewed as a precursor of the simplex method. Beginning in the 1930s, such notable mathematicians as von Neumann, Kantorovich, and Koopmans studied mathematical economics based on linear inequalities. During World War II, it was observed that decisions involving the best movement of personnel and optimal allocation of resources could be posed and solved as linear programs. Linear programming began to assume its current popularity.

An optimal solution of a linear program always lies at a vertex of the feasible region, which itself is a polyhedron. Unfortunately, the number of vertices associated with a set of n inequalities in m variables can be exponential in the dimensions—in this case, up to $n!/m!(n-m)!$. Except for small values of m and n, this number is so large as to prevent examining all possible vertices for searching an optimal vertex.

The simplex method, invented in the mid-1940s by George Dantzig, is a procedure for examining optimal candidate vertices in an intelligent fashion. It constructs a sequence of adjacent vertices with improving values of the objective function. Thus, the method travels along edges of the polyhedron until it hits an optimal vertex. Improved in various way in the intervening four decades, the simplex method continues to be the workhorse algorithm for solving linear programming problems. On average, the number of vertices or iterations visited by the simplex method seems to be roughly linear in m and perhaps logarithmic n.

Although it performs well on average, the simplex method will indeed examine every vertex when applied to certain linear programs. Klee and Minty in 1972 gave such an example. These examples confirm that, in the worst case, the simplex method needs an exponential number of iterations to find the optimal solution. As interest in complexity theory grew, many researchers believed that a good algorithm should be polynomial—i.e., broadly speaking, the running time required to compute the solution should be bounded above by a polynomial in the "size," or the total data length, of the problem. Thus, the simplex method is not a polynomial algorithm.

In 1979, a new approach to linear programming, Khachiyan's ellipsoid method, received dramatic and widespread coverage in the international press. Khachiyan proved that the ellipsoid method, developed during the 1970s by other mathematicians, is a polynomial algorithm for linear programming under a certain computational model. It constructs a sequence of shrinking ellipsoids with two properties: the current ellipsoid always contains the optimal solution set, and each member of the sequence undergoes a guaranteed reduction in volume, so that the solution set is squeezed more tightly at each iteration.

The ellipsoid method was studied intensively by practitioners as well as theoreticians. Based on the expectation that a polynomial linear program-

1.1. INTRODUCTION

ming algorithm would be faster than the simplex method, it was a great disappointment that the best implementations of the ellipsoid method were not even close to being competitive. In contrast to the simplex method, the number of steps required for the ellipsoid method to terminate was almost always close to the worst case bound—whose value, although defined by a polynomial, is typically very large. Thus, after the dust eventually settled, the prevalent view among linear programming researchers was that Khachiyan had answered a major open question on the polynomiality of solving linear programs, but the simplex method remained the clear winner in practice.

This contradiction, the fact that an algorithm with the desirable theoretical property of polynomiality might nonetheless compare unfavorably with the (worst-case exponential) simplex method, set the stage for exciting new developments. It was no wonder, then, that the announcement by Karmarkar in 1984 of a new polynomial interior-point algorithm with the potential to dramatically improve the practical effectiveness of the simplex method made front-page news in major newspapers and magazines throughout the world.

Interior-point algorithms are continuous iterative algorithms. Computation experience with sophisticated procedures suggests that the number of iterations necessarily grows much more slowly than the dimension grows. Furthermore, they have an established worst-case polynomial iteration bound, providing the potential for dramatic improvement in computation effectiveness. The success of interior-point algorithms also brought much attention to complexity theory itself.

The goal of the book is to describe some of these recent developments and to suggest a few directions in which future progress might be made. The book is organized as follows. In Chapter 1, we discuss some necessary mathematical preliminaries. We also present several decision and optimization problems, models of computation, and several basic numerical procedures used throughout the text.

Chapter 2 is devoted to studying the geometry of inequalities and interior-point algorithms. At first glance, interior-point algorithms seem less geometric than the simplex or the ellipsoid methods. Actually, they also possess many rich geometric concepts. These concepts, such as "center," "volume," and "potential" of a polytope, are generally "non-combinatorial." These geometries are always helpful for teaching, learning, and research.

In Chapter 3 we present some basic algorithms to compute a so-called analytic center, or, equivalently, to maximize a potential or minimize a barrier function for a polytope. They are key elements underlying interior-point algorithms. Then, we present several interior-point linear programming algorithms in Chapter 4. It is impossible to list all the literature in this field. Here, we select five algorithms: Karmarkar's projective algorithm, the path-following algorithm, the potential reduction algorithm, the primal-dual algorithm including the predictor-corrector algorithm, and the affine scaling algorithm,

We analyze the worst-case complexity bound for interior-point algorithms in Chapter 5. The main issues are arithmetic operation, termination, and initialization techniques. We will use the real number computation model in our analysis because of the continuous nature of interior-point algorithms. We also compare the complexity theory with the convergence rate used in numerical analysis.

The worst-case complexity bound alone hardly serves as a practical criterion for judging the efficiency of algorithms. We will discuss a common phenomenon arising from using interior-point algorithms for solving optimization problems. It is often observed that effectiveness of an algorithm is dependent on the dimension or size of a problem instance as well as a parameter, called the "condition number," inherited in the problem. This condition number represents the degree of difficulty of the problem instance. For two problems having the same dimension but different condition numbers, an algorithm may have drastically different performances. This classification will help us to understand algorithm efficiency and possibly improve the condition and, therefore, improve the complexity of the problem. We present some condition-based complexity results for LP interior-point algorithms in Chapter 5.

While most of research has been focused on the worst-case performance of interior-point algorithms, many other complexity results were quietly established during the past several years. We try to cover these less-noticeable but significant results. In particular, we present some average and probabilistic complexity results in Chapter 6 and some asymptotic complexity (local convergence) results in Chapter 7. Average complexity bounds have been successfully established for the simplex method, and asymptotic or local convergence rates have been widely accepted by the numerical and continuous optimization community as major criteria in judging efficiency of iterative procedures.

Not only has the complexity of LP algorithms been significantly improved during the last decade, but also the problem domain solvable by interior-point algorithms has dramatically widened. We present complexity results for multi-objective programming, non-smooth convex programming, positive semi-definite programming and non-polyhedron optimization, and monotone complementarity in Chapter 8. We also discuss some complexity results for fractional programming and non-monotone linear complementarity, and approximation results for solving nonconvex optimization problems in Chapter 9.

1.2. MATHEMATICAL PRELIMINARIES

Finally, we discuss major implementation issues in Chapter 10. It is common to have a gap between a theoretical algorithm and its practical implementation: theoretical algorithm makes sure that it works for all instances and never fails, while practical implementation emphasizes average performance and uses many clever "tricks" and ingenious techniques. In this chapter we discuss several effective implementation techniques frequently used in interior-point linear programming software, such as the presolver process, the sparse linear system, the high-order predictor-corrector, the homogeneous and self-dual formulation, and the optimal basis identification. Our objective is to provide *theoretical* justification for these techniques and to explain their practical pros and cons.

1.2 Mathematical Preliminaries

This section summarizes mathematical background material for linear algebra, linear programming, and nonlinear optimization.

1.2.1 Basic notations

The notation described below will be followed in general. There may be some deviation where appropriate.

By $\mathcal{R}$ we denote the set of real numbers. $\mathcal{R}_+$ denotes the set of nonnegative real numbers, and $\overset{\circ}{\mathcal{R}}_+$ denotes the set of positive numbers. For a natural number n, the symbol $\mathcal{R}^n$ ($\mathcal{R}^n_+$, $\overset{\circ}{\mathcal{R}}^n_+$) denotes the set of vectors with n components in $\mathcal{R}$ ($\mathcal{R}_+$, $\overset{\circ}{\mathcal{R}}_+$). We call $\overset{\circ}{\mathcal{R}}^n_+$ the interior of $\mathcal{R}^n_+$.

Addition of vectors and multiplication of vectors with scalars are standard. The vector inequality $x \geq y$ means $x_j \geq y_j$ for $j = 1, 2, ..., n$. Zero represents a vector whose entries are all zeros and e represents a vector whose entries are all ones, where their dimensions may vary according to other vectors in expressions. A vector is always considered as a column vector, unless otherwise stated. Upper-case letters will be used to represent matrices. Greek letters will typically be used to represent scalars.

The superscript "T" denotes transpose operation. The inner product in $\mathcal{R}^n$ is defined as follows:

$$\langle x, y \rangle := x^T y = \sum_{j=1}^{n} x_j y_j \quad \text{for} \quad x, y \in \mathcal{R}^n.$$

The l_2 norm of a vector x is given by

$$\|x\|_2 = \sqrt{x^T x},$$

and the l_∞ norm is
$$\|x\|_\infty = \max\{|x_1|, |x_2|, ..., |x_n|\}.$$
In general, the p norm is
$$\|x\|_p = \left(\sum_1^n |x_j|^p\right)^{1/p}, \quad p = 1, 2, ...$$
The dual of the p norm, denoted by $\|.\|^*$, is the q norm, where
$$\frac{1}{p} + \frac{1}{q} = 1.$$
In this book, $\|.\|$ generally represents the l_2 norm.

For convenience, we sometime write a column vector x as
$$x = (x_1; x_2; \ldots; x_n)$$
and a row vector as
$$x = (x_1, x_2, \ldots, x_n).$$

For natural numbers m and n, $\mathcal{R}^{m \times n}$ denotes the set of real matrices with m rows and n columns. For $A \in \mathcal{R}^{m \times n}$, we assume that the row index set of A is $\{1, 2, ..., m\}$ and the column index set is $\{1, 2, ..., n\}$. The ith row of A is denoted by $a_{i.}$ and the jth column of A is denoted by $a_{.j}$; the i and jth component of A is denoted by a_{ij}. If I is a subset of the row index set and J is a subset of the column index set, then A_I denotes the submatrix of A whose rows belong to I, A_J denotes the submatrix of A whose columns belong to J, A_{IJ} denotes the submatrix of A induced by those components of A whose indices belong to I and J, respectively.

The identity matrix is denoted by I. The null space of A is denoted $\mathcal{N}(A)$ and the range of A is $\mathcal{R}(A)$. The determinant of an $n \times n$-matrix A is denoted by $\det(A)$. The trace of A, denoted by $\text{tr}(A)$, is the sum of the diagonal entries in A. The operator norm of A, denoted by $\|A\|$, is
$$\|A\|^2 := \max_{0 \neq x \in \mathcal{R}^n} \frac{\|Ax\|^2}{\|x\|^2}.$$

For a vector $x \in \mathcal{R}^n$, the upper-case X represents a diagonal matrix in $\mathcal{R}^{n \times n}$ whose diagonal entries are the entries of x, i.e.,
$$X = \text{diag}(x).$$

1.2. MATHEMATICAL PRELIMINARIES

A matrix $Q \in \mathcal{R}^{n\times n}$ is said to be positive definite (PD), denoted by $Q \succ 0$, if
$$x^T Q x > 0, \quad \text{for all} \quad x \neq 0,$$
and positive semi-definite (PSD), denoted by $Q \succeq 0$, if
$$x^T Q x \geq 0, \quad \text{for all} \quad x.$$
If $Q \succ 0$, then $-Q$ is called negative definite (ND), denoted by $Q \prec 0$; if $Q \succeq 0$, then $-Q$ is called negative semi-definite (NSD), denoted by $Q \preceq 0$. If Q is symmetric, then its eigenvalues are all real numbers; furthermore, Q is PSD if and only if all its eigenvalues are non-negative, and Q is PD if and only if all its eigenvalue are positive. Given a PD matrix Q we can define a Q-norm, $\|.\|_Q$, for vector x as
$$\|x\|_Q = \sqrt{x^T Q x}.$$

$\mathcal{M}^n$ denotes the space of symmetric matrices in $\mathcal{R}^{n\times n}$. The inner product in $\mathcal{M}^n$ is defined as follows:
$$\langle X, Y \rangle := X \bullet Y = \text{tr} X^T Y = \sum_{i,j} X_{i,j} Y_{i,j} \quad \text{for} \quad X, Y \in \mathcal{M}^n.$$
This is a generalization of the vector inner product to matrices. The matrix norm associated with the inner product is called *Frobenius norm*:
$$\|X\|_f = \sqrt{\text{tr} X^T X}.$$

$\mathcal{M}_+^n$ denote the set of positive semi-definite matrices in $\mathcal{M}^n$. $\overset{\circ}{\mathcal{M}_+^n}$ denotes the set of positive definite matrices in $\mathcal{M}^n$. We call $\overset{\circ}{\mathcal{M}_+^n}$ the interior of $\mathcal{M}_+^n$.

$\{x^k\}_0^\infty$ is an ordered sequence $x^0, x^1, x^2, ..., x^k, ...$ A sequence $\{x^k\}_0^\infty$ is convergent to $\bar{x}$, denoted $x^k \to \bar{x}$, if
$$\|x^k - \bar{x}\| \to 0.$$
A point x is a limit point of $\{x^k\}_0^\infty$ if there is a subsequence of $\{x^k\}$ convergent to x.

If $g(x) \geq 0$ is a real valued function of a real nonnegative variable, the notation $g(x) = O(x)$ means that $g(x) \leq \bar{c} x$ for some constant $\bar{c}$; the notation $g(x) = \Omega(x)$ means that $g(x) \geq \underline{c} x$ for some constant $\underline{c}$; the notation $g(x) = \theta(x)$ means that $\underline{c} x \leq g(x) \leq \bar{c} x$. Another notation is $g(x) = o(x)$, which means that $g(x)$ goes to zero faster than x does:
$$\lim_{x \to 0} \frac{g(x)}{x} = 0.$$

1.2.2 Convex sets

If x is a member of the set Ω, we write $x \in \Omega$; if y is not a member of Ω, we write $y \notin \Omega$. The union of two sets S and T is denoted $S \cup T$; the intersection of them is denoted $S \cap T$. A set can be specified in the form $\Omega = \{x : P(x)\}$ as the set of all elements satisfying property P.

For $y \in \mathcal{R}^n$ and $\epsilon > 0$, $B(y, \epsilon) = \{x : \|x - y\| \leq \epsilon\}$ is the ball or sphere of radius ϵ with center y. In addition, for a positive definite matrix Q of dimension n, $E(y, Q) = \{x : (x - y)^T Q (x - y) \leq 1\}$ is called an *ellipsoid*. The vector y is the center of $E(y, Q)$. When Q is diagonal, $E(y, Q)$ is called a *coordinate-aligned ellipsoid* (Figure 1.1).

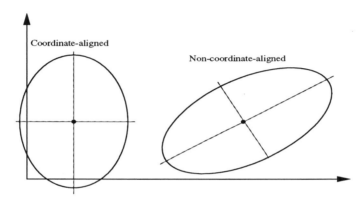

Figure 1.1. Coordinate-aligned and non-coordinate-aligned ellipsoids.

A set Ω is closed if $x^k \to x$, where $x^k \in \Omega$, implies $x \in \Omega$. A set Ω is open if around every point $y \in \Omega$ there is a ball that is contained in Ω, i.e., there is an $\epsilon > 0$ such that $B(y, \epsilon) \subset \Omega$. A set is bounded if it is contained within a ball with finite radius. A set is compact if it is both closed and bounded. The (topological) interior of any set Ω, denoted $\overset{\circ}{\Omega}$, is the set of points in Ω which are the centers of some balls contained in Ω. The closure of Ω, denoted $\hat{\Omega}$, is the smallest closed set containing Ω. The boundary of Ω is the part of $\hat{\Omega}$ that is not in $\overset{\circ}{\Omega}$.

A set C is said to be convex if for every $x^1, x^2 \in C$ and every real number α, $0 < \alpha < 1$, the point $\alpha x^1 + (1 - \alpha) x^2 \in C$. The convex hull of a set Ω is the intersection of all convex sets containing Ω.

A set C is a cone if $x \in C$ implies $\alpha x \in C$ for all $\alpha > 0$. A cone that is also convex is a convex cone. For a cone $C \subset \mathcal{E}$, the dual of C is the cone

$$C^* := \{y : \langle x, y \rangle \geq 0 \quad \text{for all} \quad x \in C\},$$

1.2. MATHEMATICAL PRELIMINARIES

where $\langle \cdot, \cdot \rangle$ is an inner product operation for space $\mathcal{E}$.

Example 1.1 *The n-dimensional non-negative orthant, $\mathcal{R}_+^n = \{x \in \mathcal{R}^n : x \geq 0\}$, is a convex cone. The dual of the cone is also $\mathcal{R}_+^n$; it is self-dual.*

Example 1.2 *The set of all positive semi-definite matrices in $\mathcal{M}^n$, $\mathcal{M}_+^n$, is a convex cone, called the* positive semi-definite matrix cone. *The dual of the cone is also $\mathcal{M}_+^n$; it is self-dual.*

Example 1.3 *The set $\{(t; x) \in \mathcal{R}^{n+1} : t \geq \|x\|\}$ is a convex cone in $\mathcal{R}^{n+1}$, called the* second-order cone. *The dual of the cone is also the second-order cone in $\mathcal{R}^{n+1}$; it is self-dual.*

A cone C is (convex) polyhedral if C can be represented by

$$C = \{x : Ax \leq 0\}$$

for some matrix A (Figure 1.2).

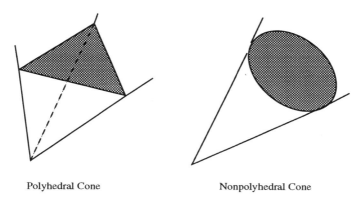

Polyhedral Cone   Nonpolyhedral Cone

Figure 1.2. Polyhedral and nonpolyhedral cones.

Example 1.4 *The non-negative orthant is a polyhedral cone, and neither the positive semi-definite matrix cone nor the second-order cone is polyhedral.*

It has been proved that for cones the concepts of "polyhedral" and "finitely generated" are equivalent according to the following theorem.

Theorem 1.1 *A convex cone C is polyhedral if and only if it is finitely generated, that is, the cone is generated by a finite number of vectors $b_1,...,b_m$:*

$$C = cone(b_1,...,b_m) := \left\{ \sum_{i=1}^{m} b_i y_i : y_i \geq 0, \ i = 1,...,m \right\}.$$

The following theorem states that a polyhedral cone can be generated by a set of basic directional vectors.

Theorem 1.2 *(Carathéodory's theorem) Let convex polyhedral cone $C = cone(b_1,...,b_m)$ and $x \in C$. Then, $x \in cone(b_{i_1},...,b_{i_d})$ for some linearly independent vectors $b_{i_1},...,b_{i_d}$ chosen from $b_1,...,b_m$.*

Example 1.5 *The following polyhedral cone of $\mathcal{R}^2$ in Figure 1.3 can be represented as either*

$$C = \left\{ \begin{pmatrix} 1 \\ 2 \end{pmatrix} y_1 + \begin{pmatrix} 2 \\ 1 \end{pmatrix} y_2 : y_1, y_2 \geq 0 \right\}$$

or

$$C = \left\{ x \in \mathcal{R}^2 : \begin{pmatrix} -2 & 1 \\ 1 & -2 \end{pmatrix} x \leq 0 \right\}.$$

The two vectors in the former representation are two basic directional vectors.

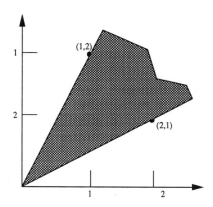

Figure 1.3. Representations of a polyhedral cone.

The most important type of convex set is a hyperplane. Hyperplanes dominate the entire theory of optimization. Let a be a nonzero n-dimensional vector, and let b be a real number. The set

$$H = \{x \in \mathcal{R}^n : a^T x = b\}$$

1.2. MATHEMATICAL PRELIMINARIES

is a hyperplane in $\mathcal{R}^n$ (Figure 1.4). Relating to hyperplane, positive and negative closed half spaces are given by

$$H_+ = \{x : a^T x \geq b\}$$
$$H_- = \{x : a^T x \leq b\}.$$

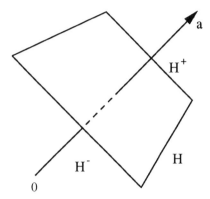

Figure 1.4. A hyperplane and half-spaces.

A set which can be expressed as the intersection of a finite number of closed half spaces is said to be a convex *polyhedron*:

$$P = \{x : Ax \leq b\}.$$

A bounded polyhedron is called *polytope*.

Let P be a polyhedron in $\mathcal{R}^n$, F is a face of P if and only if there is a vector c for which F is the set of points attaining max $\{c^T x : x \in P\}$ provided the this maximum is finite. A polyhedron has only finite many faces; each face is a nonempty polyhedron.

The most important theorem about the convex set is the following separating theorem (Figure 1.5).

Theorem 1.3 *(Separating hyperplane theorem) Let $C \subset \mathcal{E}$, where $\mathcal{E}$ is either $\mathcal{R}^n$ or $\mathcal{M}^n$, be a closed convex set and let y be a point exterior to C. Then there is a vector $a \in \mathcal{E}$ such that*

$$\langle a, y \rangle < \inf_{x \in C} \langle a, x \rangle.$$

The geometric interpretation of the theorem is that, given a convex set C and a point y outside of C, there is a hyperplane containing y that contains C in one of its open half spaces.

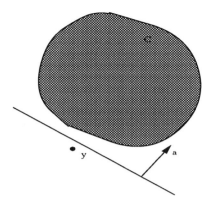

Figure 1.5. Illustration of the separating hyperplane theorem; an exterior point y is separated by a hyperplane from a convex set C.

Example 1.6 *Let C be a unit circle centered at the point $(1;1)$. That is, $C = \{x \in \mathcal{R}^2 : (x_1 - 1)^2 + (x_2 - 1)^2 \leq 1\}$. If $y = (2;0)$, $a = (-1;1)$ is a separating hyperplane vector. If $y = (0;-1)$, $a = (0;1)$ is a separating hyperplane vector. It is worth noting that these separating hyperplanes are not unique.*

We use the notation $\mathcal{E}$ to represent either $\mathcal{R}^n$ or $\mathcal{M}^n$, depending on the context, throughout this book, because all our decision and optimization problems take variables from one or both of these two vector spaces.

1.2.3 Real functions

The real function $f(x)$ is said to be continuous at x if $x^k \to x$ implies $f(x^k) \to f(x)$. The real function $f(x)$ is said to be continuous on set $\Omega \subset \mathcal{E}$, where recall that $\mathcal{E}$ is either $\mathcal{R}^n$ or $\mathcal{M}^n$, if $f(x)$ is continuous at x for every $x \in \Omega$. A continuous function f defined on a compact set (bounded and closed) $\Omega \subset \mathcal{E}$ has a minimizer in Ω; that is, there is an $x^* \in \Omega$ such that for all $x \in \Omega$, $f(x) \geq f(x^*)$. This result is called the *Weierstrass theorem*.

A function $f(x)$ is called homogeneous of degree k if $f(\alpha x) = \alpha^k f(x)$ for all $\alpha \geq 0$.

Example 1.7 *Let $c \in \mathcal{R}^n$ be given and $x \in \overset{\circ}{\mathcal{R}}{}^n_+$. Then $c^T x$ is homogeneous of degree 1 and*

$$\mathcal{P}(x) = n \log(c^T x) - \sum_{j=1}^n \log x_j$$

1.2. MATHEMATICAL PRELIMINARIES

is homogeneous of degree 0, where log is the natural logarithmic function. Let $C \in \mathcal{M}^n$ be given and $X \in \overset{\circ}{\mathcal{M}}{}^n_+$. Then $x^T C x$ is homogeneous of degree 2, $C \bullet X$ and $\det(X)$ are homogeneous of degree 1, and

$$\mathcal{P}(X) = n \log(C \bullet X) - \log \det(X)$$

is homogeneous of degree 0.

A set of real-valued function $f_1, f_2, ..., f_m$ defined on $\mathcal{R}^n$ can be written as a single vector function $f = (f_1, f_2, ..., f_m)^T \in \mathcal{R}^m$. If f has continuous partial derivatives of order p, we say $f \in C^p$. The gradient vector of a real-valued function $f \in C^1$ is a row vector

$$\nabla f(x) = \{\partial f / \partial x_i\}, \quad \text{for} \quad i = 1, ..., n.$$

If $f \in C^2$, we define the Hessian of f to be the n-dimensional symmetric matrix

$$\nabla^2 f(x) = \left\{ \frac{\partial^2 f}{\partial x_i \partial x_j} \right\} \quad \text{for} \quad i = 1, ..., n; \ j = 1, ..., n.$$

If $f = (f_1, f_2, ..., f_m)^T \in \mathcal{R}^m$, the Jacobian matrix of f is

$$\nabla f(x) = \begin{pmatrix} \nabla f_1(x) \\ ... \\ \nabla f_m(x) \end{pmatrix}.$$

f is a (continuous) convex function if and only if for $0 \leq \alpha \leq 1$,

$$f(\alpha x + (1-\alpha) y) \leq \alpha f(x) + (1-\alpha) f(y).$$

f is a (continuous) quasi-convex function if and only if for $0 \leq \alpha \leq 1$,

$$f(\alpha x + (1-\alpha) y) \leq \max[f(x), f(y)].$$

Thus, a convex function is a quasi-convex function. The level set of f is given by

$$L(z) = \{x : \ f(x) \leq z\}.$$

f is a quasi-convex function implies that the level set of f is convex for any given z (see Exercise 1.9).

A group of results that are used frequently in analysis are under the heading of *Taylor's theorem* or the mean-value theorem. The theorem establishes the linear and quadratic approximations of a function.

Theorem 1.4 *(Taylor expansion) Let $f \in C^1$ be in a region containing the line segment $[x, y]$. Then there is a α, $0 \leq \alpha \leq 1$, such that*

$$f(y) = f(x) + \nabla f(\alpha x + (1-\alpha)y)(y-x).$$

Furthermore, if $f \in C^2$ then there is a α, $0 \leq \alpha \leq 1$, such that

$$f(y) = f(x) + \nabla f(x)(y-x) + (1/2)(y-x)^T \nabla^2 f(\alpha x + (1-\alpha)y)(y-x).$$

We also have several propositions for real functions. The first indicates that the linear approximation of a convex function is a under-estimate.

Proposition 1.5 *Let $f \in C^1$. Then f is convex over a convex set Ω if and only if*
$$f(y) \geq f(x) + \nabla f(x)(y-x)$$
for all $x, y \in \Omega$.

The following proposition states that the Hessian of a convex function is positive semi-definite.

Proposition 1.6 *Let $f \in C^2$. Then f is convex over a convex set Ω if and only if the Hessian matrix of f is positive semi-definite throughout Ω.*

1.2.4 Inequalities

There are several important inequalities that are frequently used in algorithm design and complexity analysis.

Cauchy-Schwarz: given $x, y \in \mathcal{R}^n$, then

$$x^T y \leq \|x\| \|y\|.$$

Arithmetic-geometric mean: given $x \in \mathcal{R}^n_+$,

$$\frac{\sum x_j}{n} \geq \left(\prod x_j\right)^{1/n}.$$

Harmonic: given $x \in \overset{\circ}{\mathcal{R}}^n_+$,

$$\left(\sum x_j\right)\left(\sum 1/x_j\right) \geq n^2.$$

Hadamard: given $A \in \mathcal{R}^{m \times n}$ with columns $a_1, a_2, ..., a_n$, then

$$\sqrt{\det(A^T A)} \leq \prod \|a_j\|.$$

1.3 Decision and Optimization Problems

A decision or optimization problem has a form that is usually characterized by the decision variables and the constraints. A problem, $\mathcal{P}$, consists of two sets, data set $\mathcal{Z}_p$ and solution set $\mathcal{S}_p$. In general, $\mathcal{S}_p$ can be implicitly defined by the so-called optimality conditions. The solution set may be empty, i.e., problem $\mathcal{P}$ may have no solution.

In what follows, we list several decision and optimization problems. More problems will be listed later when we address them.

1.3.1 System of linear equations

Given $A \in \mathcal{R}^{m \times n}$ and $b \in \mathcal{R}^m$, the problem is to solve m linear equations for n unknowns:
$$Ax = b.$$
The data and solution sets are
$$\mathcal{Z}_p = \{A \in \mathcal{R}^{m \times n}, b \in \mathcal{R}^m\} \quad \text{and} \quad \mathcal{S}_p = \{x \in \mathcal{R}^n : Ax = b\}.$$
$\mathcal{S}_p$ in this case is an affine set. Given an x, one can easily check to see if x is in $\mathcal{S}_p$ by a matrix-vector multiplication and a vector-vector comparison. We say that a solution of this problem is easy to recognize.

To highlight the analogy with the theories of linear inequalities and linear programming, we list several well-known results of linear algebra. The first theorem provides two basic representations, the null and row spaces, of a linear subspaces.

Theorem 1.7 *Each linear subspace of $\mathcal{R}^n$ is generated by finitely many vectors, and is also the intersection of finitely many linear hyperplanes; that is, for each linear subspace of L of $\mathcal{R}^n$ there are matrices A and C such that $L = \mathcal{N}(A) = \mathcal{R}(C)$.*

The following theorem was observed by Gauss. It is sometimes called the *fundamental theorem* of linear algebra. It gives an example of a characterization in terms of necessary and sufficient conditions, where necessity is straightforward, and sufficiency is the key of the characterization.

Theorem 1.8 *Let $A \in \mathcal{R}^{m \times n}$ and $b \in \mathcal{R}^m$. The system $\{x : Ax = b\}$ has a solution if and only if that $A^T y = 0$ implies $b^T y = 0$.*

A vector y, with $A^T y = 0$ and $b^T y = 1$, is called an *infeasibility certificate* for the system $\{x : Ax = b\}$.

Example 1.8 *Let $A = (1; -1)$ and $b = (1; 1)$. Then, $y = (1/2; 1/2)$ is an infeasibility certificate for $\{x : Ax = b\}$.*

1.3.2 System of nonlinear equations

Given $f(x) : \mathcal{R}^n \to \mathcal{R}^m$, the problem is to solve m equations for n unknowns:
$$f(x) = 0.$$
The "data" and solution sets are
$$\mathcal{Z}_p = \{f, \nabla f, \ldots\} \quad \text{and} \quad \mathcal{S}_p = \{x \in \mathcal{R}^n : f(x) = 0\}.$$
In contrast to that of a linear system, $\mathcal{Z}_p$ is called an *oracle* or "black-box": for any given input x, it outputs the function values, the Jacobian matrix, and/or other finite numerical values about the function. In computer programs an oracle is a set of subroutines to perform function evaluation for a given input x. These subroutines might use a computer build-in function, a finite-element model, or a numerical partial-differential-equation procedure. In our computation models, we count one value evaluation one operation.

1.3.3 Linear least-squares problem

Given $A \in \mathcal{R}^{m \times n}$ and $c \in \mathcal{R}^n$, the system of equations $A^T y = c$ may be over-determined or have no solution. Such a case usually occurs when the number of equations is greater than the number of variables. Then, the problem is to find an $y \in \mathcal{R}^m$ or $s \in \mathcal{R}(A^T)$ such that $\|A^T y - c\|$ or $\|s - c\|$ is minimized. We can write the problem in the following format:

$$(LS) \quad \begin{array}{l} \text{minimize} \quad \|A^T y - c\|^2 \\ \text{subject to} \quad y \in \mathcal{R}^m, \end{array}$$

or

$$(LS) \quad \begin{array}{l} \text{minimize} \quad \|s - c\|^2 \\ \text{subject to} \quad s \in \mathcal{R}(A^T). \end{array}$$

In the former format, the term $\|A^T y - c\|^2$ is called the *objective function*, y is called the *decision variable*. Since y can be any point in $\mathcal{R}^m$, we say this (optimization) problem is *unconstrained*. The data and solution sets are
$$\mathcal{Z}_p = \{A \in \mathcal{R}^{m \times n}, c \in \mathcal{R}^n\}$$
and
$$\mathcal{S}_p = \{y \in \mathcal{R}^m : \|A^T y - c\|^2 \leq \|A^T x - c\|^2 \quad \text{for every} \quad x \in \mathcal{R}^m\}.$$
Given a y, to see if $y \in \mathcal{S}_p$ is as the same as the original problem. However, from a projection theorem in linear algebra, the solution set can be characterized and represented as
$$\mathcal{S}_p = \{y \in \mathcal{R}^m : AA^T y = Ac\},$$

1.3. DECISION AND OPTIMIZATION PROBLEMS

which becomes a system of linear equations and always has a solution. The vector $s = A^T y = A^T(AA^T)^+ Ac$ is the projection of c onto the range of A^T, where AA^T is called *normal matrix* and $(AA^T)^+$ is called *pseudo-inverse*. If A has full row rank then $(AA^T)^+ = (AA^T)^{-1}$, the standard inverse of full rank matrix AA^T. If A is not of full rank, neither is AA^T and $(AA^T)^+ AA^T x = x$ only for $x \in \mathcal{R}(A^T)$.

The vector $c - A^T y = (I - A^T(AA^T)^+ A)c$ is the projection of c onto the null space of A. It is the solution of the following least-squares problem:

$$(LS) \quad \begin{array}{ll} \text{minimize} & \|x - c\|^2 \\ \text{subject to} & x \in \mathcal{N}(A). \end{array}$$

In the full rank case, both matrices $A^T(AA^T)^{-1}A$ and $I - A^T(AA^T)^{-1}A$ are called *projection matrices*. These symmetric matrices have several desired properties (see Exercise 1.14).

The linear least-squares problem is a basic problem solved on each iteration of any interior-point algorithm.

1.3.4 System of linear inequalities

Given $A \in \mathcal{R}^{m \times n}$ and $b \in \mathcal{R}^m$, the problem is to find a solution $x \in \mathcal{R}^n$ satisfying $Ax \le b$ or prove that the solution set is empty. The inequality problem includes other forms such as finding an x that satisfies the combination of linear equations $Ax = b$ and inequalities $x \ge 0$. The data and solution sets of the latter are

$$\mathcal{Z}_p = \{A \in \mathcal{R}^{m \times n}, b \in \mathcal{R}^m\} \quad \text{and} \quad \mathcal{S}_p = \{x \in \mathcal{R}^n : Ax = b, \ x \ge 0\}.$$

Traditionally, a point in $\mathcal{S}_p$ is called a *feasible solution*, and a strictly positive point in $\mathcal{S}_p$ is called a *strictly feasible* or *interior feasible solution*.

The following results are Farkas' lemma and its variants.

Theorem 1.9 *(Farkas' lemma) Let $A \in \mathcal{R}^{m \times n}$ and $b \in \mathcal{R}^m$. Then, the system $\{x : Ax = b, \ x \ge 0\}$ has a feasible solution x if and only if that $A^T y \le 0$ implies $b^T y \le 0$.*

A vector y, with $A^T y \le 0$ and $b^T y = 1$, is called a (*primal*) infeasibility certificate for the system $\{x : Ax = b, \ x \ge 0\}$. Geometrically, Farkas' lemma means that if a vector $b \in \mathcal{R}^m$ does not belong to the cone generated by $a_{.1}, ..., a_{.n}$, then there is a hyperplane separating b from $\text{cone}(a_{.1}, ..., a_{.n})$.

Example 1.9 *Let $A = (1, 1)$ and $b = -1$. Then, $y = -1$ is an infeasibility certificate for $\{x : Ax = b, \ x \ge 0\}$.*

Theorem 1.10 *(Farkas' lemma variant) Let $A \in \mathcal{R}^{m \times n}$ and $c \in \mathcal{R}^n$. Then, the system $\{y : A^T y \leq c\}$ has a solution y if and only if that $Ax = 0$ and $x \geq 0$ imply $c^T x \geq 0$.*

Again, a vector $x \geq 0$, with $Ax = 0$ and $c^T x = -1$, is called a (*dual*) infeasibility certificate for the system $\{y : A^T y \leq c\}$.

Example 1.10 *Let $A = (1; -1)$ and $c = (1; -2)$. Then, $x = (1; 1)$ is an infeasibility certificate for $\{y : A^T y \leq c\}$.*

We say $\{x : Ax = b, \ x \geq 0\}$ or $\{y : A^T y \leq c\}$ is *approximately feasible* in the sense that we have an approximate solution to the equations and inequalities. In this case we can show that any certificate proving their infeasibility must have large norm. Conversely, if $\{x : Ax = b, \ x \geq 0\}$ or $\{y : A^T y \leq c\}$ is "approximately infeasible" in the sense that we have an approximate certificate in Farkas' lemma, then any feasible solution must have large norm. The details can be found in Exercise 5.12.

Example 1.11 *Given $\epsilon > 0$ but small. Let $A = (1, 1)$ and $b = -\epsilon$. Then, $x = (0; 0)$ is approximately feasible for $\{x : Ax = b, \ x \geq 0\}$, and the infeasibility certificate $y = -1/\epsilon$ has a large norm.*
Let $A = (1; -1)$ and $c = (1; -1 - \epsilon)$. Then, $y = 1$ is approximately feasible for $\{y : A^T y \leq c\}$, and the infeasibility certificate $x = (1/\epsilon; 1/\epsilon)$ has a large norm.

1.3.5 Linear programming (LP)

Given $A \in \mathcal{R}^{m \times n}$, $b \in \mathcal{R}^m$ and $c, l, u \in \mathcal{R}^n$, the linear programming (LP) problem is the following optimization problem:

$$\begin{aligned} \text{minimize} \quad & c^T x \\ \text{subject to} \quad & Ax = b, \ l \leq x \leq u, \end{aligned}$$

where some elements in l may be $-\infty$ meaning that the associated variables are unbounded from below, and some elements in u may be ∞ meaning that the associated variables are unbounded from above. If a variable is unbounded either from below or above, then it is called a *"free"* variable

The standard form linear programming problem is given below, which we will use throughout this book:

$$(LP) \quad \begin{aligned} \text{minimize} \quad & c^T x \\ \text{subject to} \quad & Ax = b, \ x \geq 0. \end{aligned}$$

The linear function $c^T x$ is called the *objective function*, and x is called the *decision variables*. In this problem, $Ax = b$ and $x \geq 0$ enforce *constraints*

1.3. DECISION AND OPTIMIZATION PROBLEMS

on the selection of x. The set $\mathcal{F}_p = \{x : Ax = b, x \geq 0\}$ is called *feasible set* or *feasible region*. A point $x \in \mathcal{F}_p$ is called a *feasible point*, and a feasible point x^* is called an *optimal solution* if $c^T x^* \leq c^T x$ for all feasible points x. If there is a sequence $\{x^k\}$ such that x^k is feasible and $c^T x^k \to -\infty$, then (LP) is said to be *unbounded*.

The data and solution sets for (LP), respectively, are

$$\mathcal{Z}_p = \{A \in \mathcal{R}^{m \times n}, b \in \mathcal{R}^m, c \in \mathcal{R}^n\}$$

and

$$\mathcal{S}_p = \{x \in \mathcal{F}_p : c^T x \leq c^T y, \quad \text{for every} \quad y \in \mathcal{F}_p\}.$$

Again, given an x, to see if $x \in \mathcal{S}_p$ is as difficult as the original problem. However, due to the duality theorem, we can simplify the representation of the solution set significantly.

With every (LP), another linear program, called the dual (LD), is the following problem:

$$(LD) \quad \begin{array}{ll} \text{maximize} & b^T y \\ \text{subject to} & A^T y + s = c, \ s \geq 0, \end{array}$$

where $y \in \mathcal{R}^m$ and $s \in \mathcal{R}^n$. The components of s are called *dual slacks*. Denote by $\mathcal{F}_d$ the sets of all (y, s) that are feasible for the dual. We see that (LD) is also a linear programming problem where y is a "free" vector.

The following theorems give us an important relation between the two problems.

Theorem 1.11 *(Weak duality theorem) Let $\mathcal{F}_p$ and $\mathcal{F}_d$ be non-empty. Then,*

$$c^T x \geq b^T y \quad \text{where} \quad x \in \mathcal{F}_p, \ (y, s) \in \mathcal{F}_d.$$

This theorem shows that a feasible solution to either problem yields a bound on the value of the other problem. We call $c^T x - b^T y$ the *duality gap*. From this we have important results.

Theorem 1.12 *(Strong duality theorem) Let $\mathcal{F}_p$ and $\mathcal{F}_d$ be non-empty. Then, x^* is optimal for (LP) if and only if the following conditions hold:*

i) $x^* \in \mathcal{F}_p$;

ii) *there is* $(y^*, s^*) \in \mathcal{F}_d$;

iii) $c^T x^* = b^T y^*$.

Theorem 1.13 *(LP duality theorem) If (LP) and (LD) both have feasible solutions then both problems have optimal solutions and the optimal objective values of the objective functions are equal.*

If one of (LP) or (LD) has no feasible solution, then the other is either unbounded or has no feasible solution. If one of (LP) or (LD) is unbounded then the other has no feasible solution.

The above theorems show that if a pair of feasible solutions can be found to the primal and dual problems with equal objective values, then these are both optimal. The converse is also true; there is no "gap." From this condition, the solution set for (LP) and (LD) is

$$\mathcal{S}_p = \left\{ (x,y,s) \in (\mathcal{R}_+^n, \mathcal{R}^m, \mathcal{R}_+^n) : \begin{array}{rcl} c^T x - b^T y & = & 0 \\ Ax & = & b \\ -A^T y - s & = & -c \end{array} \right\}, \quad (1.1)$$

which is a system of linear inequalities and equations. Now it is easy to verify whether or not a pair (x, y, s) is optimal.

For feasible x and (y, s), $x^T s = x^T(c - A^T y) = c^T x - b^T y$ is called the *complementarity gap*. If $x^T s = 0$, then we say x and s are complementary to each other. Since both x and s are nonnegative, $x^T s = 0$ implies that $x_j s_j = 0$ for all $j = 1, \ldots, n$. Thus, one equation plus nonnegativity are transformed into n equations. Equations in (1.1) become

$$\begin{array}{rcl} Xs & = & 0 \\ Ax & = & b \\ -A^T y - s & = & -c. \end{array} \quad (1.2)$$

This system has total $2n + m$ unknowns and $2n + m$ equations including n nonlinear equations.

The following theorem plays an important role in analyzing LP interior-point algorithms. It give a unique partition of the LP variables in terms of complementarity.

Theorem 1.14 *(Strict complementarity theorem) If (LP) and (LD) both have feasible solutions then both problems have a pair of strictly complementary solutions $x^* \geq 0$ and $s^* \geq 0$ meaning*

$$X^* s^* = 0 \quad \text{and} \quad x^* + s^* > 0.$$

Moreover, the supports

$$P^* = \{j : x_j^* > 0\} \quad \text{and} \quad Z^* = \{j : s_j^* > 0\}$$

are invariant for all pairs of strictly complementary solutions.

1.3. DECISION AND OPTIMIZATION PROBLEMS

Given (LP) or (LD), the pair of P^* and Z^* is called the (strict) *complementarity partition*. $\{x : A_{P^*}x_{P^*} = b,\ x_{P^*} \geq 0,\ x_{Z^*} = 0\}$ is called the *primal optimal face*, and $\{y : c_{Z^*} - A_{Z^*}^T y \geq 0,\ c_{P^*} - A_{P^*}^T y = 0\}$ is called the *dual optimal face*.

Select m linearly independent columns, denoted by the index set B, from A. Then matrix A_B is nonsingular and we may uniquely solve

$$A_B x_B = b$$

for the m-vector x_B. By setting the variables, x_N, of x corresponding to the remaining columns of A equal to zero, we obtain a solution x such that

$$Ax = b.$$

Then, x is said to be a (*primal*) *basic solution* to (LP) with respect to the *basis* A_B. The components of x_B are called *basic variables*. A dual vector y satisfying

$$A_B^T y = c_B$$

is said to be the corresponding *dual basic solution*. If a basic solution $x \geq 0$, then x is called a *basic feasible solution*. If the dual solution is also feasible, that is

$$s = c - A^T y \geq 0,$$

then x is called an *optimal basic solution* and A_B an *optimal basis*. A basic feasible solution is a vertex on the boundary of the feasible region. An optimal basic solution is an optimal vertex of the feasible region.

If one or more components in x_B has value zero, that basic solution x is said to be (*primal*) *degenerate*. Note that in a nondegenerate basic solution the basic variables and the basis can be immediately identified from the nonzero components of the basic solution. If all components, s_N, in the corresponding dual slack vector s, except for s_B, are non-zero, then y is said to be (*dual*) *nondegenerate*. If both primal and dual basic solutions are nondegenerate, A_B is called a *nondegenerate basis*.

Theorem 1.15 (*LP fundamental theorem*) *Given (LP) and (LD) where A has full row rank m,*

i) *if there is a feasible solution, there is a basic feasible solution;*

ii) *if there is an optimal solution, there is an optimal basic solution.*

The above theorem reduces the task of solving a linear program to that searching over basic feasible solutions. By expanding upon this result, the simplex method, a finite search procedure, is derived. The simplex method is to proceed from one basic feasible solution (an extreme point of the feasible region) to an adjacent one, in such a way as to continuously decrease the value of the objective function until a minimizer is reached. In contrast, interior-point algorithms will move in the interior of the feasible region and reduce the value of the objective function, hoping to by-pass many extreme points on the boundary of the region.

1.3.6 Quadratic programming (QP)

Given $Q \in \mathcal{R}^{n \times n}$, $A \in \mathcal{R}^{m \times n}$, $b \in \mathcal{R}^m$ and $c \in \mathcal{R}^n$, the quadratic programming (QP) problem is the following optimization problem:

$$(QP) \quad \begin{array}{ll} \text{minimize} & q(x) := (1/2)x^T Q x + c^T x \\ \text{subject to} & Ax = b, \; x \geq 0. \end{array}$$

We may denote the feasible set by $\mathcal{F}_p$. The data and solution sets for (QP) are

$$\mathcal{Z}_p = \{Q \in \mathcal{R}^{n \times n}, A \in \mathcal{R}^{m \times n}, b \in \mathcal{R}^m, c \in \mathcal{R}^n\}$$

and

$$\mathcal{S}_p = \{x \in \mathcal{F}_p : q(x) \leq q(y) \quad \text{for every} \quad y \in \mathcal{F}_p\}.$$

A feasible point x^* is called a *KKT* point, where KKT stands for Karush-Kuhn-Tucker, if the following KKT conditions hold: there exists $(y^* \in \mathcal{R}^m, s^* \in \mathcal{R}^n)$ such that (x^*, y^*, s^*) is feasible for the following dual problem:

$$(QD) \quad \begin{array}{ll} \text{maximize} & d(x,y) := b^T y - (1/2) x^T Q x \\ \text{subject to} & A^T y + s - Qx = c, \; x, \; s \geq 0, \end{array}$$

and satisfies the complementarity condition

$$(x^*)^T s^* = (1/2)(x^*)^T Q x^* + c^T x^* - (b^T y^* - (1/2)(x^*)^T Q x^*) = 0.$$

Similar to LP, we can write the KKT condition as:

$$(x, y, s) \in (\mathcal{R}_+^n, \mathcal{R}^m, \mathcal{R}_+^n)$$

and

$$\begin{array}{rcl} Xs & = & 0 \\ Ax & = & b \\ -A^T y + Qx - s & = & -c. \end{array} \quad (1.3)$$

Again, this system has total $2n + m$ unknowns and $2n + m$ equations including n nonlinear equations.

1.3. DECISION AND OPTIMIZATION PROBLEMS

The above condition is also called the *first-order necessary condition*. If Q is positive semi-definite, then x^* is an optimal solution for (QP) if and only if x^* is a KKT point for (QP). In this case, the solution set for (QP) is characterized by a system of linear inequalities and equations. One can see (LP) is a special case of (QP).

1.3.7 Linear complementarity problem (LCP)

Given $M \in \mathcal{R}^{n \times n}$ and $q \in \mathcal{R}^n$, the linear complementarity problem (LCP) is to find a pair $x, s \in \mathcal{R}^n$ such that

$$s = Mx + q, \quad (x, s) \geq 0 \quad \text{and} \quad x^T s = 0.$$

A pair $(x, s) \geq 0$ satisfying $s = Mx + q$ is called a *feasible pair*. The LCP problem can be written as an optimization problem:

$$(LCP) \quad \begin{array}{ll} \text{minimize} & x^T s \\ \text{subject to} & Mx - s = -q, \ x, s \geq 0. \end{array}$$

The LCP is a fundamental decision and optimization problem. It also arises from economic equilibrium problems, noncooperative games, traffic assignment problems, and optimization problems.

The data and solution sets for LCP are

$$\mathcal{Z}_p = \{M \in \mathcal{R}^{n \times n}, q \in \mathcal{R}^n\}$$

and

$$\mathcal{S}_p = \left\{ (x, s) \in (\mathcal{R}^n_+, \mathcal{R}^n_+) : \begin{array}{rl} Xs &= 0 \\ Mx - s &= -q \end{array} \right\}. \quad (1.4)$$

(Again, $x^T s = 0$ plus nonnegativity on x and s imply n equations $Xs = 0$.)

There are extensions from LCP. One extension is some variables can be "free" (unrestricted in sign) and their counter-parts are zeros:

$$\begin{array}{ll} \text{minimize} & x^T s \\ \text{subject to} & M \begin{pmatrix} x \\ y \end{pmatrix} - \begin{pmatrix} s \\ 0 \end{pmatrix} = -q, \ x, s \geq 0. \end{array} \quad (1.5)$$

One can verify that the solution set of (LP) and the KKT set of (QP) can be formulated as this problem, where

$$M = \begin{pmatrix} Q & -A^T \\ A & 0 \end{pmatrix} \quad \text{and} \quad q = \begin{pmatrix} c \\ -b \end{pmatrix}. \quad (1.6)$$

Another extension is called the *generalized linear complementarity problem*:

(GLCP) minimize $x^T s$
subject to $Ax + Bs + Cz = q$, $(x \in \mathcal{R}^n, s \in \mathcal{R}^n, z \in \mathcal{R}^d) \geq 0$,

where matrices $A, B \in \mathcal{R}^{m \times n}$, $C \in \mathcal{R}^{m \times d}$ and $q \in \mathcal{R}^m$. We will discuss on solving GLCP in Chapter 9.

If M (may not be symmetric) is monotone or positive semi-definite, i.e.,

$$x^T M x \geq 0, \quad \text{for all} \quad x,$$

then the LCP is called a *monotone LCP*. The following theorem characterizes the solution set of a monotone linear complementarity problem.

Theorem 1.16 *(Monotone linear complementarity theorem) Consider the monotone linear complementarity problem. We have*

i) *if the problem is feasible, then it has a solution with $x^T s = 0$;*

ii) *the solution set is convex;*

iii) *it has a maximal complementary solution pair x^* and s^*, meaning that the number of the positive components in vector $x^* + s^*$ is maximized. Moreover, the supports*

$$P^* = \{j : x_j^* > 0\} \quad and \quad Z^* = \{j : s_j^* > 0\}$$

are invariant for all pairs of maximal complementary solutions.

Note that, in general, a monotone complementarity problem, unlike linear programming, may not have a strictly complementary solution.

1.3.8 Positive semi-definite programming (PSP)

Given $C \in \mathcal{M}^n$, $A_i \in \mathcal{M}^n$, $i = 1, 2, ..., m$, and $b \in \mathcal{R}^m$, the positive semi-definite programming problem is to find a matrix $X \in \mathcal{M}^n$ for the optimization problem:

(PSP) inf $C \bullet X$
subject to $A_i \bullet X = b_i, i = 1, 2, ..., m, \ X \succeq 0.$

Recall that the $\bullet$ operation is the matrix inner product

$$A \bullet B := \operatorname{tr} A^T B.$$

1.3. DECISION AND OPTIMIZATION PROBLEMS

The notation $X \succeq 0$ means that X is a positive semi-definite matrix, and $X \succ 0$ means that X is a positive definite matrix. If a point $X \succ 0$ and satisfies all equations in (PSP), it is called a (*primal*) strictly or interior feasible solution. .

The data set of (PSP) is

$$\mathcal{Z}_p = \{A_i \in \mathcal{M}^n,\ i=1,\ldots,m,\ b \in \mathcal{R}^m, c \in \mathcal{M}^n\}.$$

The dual problem to (PSP) can be written as:

(PSD) $\quad$ sup $\quad b^T y$
$\qquad$ subject to $\quad \sum_i^m y_i A_i + S = C,\ S \succeq 0,$

which is analogous to the dual (LD) of LP. Here $y \in \mathcal{R}^m$ and $S \in \mathcal{M}^n$. If a point $(y, S \succ 0)$ satisfies all equations in (PSD), it is called a *dual interior feasible solution*.

Example 1.12 Let $P(y \in \mathcal{R}^m) = -C + \sum_i^m y_i A_i$, where C and A_i, $i = 1,\ldots,m$, are given symmetric matrices. The problem of minimizing the max-eigenvalue of $P(y)$ can be cast as a (PSD) problem.

In positive semi-definite programming, we minimize a linear function of a matrix in the positive semi-definite matrix cone subject to affine constraints. In contrast to the positive orthant cone of linear programming, the positive semi-definite matrix cone is non-polyhedral (or "non-linear"), but convex. So positive semi-definite programs are convex optimization problems. Positive semi-definite programming unifies several standard problems, such as linear programming, quadratic programming, and convex quadratic minimization with convex quadratic constraints, and finds many applications in engineering, control, and combinatorial optimization.

We have several theorems analogous to Farkas' lemma.

Theorem 1.17 *(Farkas' lemma in PSP) Let $A_i \in \mathcal{M}^n$, $i = 1,\ldots,m$, have rank m (i.e., $\sum_i^m y_i A_i = 0$ implies $y = 0$) and $b \in \mathcal{R}^m$. Then, there exists a symmetric matrix $X \succ 0$ with*

$$A_i \bullet X = b_i, \quad i = 1,\ldots,m,$$

if and only if $\sum_i^m y_i A_i \preceq 0$ and $\sum_i^m y_i A_i \neq 0$ implies $b^T y < 0$.

Note the difference between the above theorem and Theorem 1.9.

Theorem 1.18 *(Weak duality theorem in PSP) Let $\mathcal{F}_p$ and $\mathcal{F}_d$, the feasible sets for the primal and dual, be non-empty. Then,*

$$C \bullet X \geq b^T y \quad \text{where} \quad X \in \mathcal{F}_p,\ (y, S) \in \mathcal{F}_d.$$

The weak duality theorem is identical to that of (LP) and (LD).

Corollary 1.19 *(Strong duality theorem in PSP) Let $\mathcal{F}_p$ and $\mathcal{F}_d$ be non-empty and have an interior. Then, X is optimal for (PS) if and only if the following conditions hold:*

i) $X \in \mathcal{F}_p$;

ii) *there is* $(y, S) \in \mathcal{F}_d$;

iii) $C \bullet X = b^T y$ *or* $X \bullet S = 0$.

Again note the difference between the above theorem and the strong duality theorem for LP.

Two positive semi-definite matrices are complementary to each other, $X \bullet S = 0$, if and only if $XS = 0$ (Exercise 1.24). From the optimality conditions, the solution set for certain (PSP) and (PSD) is

$$\mathcal{S}_p = \{X \in \mathcal{F}_p,\ (y, S) \in \mathcal{F}_d : C \bullet X - b^T y = 0\},$$

or

$$\mathcal{S}_p = \{X \in \mathcal{F}_p,\ (y, S) \in \mathcal{F}_d : XS = 0\},$$

which is a system of linear matrix inequalities and equations.

In general, we have

Theorem 1.20 *(PSP duality theorem) If one of (PSP) or (PSD) has a strictly or interior feasible solution and its optimal value is finite, then the other is feasible and has the same optimal value. If one of (PSP) or (PSD) is unbounded then the other has no feasible solution.*

Note that a duality gap may exists if neither (PSP) nor (PSD) has a strictly feasible point. This is in contrast to (LP) and (LD) where no duality gap exists if both are feasible.

Although positive semi-definite programs are much more general than linear programs, they are not much harder to solve. It has turned out that most interior-point methods for LP have been generalized to positive semi-definite programs. As in LP, these algorithms possess polynomial worst-case complexity under certain computation models. They also perform well in practice. We will describe such extensions later in this book.

1.3. DECISION AND OPTIMIZATION PROBLEMS

1.3.9 Nonlinear programming (NP)

Given $f : \mathcal{R}^n \to \mathcal{R}$, $h : \mathcal{R}^n \to \mathcal{R}^m$, $g : \mathcal{R}^n \to \mathcal{R}^d$, the nonlinear programming problem is the following optimization problem:

$$(NP) \quad \text{minimize} \quad f(x)$$
$$\text{subject to} \quad h(x) = b, \ g(x) \geq 0.$$

We may denote the feasible set by $\mathcal{F}$. The data and solution sets for (NP) are

$$\mathcal{Z}_p = \{f, h, g, \nabla f, \nabla h, \nabla g, \ldots\}$$

and

$$\mathcal{S}_p = \{x \in \mathcal{F} : f(x) \leq f(y) \text{ for every } y \in \mathcal{F}\}.$$

A feasible point x^* is called a *KKT point* if the following KKT conditions hold: there exists $(y^* \in \mathcal{R}^m, s^* \in \mathcal{R}^d)$ such that (x^*, y^*, s^*) are satisfying

$$\nabla^T f(x^*) - \nabla^T h(x^*) y^* - \nabla^T g(x^*) s^* = 0, \ s^* \geq 0,$$

and

$$g(x^*)^T s^* = 0.$$

Here, y^* and s^* are called the *Lagrange* or *dual multipliers*. We have the following necessary condition for a point to be a local minimizer.

Theorem 1.21 *(Karush-Kuhn-Tucker theorem) Let x^* be a relative (local) minimum solution for (NP) and suppose x^* is a regular point for the constraints, meaning that the Jacobian matrix $\nabla h(x^*)$ and the gradient vectors $\nabla g_j(x^*)$ for all j with $g_j(x^*) = 0$ (they called active inequality constraints) are linearly independent. Then, x^* must be a KKT point.*

If f is convex, h is affine, and g is concave, then x^* is optimal if and only if x^* is a KKT point for (NP). Thus, the necessary condition becomes sufficient.

1.3.10 Nonlinear complementarity problem (NCP)

Let $f(x) : \mathcal{R}^n_+ \to \mathcal{R}^n$ be a continuous real function. Then, the nonlinear complementarity problem (NCP) is to find a pair $x, s \in \mathcal{R}^n$ such that

$$s = f(x), \quad (x, s) \geq 0 \quad \text{and} \quad Xs = 0.$$

We say (NCP) is feasible if there exists an $x \geq 0$ and $s = f(x) \geq 0$.

If $f(x)$ is a monotone function over $\mathcal{R}^n_+$, that is,

$$(x^1 - x^2)^T (f(x^1) - f(x^2)) \geq 0 \quad \text{for all} \quad x^1, x^2 \in \mathcal{R}^n_+,$$

then the problem is called the *monotone complementarity problem*. It can be written as an optimization problem:

$$(MCP) \quad \begin{array}{ll} \text{minimize} & x^T s \\ \text{subject to} & s = f(x), \ (x, s) \geq 0. \end{array}$$

The Jacobian matrix of f, ∇f, is positive semi-definite in $\mathcal{R}^n_+$.

Denote the feasible set by $\mathcal{F}$ and the solution set by $\mathcal{S}$. Note that (MCP) being feasible does not imply that (MCP) has a solution. The following theorem characterizes the solution set of a monotone complementarity problem.

Theorem 1.22 *(Monotone complementarity theorem) Consider the monotone nonlinear complementarity problem. We have*

i) *If (MCP) has an interior feasible point, i.e., $x > 0$ and $s = f(x) > 0$, then it has a solution with $x^T s = 0$;*

ii) *the solution set is convex;*

iii) *it has a maximal complementary solution pair x^* and s^*, meaning that the number of the positive components in vector $x^* + s^*$ is maximized. Moreover, the supports*

$$P^* = \{j : x^*_j > 0\} \quad and \quad Z^* = \{j : s^*_j > 0\}$$

are invariant for all pairs of maximal complementary solutions.

Note that, in general, a monotone complementarity problem, unlike linear programming, may not have a strictly complementary solution.

Similar to our discussion on linear complementarity, finding a KKT point or proving infeasibility or unboundedness of a convex nonlinear programming problem can be reduced to solving a monotone complementarity problem with possible "free" variables.

1.4 Algorithms and Computation Models

An algorithm is a list of instructions to solve a problem. For every instance of problem $\mathcal{P}$, i.e., for every given data $Z \in \mathcal{Z}_p$, an algorithm for solving $\mathcal{P}$ either determines that $\mathcal{S}_p$ is empty or generates an output x such that $x \in \mathcal{S}_p$ or x is close to $\mathcal{S}_p$ in certain measure. The latter x is called an *approximate* solution.

Let us use $\mathcal{A}_p$ to denote the collection of all possible algorithm for solving every instance in $\mathcal{P}$. Then, the (operation) complexity of an algorithm

1.4. ALGORITHMS AND COMPUTATION MODELS

$A \in \mathcal{A}_p$ for solving an instance $Z \in \mathcal{Z}_p$ is defined as the total arithmetic operations: $+$, $-$, $*$, $/$, and comparison on real numbers. Denote it by $c_o(A, Z)$. Sometimes it is convenient to define the iteration complexity, denoted by $c_i(A, Z)$, where we assume that each iteration costs a polynomial number (in m and n) of arithmetic operations. In most iterative algorithms, each iteration can be performed efficiently both sequentially and in parallel, such as solving a system of linear equations, rank-one updating the inversion of a matrix, pivoting operation of a matrix, multiplying a matrix by a vector, etc.

We present four algorithm efficiency or complexity measures frequently used in algorithm analysis and optimization theory. The first two are concerned by the worst-case performance, the third one is on the "average" performance, and the fourth one is about the "local" performance. They will all be used in our analysis of interior-point algorithms.

1.4.1 Worst-case complexity

The worst-case complexity of algorithm A for problem $\mathcal{P}$ is defined as

$$c(A) := \sup_{Z \in \mathcal{Z}_p} c(A, Z).$$

It is better to distinguish the worst-case complexity of an algorithm, A, from that of a problem $\mathcal{P}$. The worst-case complexity of the problem is

$$c^p := \inf_{A \in \mathcal{A}_p} c(A).$$

Analyzing the worst-case complexity of a problem is challenging since $\mathcal{A}_p$ is an unknown domain, and the analysis of the complexity of the algorithm is equally difficult since $\mathcal{P}$ is also immense. However, the complexity theory does not directly attack the algorithm complexity for every instance. Instead, it classifies $\mathcal{P}$ using its data bit-size L, where the data are assumed rational. This is the Turing Machine Model for computation. We may call this type of complexity *size-based*. Then, we express an upper bound $f_A(m, n, L)$, in terms of the parameters m, n, and L, for the size-based complexity of algorithm A as

$$c(A, L) := \sup_{Z \in \mathcal{Z}_p,\ \text{size}(Z) \leq L} c(A, Z) \leq f_A(m, n, L).$$

(Recall that integer m and n represent the problem dimensions.) Then, the size-based complexity of problem $\mathcal{P}$ has a relation

$$c^p(L) := \inf_{A \in \mathcal{A}_p} c(A, L) \leq f_A(m, n, L).$$

We see that the complexity of algorithms is an upper bound for the complexity of the problem. Another active pursuit in computer science is the analysis of a lower bound for the problem's complexity, which is outside of the scope of this monograph.

If $f_A(m, n, L)$ is a polynomial in m, n, and L, then we say algorithm A is a *polynomial-time* or *polynomial* algorithm and problem $\mathcal{P}$ is polynomially solvable. If $f_A(m, n, L)$ is independent of L and polynomial in m and n, then we say algorithm A is a *strongly polynomial algorithm*.

In the real number model, the use of L is not suitable. We may use ϵ, the error for an approximate solution as a parameter. Let $c(A, Z, \epsilon)$ be the total number of operations of algorithm A for generating an ϵ-approximate solution, with a well-defined measure, to problem $\mathcal{P}$. Then,

$$c(A, \epsilon) := \sup_{Z \in \mathcal{Z}_p} c(A, Z, \epsilon) \leq f_A(m, n, \epsilon) \quad \text{for any} \quad \epsilon > 0.$$

We call this complexity model *error-based*. One may also view an approximate solution an exact solution to a problem ϵ-near to $\mathcal{P}$ with a well-defined measure in the data space. This is the so-called *backward analysis* model in numerical analysis.

If $f_A(m, n, \epsilon)$ is a polynomial in m, n, and $\log(1/\epsilon)$, then algorithm A is a polynomial algorithm and problem $\mathcal{P}$ is polynomially solvable. Again, if $f_A(m, n, \epsilon)$ is independent of ϵ and polynomial in m and n, then we say algorithm A is a strongly polynomial algorithm. If $f_A(m, n, \epsilon)$ is a polynomial in m, n, and $(1/\epsilon)$, then algorithm A is a polynomial approximation scheme or pseudo-polynomial algorithm . For some optimization problems, the complexity theory can be applied to prove not only that they cannot be solved in polynomial-time, but also that they do not have polynomial approximation schemes. In practice, approximation algorithms are widely used and accepted in practice.

Example 1.13 *There is a strongly polynomial algorithm for sorting a vector in descending or ascending order, for matrix-vector multiplication, and for computing the norm of a vector.*

Example 1.14 *Consider the bisection method to locate a root of a continuous function $f(x) : \mathcal{R} \to \mathcal{R}$ within interval $[0, 1]$, where $f(0) > 0$ and $f(1) < 0$. The method calls the oracle to evaluate $f(1/2)$ (counted as one operation). If $f(1/2) > 0$, we throw away $[0, 1/2]$; if $f(1/2) < 0$, we throw away $(1/2, 1]$. Then we repeat this process on the remaining half interval. Each step of the method halves the interval that contains the root. Thus, in $\log(1/\epsilon)$ steps, we must have an approximate root whose distance to the root is less than ϵ. Therefore, the bisection method is a polynomial algorithm.*

1.4. ALGORITHMS AND COMPUTATION MODELS

We have to admit that the criterion of polynomiality is somewhat controversial. Many algorithms may not be polynomial but work fine in practice. This is because polynomiality is built upon the worst-case analysis. However, this criterion generally provides a qualitative statement: if a problem is polynomial solvable, then the problem is indeed relatively easy to solve regardless of the algorithm used. Furthermore, it is ideal to develop an algorithm with both polynomiality and practical efficiency.

1.4.2 Condition-based complexity

As we discussed before, in the Turing Machine Model the parameters are selected as the number of variables, the number of constraints, and the bit-size of the data of an instance. In fact, expressing the algorithm complexity in terms of the size of the problem does not really measure the difficulty of an instance of the problem. Two instances with the same size may result in drastically different performances by the same algorithm.

Example 1.15 *Consider the steepest descent algorithm for solving*

$$minimize \quad q(x) = (1/2)x^T Q x + c^T x, \qquad (1.7)$$

where $Q \in \mathcal{R}^{n \times n}$ is positive definite. Denote by $x^* = -Q^{-1}c \in \mathcal{R}^n$ the minimizer of the problem. Starting from an $x^0 \in \mathcal{R}^n$, the method uses iterative formula

$$x^{k+1} = x^k - \alpha^k d^k, \quad where \quad d^k = Qx^k + c$$

and

$$\alpha^k = \frac{\|d^k\|^2}{\|d^k\|_Q^2},$$

which minimizes $q(x^k - \alpha d^k)$. (Recall that the norm $\|d\|_Q = \sqrt{d^T Q d}$ for positive definite Q.) It is well known that the algorithm generates a sequence of $\{x^k\}$ such that

$$\frac{\|x^{k+1} - x^*\|_Q}{\|x^k - x^*\|_Q} \le \frac{\lambda_n - \lambda_1}{\lambda_n + \lambda_1},$$

where $\lambda_1 \le \lambda_2, ... \le \lambda_n$ are n eigenvalues of Q with ascending order. Let

$$\xi(Q) := \frac{\lambda_n - \lambda_1}{\lambda_n + \lambda_1} < 1.$$

Then, to reach $\frac{\|x^{k+1} - x^*\|_Q}{\|x^k - x^*\|_Q} \le \epsilon$ we need no more than $\log(1/\epsilon)/\log(1/\xi(Q))$ iterations; the smaller $\xi(Q)$, the faster of the algorithm.

In this example two Q matrices with the same size but different eigenvalue structure will possess quite different convergence speed. This phenomenon is surprisingly common in optimization, due to mathematical bases upon which algorithms are designed. Thus, the upper bound for the complexity of an algorithm may be expressed as $f_A(m,n,\eta(Z))$ or $f_A(m,n,\epsilon,\eta(Z))$ in both the rational-number and the real number models, where $\eta(Z)$ can be viewed as a condition number for the instance Z. ($\eta(Z) = \frac{1}{\log(1/\xi(Q))}$ in this example.) The better the condition number, the less difficult the instance. It is our goal to study this phenomenon and to improve the condition number and, thereby, the performance of an algorithm.

1.4.3 Average complexity

Let $\mathcal{Z}_p$ be a random sample space, then one can define the average or expected complexity of the algorithm for the problem as

$$c^a(A) = E_{Z \in \mathcal{Z}_p}(c(A,Z)).$$

If we know the condition-based complexity of an algorithm for $\mathcal{P}$, then the average complexity of the algorithm is

$$c^a(A) \le E_{Z \in \mathcal{Z}}(f_A(m,n,\eta(Z))).$$

In many cases, $f_A(m,n,\eta(Z))$ can be expressed as

$$f_A(m,n,\eta(Z)) = f^1(m,n)f^2(\eta(Z)).$$

Thus,
$$c^a(A) \le f^1(m,n) E_{Z \in \mathcal{Z}}(f^2(\eta(Z))),$$

which will simplify analysis a great deal.

Example 1.16 *Let us randomly generate positive definite matrix Q in Example 1.15. Then, $\xi(Q)$ is a random number, and the expected number of iterations to reach $\frac{\|x^{k+1}-x^*\|_Q}{\|x^k-x^*\|_Q} \le \epsilon$ is bounded by*

$$\log(1/\epsilon) E\left(\frac{1}{\log(1/\xi(Q))}\right).$$

Another probabilistic model is called *high-probability analysis*. We say that a problem $\mathcal{P}$ can be solved by algorithm A in $f_A(m,n)$ time with high probability if

$$\Pr\{c(A,Z) \le f_A(m,n)\} \to 1$$

1.4. ALGORITHMS AND COMPUTATION MODELS

as $m, n \to \infty$. Again, if we have a condition-based complexity and if we have
$$\Pr\{f^2(\eta(Z)) \le f^3(m,n)\} \to 1$$
as $m, n \to \infty$, then the algorithm solves P in
$$f_A(m,n) = f^1(m,n) f^3(m,n)$$
operations with high probability.

1.4.4 Asymptotic complexity

Most algorithms are iterative in nature. They generate a sequence of ever-improving points $x^0, x^1, ..., x^k, ...$ approaching the solution set. For many optimization problems and/or algorithms, the sequence will never exactly reach the solution set. One theory of iterative algorithms, referred to as local or asymptotic convergence analysis, is concerned with the rate at which the optimality error of the generated sequence converges to zero.

Obviously, if each iteration of competing algorithms requires the same amount of work, the speed of the convergence of the error reflects the speed of the algorithm. This convergence rate, although it holds locally or asymptotically, provides evaluation and comparison of different algorithms. It has been widely used by the nonlinear optimization and numerical analysis community as an efficiency criterion. In many cases, this criterion does explain practical behavior of iterative algorithms.

Consider a sequence of real numbers $\{r^k\}$ converging to zero. One can define several notions related to the speed of convergence of such a sequence.

Definition 1.1 . *Let the sequence $\{r^k\}$ converge to zero. The order of convergence of $\{r^k\}$ is defined as the supremum of the nonnegative numbers p satisfying*
$$0 \le \limsup_{k \to \infty} \frac{|r^{k+1}|}{|r^k|^p} < \infty.$$

Definition 1.2 . *Let the sequence $\{r^k\}$ converge to zero such that*
$$\limsup_{k \to \infty} \frac{|r^{k+1}|}{|r^k|^2} < \infty.$$
Then, the sequence is said to converge quadratically to zero.

It should be noted that the order of convergence is determined only by the properties of the sequence that holds as $k \to \infty$. In this sense we might say that the order of convergence is a measure of how good the tail of $\{r^k\}$ is. Large values of p imply the faster convergence of the tail.

Definition 1.3 . *Let the sequence $\{r^k\}$ converge to zero such that*

$$\limsup_{k \to \infty} \frac{|r^{k+1}|}{|r^k|} = \beta < 1.$$

Then, the sequence is said to converge linearly *to zero with convergence ratio β.*

Linear convergence is the most important type of convergence behavior. A linearly convergence sequence, with convergence ratio β, can be said to have a tail that converges to zero at least as fast as the geometric sequence $c\beta^k$ for a fixed number c. Thus, we also call linear convergence *geometric* convergence.

As a rule, when comparing the relative effectiveness of two competing algorithms both of which produce linearly convergent sequences, the comparison is based on their corresponding convergence ratio—the smaller the ratio, the faster the algorithm. The ultimate case where $\beta = 0$ is referred to as *superlinear* convergence.

Example 1.17 *The steepest descent algorithm for solving problem (1.7) in Example 1.15 has a linear convergence with convergence ratio $\xi(Q)$. If Q has equal eigenvalues, then the algorithm is superlinearly convergent.*

Example 1.18 *Consider the conjugate gradient algorithm for solving problem (1.7). Starting from an $x^0 \in \mathcal{R}^n$ and $d^0 = Qx^0 + c$, the method uses iterative formula*

$$x^{k+1} = x^k - \alpha^k d^k$$

where

$$\alpha^k = \frac{(d^k)^T(Qx^k + c)}{\|d^k\|_Q^2},$$

and

$$d^{k+1} = Qx^{k+1} - \theta^k d^k$$

where

$$\theta^k = \frac{(d^k)^T Q(Qx^{k+1} + c)}{\|d^k\|_Q^2}.$$

This algorithm is superlinearly convergent (in fact, it converges in finite number of steps).

1.5 Basic Computational Procedures

There are several basic numerical problems frequently solved by interior-point algorithms.

1.5.1 Gaussian elimination method

Probably the best-known algorithm for solving a system of linear equations is the Gaussian elimination method. Suppose we want to solve

$$Ax = b.$$

We may assume $a_{11} \neq 0$ after some row switching, where a_{ij} is the component of A in row i and column j. Then we can subtract appropriate multiples of the first equation from the other equations so as to have an equivalent system:

$$\begin{pmatrix} a_{11} & A_{1.} \\ 0 & A' \end{pmatrix} \begin{pmatrix} x_1 \\ x' \end{pmatrix} = \begin{pmatrix} b_1 \\ b' \end{pmatrix}.$$

This is a pivot step, where a_{11} is called a *pivot*, and A' is called a *Schur complement*. Now, recursively, we solve the system of the last $m-1$ equations for x'. Substituting the solution x' found into the first equation yields a value for x_1. The last process is called *back-substitution*.

In matrix form, the Gaussian elimination method transforms A into the form

$$\begin{pmatrix} U & C \\ 0 & 0 \end{pmatrix}$$

where U is a nonsingular, upper-triangular matrix,

$$A = L \begin{pmatrix} U & C \\ 0 & 0 \end{pmatrix},$$

and L is a nonsingular, lower-triangular matrix. This is called the *LU-decomposition*.

Sometimes, the matrix is transformed further to a form

$$\begin{pmatrix} D & C \\ 0 & 0 \end{pmatrix}$$

where D is a nonsingular, diagonal matrix. This whole procedure uses about nm^2 arithmetic operations. Thus, it is a strong polynomial-time algorithm.

1.5.2 Choleski decomposition method

Another useful method is to solve the least squares problem:

$$(LS) \quad \text{minimize} \quad \|A^T y - c\|.$$

The theory says that y^* minimizes $\|A^T y - c\|$ if and only if

$$AA^T y^* = Ac.$$

So the problem is reduced to solving a system of linear equations with a symmetric semi-positive definite matrix.

One method is Choleski's decomposition. In matrix form, the method transforms AA^T into the form

$$AA^T = L\Lambda L^T,$$

where L is a lower-triangular matrix and Λ is a diagonal matrix. (Such a transformation can be done in about nm^2 arithmetic operations as indicated in the preceding section.) L is called the *Choleski factor* of AA^T. Thus, the above linear system becomes

$$L\Lambda L^T y^* = Ac,$$

and y^* can be obtained by solving two triangle systems of linear equations.

1.5.3 The Newton method

The Newton method is used to solve a system of nonlinear equations: given $f(x) : \mathcal{R}^n \to \mathcal{R}^n$, the problem is to solve n equations for n unknowns such that

$$f(x) = 0.$$

The idea behind Newton's method is to use the Taylor linear approximation at the current iterate x^k and let the approximation be zero:

$$f(x) \simeq f(x^k) + \nabla f(x^k)(x - x^k) = 0.$$

The Newton method is thus defined by the following iterative formula:

$$x^{k+1} = x^k - \alpha(\nabla f(x^k))^{-1} f(x^k),$$

where scalar $\alpha \geq 0$ is called *step-size*. Rarely, however, is the Jacobian matrix ∇f inverted. Generally the system of linear equations

$$\nabla f(x^k) d_x = -f(x^k)$$

is solved and $x^{k+1} = x^k + \alpha d_x$ is used. The direction vector d_x is called a *Newton step*, which can be carried out in strongly polynomial time.

A modified or quasi Newton method is defined by

$$x^{k+1} = x^k - \alpha M^k f(x^k),$$

1.5. BASIC COMPUTATIONAL PROCEDURES

where M^k is an $n \times n$ symmetric matrix. In particular, if $M^k = I$, the method is called the *gradient method*, where f is viewed as the gradient vector of a real function.

The Newton method has a superior asymptotic convergence order equal 2 for $\|f(x^k)\|$. It is frequently used in interior-point algorithms, and believed to be the key to their effectiveness.

1.5.4 Solving ball-constrained linear problem

The ball-constrained linear problem has the following form:

$$(BP) \quad \text{minimize} \quad c^T x$$
$$\text{subject to} \quad Ax = 0, \ \|x\|^2 \leq 1,$$

or

$$(BD) \quad \text{minimize} \quad b^T y$$
$$\text{subject to} \quad \|A^T y\|^2 \leq 1.$$

x^* minimizes (BP) if and only if there always exists a y such that they satisfy
$$AA^T y = Ac,$$
and if $c - A^T y \neq 0$ then
$$x^* = -(c - A^T y)/\|c - A^T y\|;$$

otherwise any feasible x is a solution. The solution y^* for (BD) is given as follows: Solve
$$AA^T \bar{y} = b,$$
and if $\bar{y} \neq 0$ then set
$$y^* = -\bar{y}/\|A^T \bar{y}\|;$$

otherwise any feasible y is a solution. So these two problems can be reduced to solving a system of linear equations.

1.5.5 Solving ball-constrained quadratic problem

The ball-constrained quadratic problem has the following form:

$$(BP) \quad \text{minimize} \quad (1/2)x^T Q x + c^T x$$
$$\text{subject to} \quad Ax = 0, \ \|x\|^2 \leq 1,$$

or simply

$$(BD) \quad \text{minimize} \quad (1/2)y^T Q y + b^T y$$
$$\text{subject to} \quad \|y\|^2 \leq 1.$$

This problem is used by the classical *trust region* method for nonlinear optimization. The optimality conditions for the minimizer y^* of (BD) are

$$(Q + \mu^* I)y^* = -b, \quad \mu^* \geq 0, \quad \|y^*\|^2 \leq 1, \quad \mu^*(1 - \|y^*\|^2) = 0,$$

and

$$(Q + \mu^* I) \succeq 0.$$

These conditions are necessary and sufficient. This problem can be solved in polynomial time $\log(1/\epsilon)$ and $\log(\log(1/\epsilon))$ by the bisection method or a hybrid of the bisection and Newton methods, respectively. In practice, several trust region procedures have been very effective in solving this problem.

The ball-constrained quadratic problem will be used an a sub-problem by several interior-point algorithms in solving complex optimization problems. We will discuss them later in the book.

1.6 Notes

The term "complexity" was introduced by Hartmanis and Stearns [180]. Also see Garey and Johnson [131] and Papadimitriou and Steiglitz [337]. The NP theory was due to Cook [88] and Karp [219]. The importance of P was observed by Edmonds [106].

Linear programming and the simplex method were introduced by Dantzig [93]. Other inequality problems and convexity theories can be seen in Gritzmann and Klee [169], Grötschel, Lovász and Schrijver [170], Grünbaum [171], Rockafellar [364], and Schrijver [373]. Various complementarity problems can be found found in Cottle, Pang and Stone [91]. The positive semi-definite programming, an optimization problem in nonpolyhedral cones, and its applications can be seen in Nesterov and Nemirovskii [327], Alizadeh [9], and Boyd, Ghaoui, Feron and Balakrishnan [72]. Recently, Goemans and Williamson [141] obtained several breakthrough results on approximation algorithms using positive semi-definite programming. The KKT condition for nonlinear programming was given by Karush, Kuhn and Tucker [240].

It was shown by Klee and Minty [223] that the simplex method is not a polynomial-time algorithm. The ellipsoid method, the first polynomial-time algorithm for linear programming with rational data, was proven by Khachiyan [221]; also see Bland, Goldfarb and Todd [63]. The method was devised independently by Shor [380] and by Nemirovskii and Yudin [321]. The interior-point method, another polynomial-time algorithm for linear programming, was developed by Karmarkar. It is related to the classical barrier-function method studied by Frisch [126] and Fiacco and McCormick [116]; see Gill, Murray, Saunders, Tomlin and Wright [139], and Anstreicher [23]. For a brief LP history, see the excellent article by Wright [457].

The real computation model was developed by Blum, Shub and Smale [66] and Nemirovskii and Yudin [321]. The average setting can be seen in Traub, Wasilkowski and Wozniakowski [420]. The asymptotic convergence rate and ratio can be seen in Luenberger [248], Ortega and Rheinboldt [334], and Traub [419]. Other complexity issues in numerical optimization were discussed in Vavasis [450].

Many basic numerical procedures listed in this chapter can be found in Golub and Van Loan [154]. The ball-constrained quadratic problem and its solution methods can be seen in Moré [308], Sorenson [387], and Dennis and Schnable [96]. The complexity result of the ball-constrained quadratic problem was proved by Vavasis [450] and Ye [469, 473].

1.7 Exercises

1.1 *Let $Q \in \mathcal{R}^{n \times n}$ be a given nonsingular matrix, and a and b be given $\mathcal{R}^n$ vectors. Show*

$$(Q + ab^T)^{-1} = Q^{-1} - \frac{1}{1 + b^T Q^{-1} a} Q^{-1} ab^T Q^{-1}.$$

This formula is called the **Sherman-Morrison-Woodbury formula**.

1.2 *Prove that the eigenvalues of all matrices $Q \in \mathcal{M}^{n \times n}$ are real. Furthermore, show that Q is PSD if and only if all its eigenvalues are nonnegative, and Q is PD if and only if all its eigenvalues are positive.*

1.3 *Using the ellipsoid representation in Section 1.2.2, find the matrix Q and vector y that describes the following ellipsoids:*

1. *The 3-dimensional sphere of radius 2 centered at the origin;*

2. *The 2-dimensional ellipsoid centered at $(1; 2)$ that passes the points $(0; 2)$, $(1; 0)$, $(2; 2)$, and $(1; 4)$;*

3. *The 2-dimensional ellipsoid centered at $(1; 2)$ with axes parallel to the line $y = x$ and $y = -x$, and passing through $(-1; 0)$, $(3; 4)$, $(0; 3)$, and $(2; 1)$.*

1.4 *Show that the biggest coordinate-aligned ellipsoid that is entirely contained in $\mathcal{R}^n_+$ and has its center at $x^a \in \overset{\circ}{\mathcal{R}}{}^n_+$ can be written as:*

$$E(x^a) = \{x \in \mathcal{R}^n : \ \|(X^a)^{-1}(x - x^a)\| \leq 1\}.$$

1.5 Show that the non-negative orthant, the positive semi-definite cone, and the second-order cone are all self-dual.

1.6 Consider the convex set $C = \{x \in \mathcal{R}^2 : (x_1 - 1)^2 + (x_2 - 1)^2 \leq 1\}$ and let $y \in \mathcal{R}^2$. Assuming $y \notin C$,

1. Find the point in C that is closest to y;

2. Find a separating hyperplane vector as a function of y.

1.7 Using the idea of Exercise 1.6, prove the separating hyperplane theorem 1.3.

1.8 Given an $m \times n$ matrix A and a vector $c \in \mathcal{R}^n$, consider the function $\mathcal{B}(y) = \sum_{j=1}^n \log s_j$ where $s = c - A^T y > 0$. Find $\nabla \mathcal{B}(y)$ and $\nabla^2 \mathcal{B}(y)$ in terms of s.

1.9 Prove that the level set of a quasi-convex function is convex.

1.10 Prove Propositions 1.5 and 1.6 for convex functions in Section 1.2.3.

1.11 Prove the Harmonic inequality described in Section 1.2.4.

1.12 Prove Farkas' lemma 1.8 for linear equations.

1.13 Prove the linear least-squares problem always has a solution.

1.14 Let $P = A^T(AA^T)^{-1}A$ or $P = I - A^T(AA^T)^{-1}A$. Then prove

1. $P = P^2$.

2. P is positive semi-definite.

3. The eigenvalues of P are either 0 or 1.

1.15 Using the separating theorem, prove Farkas' lemmas 1.9 and 1.10.

1.16 If a system $A^T y \leq c$ of linear inequalities in m variables has no solution, show that $A^T y \leq c$ has a subsystem $(A')^T y \leq c'$ of at most $m+1$ inequalities having no solution.

1.17 Prove the LP fundamental theorem 1.15.

1.18 If (LP) and (LD) have a nondegenerate optimal basis A_B, prove that the strict complementarity partition in Theorem 1.14 is

$$P^* = B.$$

1.7. EXERCISES

1.19 If Q is positive semi-definite, prove that x^* is an optimal solution for (QP) if and only if x^* is a KKT point for (QP).

1.20 Show that M of (1.6) is monotone if and only if Q is positive semi-definite.

1.21 Prove the monotone linear complementarity theorem 1.16.

1.22 Let $P(y) = -C + \sum_i^m y_i A_i$, where C and A_i, $i = 1, \ldots, m$, are given symmetric matrices. Formulate the minimization of the max-eigenvalue of $P(y)$ as a (PSD) problem. What does its primal problem look like?

1.23 Prove $X \bullet S \geq 0$ if both X and S are positive semi-definite matrices.

1.24 Prove that two positive semi-definite matrices are complementary to each other, $X \bullet S = 0$, if and only if $XS = 0$.

1.25 Prove Farkas' lemma 1.17 for positive semi-definite programming.

1.26 Let both (LP) and (LD) for a given data set (A, b, c) have interior feasible points. Then consider the level set

$$\Omega(z) = \{y : c - A^T y \geq 0, -z + b^T y \geq 0\}$$

where $z < z^*$ and z^* designates the optimal objective value. Prove that $\Omega(z)$ is bounded and has an interior for any finite $z < z^*$, even $\mathcal{F}_d$ is unbounded.

1.27 Given an (LP) data set (A, b, c) and an interior feasible point x^0, find the feasible direction d_x ($Ad_x = 0$) that achieves the steepest decrease in the objective function.

1.28 Given an (LP) data set (A, b, c) and a feasible point $(x^0, y^0, s^0) \in (\mathcal{R}_+^n, \mathcal{R}^m, \mathcal{R}_+^n)$ for the primal and dual, and ignoring the nonnegativity condition, write the systems of linear equations used to calculate the Newton steps for finding points that satisfy the optimality equations (1.2), (1.3), and (1.4), respectively.

1.29 Similar to our discussion on quadratic programming and linear complementarity, demonstrate that finding a KKT point of a convex nonlinear programming problem can be reduced to solving a monotone complementarity problem with possible "free" variables.

1.30 Show that the ball-constrained linear problem (BP) can be written in the (BD) form and write the KKT conditions for both of them.

1.31 *Given a scalar $\alpha > 0$, a positive diagonal matrix S, and an $m \times n$ matrix A, find the formula for $y \in \mathcal{R}^m$ such that*

$$\begin{array}{ll} \text{minimize} & e^T S^{-1} A^T y \\ \text{subject to} & \|S^{-1} A^T y\|^2 \leq \alpha^2. \end{array}$$

1.32 *Show the optimality conditions for the minimizer y^* of (BD) in Section 1.5.5:*

$$(Q + \mu^* I) y^* = -b, \quad \mu^* \geq 0, \quad \|y^*\| \leq 1, \quad \mu^*(1 - \|y^*\|) = 0,$$

and

$$(Q + \mu^* I) \succeq 0,$$

are necessary and sufficient.

Chapter 2

Geometry of Convex Inequalities

Most optimization algorithms are iterative in nature, that is, they generate a sequence of improved points. Algorithm design is closely related to how the improvement is measured. Most optimization algorithms use a merit or descent function to measure the progress. Some merit or descent functions are based on the objective function. For example, if we know a lower bound z of the optimal objective value, $f(x) - z$ is a measure of how far x is from the solution set. Another example measures the residual or error of the optimality conditions represented by a system of equations and inequalities involving the derivatives of the objective and constraint functions, as discussed in the preceding chapter.

One particular merit or descent function measures the "size" of the containing set—a set that contains a solution. A typical example is the bisection method for finding a root of a continuous function within an interval. The method measures the length of the containing interval. In each step, the middle point of the containing interval is tested, and, subsequently, a new containing interval is selected and its length is a half of the previous one. Thus, these containing intervals shrink at a constant rate $1/2$.

A generic central-section algorithm for multiple-variable problems can be described as follows: Given x^k, a "good" interior point in a containing set, we check to see if x^k is desirable. If not, we generate a separating hyperplane and place it through x^k and cut the containing set into two parts. If we can assure that the solution set lies in one of the two parts, then the other part can be deleted. This leads to a new containing set that is smaller than the previous one. A new "good" interior point in the new containing set can be tested and the process continues. Obviously, this method can be applied to any convex problems.

The question that arises is how to select the test point and where to place the cut. Ideally, we would like to select a "center" point that divides the containing set into two approximately "equal" parts with respect to certain measures of the containing set. Then, we will have a shrinking rate of about 1/2 for the sequence of the containing sets.

Various centers were considered as test points. In this chapter, we review these centers and their associated measures. We show that, similar to these central-section algorithms, interior-point algorithms use a new measure of the containing set represented by linear inequalities. This measure is "*analytic*" and is relatively easy to compute. Its associated center is called the *analytic center*.

2.1 Convex Bodies

A natural choice of the measure would be the volume of the convex body. Interest in measure of convex bodies dates back as far as the ancient Greeks and Chinese who computed centers, areas, perimeters and curvatures of circles, triangle, and polygons. Unlike the length of a line segment, the computation of volumes, even in two and three dimensional spaces, is not an easy task for a slightly complex shaped body. In order to measure cultivated lands, Chinese farmers weighted the amount of sand contained in a down-scaled body whose shape is identical to the land, then compared the weight to the amount of sand in an equally down-scaled unit-square box.

2.1.1 Center of gravity

Associated with the volume of a convex body, the center of gravity will be the choice as the test point (Figure 2.1). We have the following theorem:

Theorem 2.1 *Let Ω be a compact convex body in $\mathcal{R}^m$ with center of gravity y^g, and let Ω^+ and Ω^- be the bodies in which a hyperplane H passing through y^g divides Ω. Then the volumes $V(\Omega^+)$ and $V(\Omega^-)$ satisfy the inequality*

$$V(\Omega^*) \leq \left(1 - (1 - \frac{1}{n+1})^n\right) V(\Omega), \quad where \quad * = + \ or \ -.$$

This result shows that by successively cutting through the center of gravity, these convex bodies shrink at a constant rate of at most (1 −

2.1. CONVEX BODIES

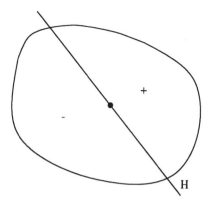

Figure 2.1. A hyperplane H cuts through the center of gravity of a convex body.

$1/\exp(1))$, where $\exp(1)$ is the natural number 2.718.... This rate is just slightly worse than $1/2$ in the bisection method.

Let the solution set contain a ball with radius r and the initial containing set be contained in a ball with radius R. Then we know that the volumes of the containing sets are bounded from below by $\pi_m r^m$ and bounded from above by $\pi_m R^m$, where π_m is the volume of the unit ball in $\mathcal{R}^m$. Thus, a solution point must be found in $O(m \log(R/r))$ central-section steps, because the volumes of the containing sets eventually become too small to contain the solution set.

The difficulty with the gravity-center section method lies in computing the center and volume of a convex body. It is well known that computing the volume of a convex polytope, either given as a list of facets or vertices, is as difficult as computing the permanent of a matrix, which is itself very hard (called *#P-Hard*). Since the computation of the center of gravity is closely related to the volume computation, it seems reasonable to conclude that no efficient algorithm can compute the center of gravity of Ω.

Although computing the center of gravity is difficult for general convex bodies, it is relatively easy for some simple convex bodies like a cube, a simplex, or an ellipsoid. This leads researchers to use some simple convex bodies to estimate Ω.

2.1.2 Ellipsoids

It is known that every convex body contains a unique ellipsoid of maximal volume and is contained in a unique ellipsoid of minimal volume (Figure

2.2). We have the following general theorem:

Theorem 2.2 *For every full dimensional (bounded) convex body $\Omega \subset \mathcal{R}^m$ there exists a unique ellipsoid $\underline{E}(\Omega)$ of maximal volume contained in Ω. Moreover, Ω is contained in the ellipsoid obtained from $\underline{E}(\Omega)$ by enlarging it from its center by a factor of m.*

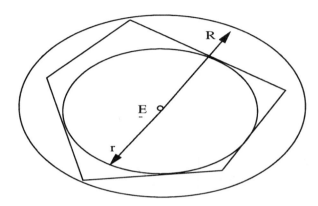

Figure 2.2. The max-volume ellipsoid inscribing a polytope and the concentric ellipsoid circumscribing the polytope ($R/r \leq m$).

Let $\underline{y}^e$ be the center of the max-volume ellipsoid inscribing Ω. Through the center we place a hyperplane H and divide Ω into two bodies Ω^+ and Ω^-. Then, we have a central-section theorem:

Theorem 2.3 *The Volumes of the new ellipsoids satisfy the inequality*

$$V(\underline{E}(\Omega^*)) \leq 0.843 V(\underline{E}(\Omega)), \quad \text{where} \quad * = + \text{ or } -.$$

Thus, one can use the max-volume inscribing ellipsoid as an estimate of Ω. These ellipsoids will shrink at a constant rate in the central-section method. The volume of $\underline{E}(\Omega)$ is also bounded from below by $\pi_m r^m$ and bounded from above by $\pi_m R^m$. Thus, a solution point must be found in $O(m \log(R/r))$ central-section steps. Apparently, to compute $\underline{y}^e$ one needs to use the structure of the convex body. If the convex body is represented by linear inequalities, there is a polynomial complexity bound for computing an approximate point of $\underline{y}^e$.

Another approach is the original ellipsoid method, which monitors the volume of an ellipsoid that contains the solution set. This is based on a similar theorem:

2.1. CONVEX BODIES

Theorem 2.4 *For every (bounded) convex body $\Omega \subset \mathcal{R}^m$ there exists a unique ellipsoid $\overline{E}(\Omega)$ of minimal volume containing Ω. Moreover, Ω contains the ellipsoid obtained from $\overline{E}(\Omega)$ by shrinking it from its center by a factor of m.*

Moreover, let y^e be the center of an ellipsoid $E \subset \mathcal{R}^m$. Through its centers we place a hyperplane H and divide E into two bodies (half ellipsoids) E^+ and E^-. Let $\overline{E}(E^+)$ and $\overline{E}(E^-)$ be the new min-volume ellipsoids containing E^+ and E^-, respectively (Figure 2.3). Then, we have the following central-section theorem:

Theorem 2.5 *The Volumes of the new ellipsoids satisfy the inequality*

$$V(\overline{E}(E^*)) \leq \exp(-.5/(m+1))V(E), \quad \text{where} \quad * = + \text{ or } -.$$

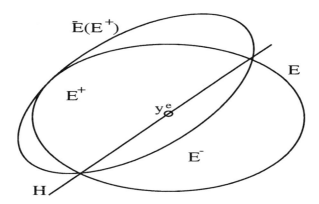

Figure 2.3. Illustration of the min-volume ellipsoid containing a half ellipsoid.

In algebra, let the ellipsoid

$$E = \{y \in \mathcal{R}^m : (y - y^e)^T Q(y - y^e) \leq 1\}.$$

Then, $\overline{E}$ containing the half ellipsoid $\{y \in E : a^T y \leq a^T y^e\}$ is given by

$$\overline{E} = \{y \in \mathcal{R}^m : (y - \bar{y}^e)^T \bar{Q}(y - \bar{y}^e) \leq 1\},$$

where

$$\bar{y}^e = y^e - \frac{1}{m+1} \frac{Qa}{\sqrt{a^T Q a}}$$

and
$$\bar{Q} = \frac{m^2}{m^2 - 1}\left(Q - \frac{2}{m+1}\frac{Qaa^TQ}{a^TQa}\right).$$

Therefore, the new containing ellipsoid can be easily constructed and its center can be computed in $O(m^2)$ arithmetic operations. Since the volumes of the ellipsoids are bounded from below by $\pi_m r^m$ and the initial one is bounded from above by $\pi_m R^m$, a solution point must be found in $O(m^2 \log(R/r))$ central-section steps. We see that the ellipsoid method does not keep the "knowledge" of the cutting plane after the new containing ellipsoid is updated.

2.2 Analytic Center

The centers discussed in the preceding section are "universal," meaning that they are invariant of the representation of a convex body. A drawback of these centers is that they generally cannot be computed cost-effectively. For the ellipsoid method, its advantage in not keeping knowledge of the cutting planes is also a disadvantage to practical efficiency for solving certain problems, such as linear programs. Thus, another type of center, called the *analytic center for a convex polyhedron given by linear inequalities*, was introduced.

2.2.1 Analytic center

Let Ω be a bounded polytope in $\mathcal{R}^m$ represented by $n\ (>m)$ linear inequalities, i.e.,
$$\Omega = \{y \in \mathcal{R}^m : c - A^T y \geq 0\},$$
where $A \in \mathcal{R}^{m \times n}$ and $c \in \mathcal{R}^n$ are given and A has rank m. Denote the interior of Ω by
$$\overset{\circ}{\Omega} = \{y \in \mathcal{R}^m : c - A^T y > 0\}.$$

Given a point in Ω, let a "position" function of y satisfy

1. $d(y, \Omega) = 0$ if y is on the boundary of Ω;

2. $d(y, \Omega) > 0$ if y is in $\overset{\circ}{\Omega}$;

3. If $c' \leq c$ and let $\Omega' = \{y \in \mathcal{R}^m : c' - A^T y \geq 0\}$ (thereby, $\Omega' \subset \Omega$), then $d(y, \Omega') \leq d(y, \Omega)$.

2.2. ANALYTIC CENTER

This function is dependent on the analytic representations of Ω, not just the set. We may have $d(y, \Omega') \leq d(y, \Omega)$ even when $\Omega \subset \Omega'$ geometrically. Thus, $d(y, \Omega)$ is really a function of point y, and the data A and c as well.

One choice of the functions is

$$d(y, \Omega) = \prod_{j=1}^{n}(c_j - a_j^T y), \quad y \in \Omega,$$

where $a_{.j}$ is the jth column of A. Traditionally, we let $s := c - A^T y$ and call it a *slack vector*. Thus, the function is the product of all slack variables. Its logarithm is called the *(dual) potential function*,

$$\mathcal{B}(y, \Omega) := \log d(y, \Omega) = \sum_{j=1}^{n} \log(c_j - a_{.j}^T y) = \sum_{j=1}^{n} \log s_j, \quad (2.1)$$

and $-\mathcal{B}(y, \Omega)$ is the classical logarithmic barrier function. For convenience, in what follows we may write $\mathcal{B}(s, \Omega)$ to replace $\mathcal{B}(y, \Omega)$ where s is always equal to $c - A^T y$.

Example 2.1 *Let $A = (1, -1)$ and $c = (1; 1)$. Then the set of Ω is the interval $[-1, 1]$. Let $A' = (1, -1, -1)$ and $c' = (1; 1; 1)$. Then the set of Ω' is also the interval $[-1, 1]$. Note that*

$$d(-1/2, \Omega) = (3/2)(1/2) = 3/4 \quad \text{and} \quad \mathcal{B}(-1/2, \Omega) = \log(3/4),$$

and

$$d(-1/2, \Omega') = (3/)(1/2)(1/2) = 3/8 \quad \text{and} \quad \mathcal{B}(-1/2, \Omega') = \log(3/8).$$

The interior point, denoted by y^a and $s^a = c - A^T y^a$, in Ω that maximizes the potential function is called the *analytic center* of Ω, i.e.,

$$\mathcal{B}(\Omega) := \mathcal{B}(y^a, \Omega) = \max_{y \in \Omega} \log d(y, \Omega).$$

(y^a, s^a) is uniquely defined, since the potential function is strictly concave in a bounded convex $\overset{\circ}{\Omega}$. Setting $\nabla \mathcal{B}(y, \Omega) = 0$ and letting $x^a = (S^a)^{-1} e$, the analytic center (y^a, s^a) together with x^a satisfy the following optimality conditions:

$$\begin{aligned} Xs &= e \\ Ax &= 0 \\ -A^T y - s &= -c. \end{aligned} \quad (2.2)$$

Note that adding or deleting a redundant inequality changes the location of the analytic center.

Example 2.2 *Consider* $\Omega = \{y \in R : -y \leq 0, \ y \leq 1\}$, *which is interval* $[0,1]$. *The analytic center is* $y^a = 1/2$ *with* $x^a = (2,2)^T$.
Consider
$$\Omega' = \{y \in R : \overbrace{-y \leq 0, \cdots, -y \leq 0}^{n \text{ times}}, \ y \leq 1\},$$
which is, again, interval $[0,1]$ *but* "$-y \leq 0$" *is copied n times. The analytic center for this system is* $y^a = n/(n+1)$ *with* $x^a = ((n+1)/n, \cdots, (n+1)/n, \ (n+1))^T$.

The analytic center can be defined when the interior is empty or equalities are presented, such as
$$\Omega = \{y \in \mathcal{R}^m : c - A^T y \geq 0, \ By = b\}.$$
Then the analytic center is chosen on the hyperplane $\{y : By = b\}$ to maximize the product of the slack variables $s = c - A^T y$. Thus, the interior of Ω is not used in the sense that the topological interior for a set is used. Rather, it refers to the interior of the positive orthant of slack variables: $\mathcal{R}^n_+ := \{s : s \geq 0\}$. When say Ω has an interior, we mean that
$$\overset{\circ}{\mathcal{R}^n_+} \cap \{s : s = c - A^T y \text{ for some } y \text{ where } By = b\} \neq \emptyset.$$
Again $\overset{\circ}{\mathcal{R}^n_+} := \{s \in \mathcal{R}^n_+ : s > 0\}$, i.e., the interior of the orthant $\mathcal{R}^n_+$. Thus, if Ω has only a single point y with $s = c - A^T y > 0$, we still say $\overset{\circ}{\Omega}$ is not empty.

Example 2.3 *Consider the system* $\Omega = \{x : Ax = 0, \ e^T x = n, \ x \geq 0\}$, *which is called* **Karmarkar's** *canonical set. If* $x = e$ *is in* Ω *then* e *is the analytic center of* Ω, *the intersection of the simplex* $\{x : e^T x = n, \ x \geq 0\}$ *and the hyperplane* $\{x : Ax = 0\}$ *(Figure 2.4)*.

2.2.2 Dual potential function

In this section we present some geometry of the dual potential function. We may describe $\Omega = \{y \in \mathcal{R}^m : c - A^T y \geq 0\}$ using only the slack variable s:
$$\mathcal{S}_\Omega := \{s \in \mathcal{R}^n : A^T y + s = c \text{ for some } y, \ s \geq 0\},$$
or
$$\mathcal{S}_\Omega := \{s \in \mathcal{R}^n : s - c \in \mathcal{R}(A^T), \ s \geq 0\},$$

2.2. ANALYTIC CENTER

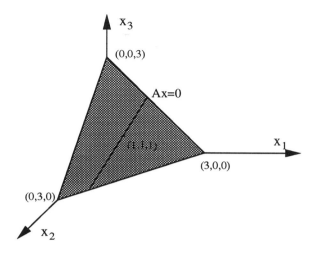

Figure 2.4. Illustration of the Karmarkar (simplex) polytope and its analytic center.

which is the intersection of the affine set

$$\mathcal{A}_\Omega = \{s \in \mathcal{R}^n : s - c \in \mathcal{R}(A^T)\}$$

and the positive cone (orthant) $\mathcal{R}^n_+$. The interior of $\mathcal{S}_\Omega$ is denoted by

$$\overset{\circ}{\mathcal{S}}_\Omega := \mathcal{A}_\Omega \cap \overset{\circ}{\mathcal{R}}^n_+.$$

Let s be an interior point in $\mathcal{S}_\Omega$. Then consider the ellipsoid

$$E_s = \{t \in \mathcal{R}^n : \|S^{-1}(t-s)\| \le 1\}.$$

This is a coordinate-aligned ellipsoid centered at s and inscribing the positive orthant $\mathcal{R}^n_+$. The volume of the coordinate-aligned ellipsoid is

$$V(E_s) = \pi_n \prod_{j=1}^n s_j.$$

Moreover, we have

$$(E_s \cap \mathcal{A}_\Omega) \subset \mathcal{S}_\Omega,$$

that is, the intersection of E_s and $\mathcal{A}_\Omega$ is contained in $\mathcal{S}_\Omega$ (Figure 2.5).

Thus, the potential function value, $\mathcal{B}(s, \Omega)$, at an $s \in \overset{\circ}{\mathcal{S}}_\Omega$ plus $\log \pi_n$ is the logarithmic volume of the coordinate-aligned ellipsoid centered at s and

CHAPTER 2. GEOMETRY OF CONVEX INEQUALITIES

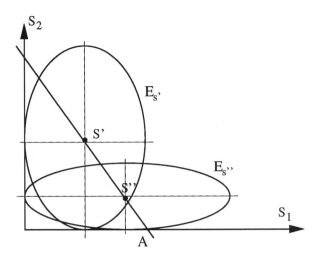

Figure 2.5. Coordinate-aligned (dual) ellipsoids centered at points s's on the intersection of an affine set A and the positive orthant; they are also contained by the positive orthant.

inscribing $\mathcal{R}_+^n$. Therefore, the inscribing coordinate-aligned ellipsoid centered at the analytic center of $\mathcal{S}_\Omega$, among all of these inscribing coordinate-aligned ellipsoids, has the maximal volume. We denote this max-potential of $\mathcal{S}_\Omega$ by $\mathcal{B}(\Omega)$.

We now argue that the exponential of $\mathcal{B}(\Omega)$ is an "analytic" measure of $\mathcal{S}_\Omega$ or Ω.

1. $\exp(\mathcal{B}(\Omega)) = 0$ if $\overset{\circ}{\Omega} = \emptyset$;

2. $\exp(\mathcal{B}(\Omega)) > 0$ if $\overset{\circ}{\Omega} \neq \emptyset$, and if Ω contains a full dimensional ball with radius r or $\{y : A^T y \leq c - re\} \neq \emptyset$ (here we assume that $\|a_{.j}\| = 1$ for $j = 1, 2, ..., n$), then $\mathcal{B}(\Omega) \geq n \log r$;

3. If $c' \leq c$ and let $\Omega' = \{y \in \mathcal{R}^m : c' - A^T y \geq 0\}$ (thereby, $\Omega' \subset \Omega$), then $\mathcal{B}(\Omega') \leq \mathcal{B}(\Omega)$.

Note that the max-potential $\mathcal{B}(\Omega)$ is now a function of data A and c.

Let s^a (or y^a) be the analytic center of $\mathcal{S}_\Omega$ (or Ω). We now consider the ellipsoid
$$nE_{s^a} = \{t \in \mathcal{R}^n : \|(S^a)^{-1}(t - s^a)\| \leq n\},$$

2.2. ANALYTIC CENTER

which is enlarged from E_{s^a} by a factor of n. The question is whether or not this enlarged coordinate-aligned ellipsoid $nE_{s^a} \cap \mathcal{A}_\Omega$ contains $\mathcal{S}_\Omega$. The answer is "yes" according to the following theorem:

Theorem 2.6 *The analytic center $s^a \in \mathcal{S}_\Omega$ is the center of the unique maximal-volume coordinate-aligned ellipsoid E_{s^a} inscribing the orthant $\mathcal{R}_+^n$. Its intersection with $\mathcal{A}_\Omega$ is contained by polytope $\mathcal{S}_\Omega$. Moreover, polytope $\mathcal{S}_\Omega$ itself is contained by the intersection of $\mathcal{A}_\Omega$ and the ellipsoid obtained from E_{s^a} by enlarging it from its center by a factor n.*

Proof. The uniqueness of the analytic center is resulted from the fact that the potential function is strictly concave in the interior of the polytope and A has a full row-rank. Let y^a be the analytic center and $s^a = c - A^T y^a$, then there is x^a such that $A x^a = 0$ and $X^a s^a = e$. Thus, we have $x^a > 0$ and $c^T x^a = n$. For all $s = c - A^T y \geq 0$ we have

$$\|(S^a)^{-1}(s-s^a)\|^2 = \|X^a s - e\|^2 = \|X^a s\|^2 - n \leq ((x^a)^T s)^2 - n = n^2 - n < n^2.$$

This completes the proof.

$\square$

2.2.3 Analytic central-section inequalities

We now develop two central-section inequalities for the analytic center. They resemble the results for the previously discussed centers in that they show the volume of a polytope containing a solution set can be reduced at a geometric rate. First, we study how translating a hyperplane in $\Omega = \{y : c - A^T y \geq 0\}$ will affect the max-potential value. More specifically, we have the following problem: If one inequality in Ω, say the first one, of $c - A^T y \geq 0$ needs to be translated, change $c_1 - a_1^T y \geq 0$ to $a_1^T y^a - a_1^T y \geq 0$; i.e., the first inequality is parallelly translated, and it cuts through the center y^a and divides Ω into two bodies (Figure 2.6). Analytically, c_1 is replaced by $a_1^T y^a$ and the rest of data are unchanged.

Let

$$\Omega^+ := \{y : a_1^T y^a - a_1^T y \geq 0, \ c_j - a_j^T y \geq 0, \ j = 2, ..., n\}$$

and let $\bar{y}^a$ be the analytic center of Ω^+. Then, the max-potential for the new convex polytope Ω^+ is

$$\exp(\mathcal{B}(\Omega^+)) = (a_1^T y^a - a_1^T \bar{y}^a) \prod_{j=2}^n (c_j - a_j^T \bar{y}^a).$$

Regarding $\mathcal{B}(\Omega)$ and $\mathcal{B}(\Omega^+)$, we prove the following theorem:

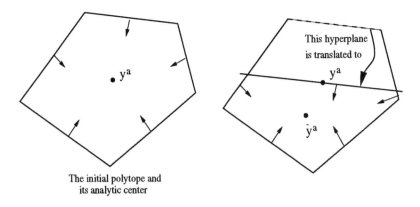

Figure 2.6. Translation of a hyperplane.

Theorem 2.7 *Let Ω and Ω^+ be defined as the above. Then*

$$\mathcal{B}(\Omega^+) \leq \mathcal{B}(\Omega) - 1,$$

or

$$\exp(\mathcal{B}(\Omega^+)) \leq \exp(-1)\exp(\mathcal{B}(\Omega)),$$

where $\exp(-1) = 1/\exp(1)$.

Proof. Since y^a is the analytic center of Ω, there exists $x^a > 0$ such that

$$X^a(c - A^T y^a) = e \quad \text{and} \quad Ax^a = 0. \qquad (2.3)$$

Recall that e is the vector of all ones and X^a designates the diagonal matrix of x^a. Thus, we have

$$s^a = (c - A^T y^a) = (X^a)^{-1} e \quad \text{and} \quad c^T x^a = n.$$

Let $\bar{c}_j = c_j$ for $j = 2, ..., n$ and $\bar{c}_1 = a_1^T y^a$, and let $\bar{s}^a = \bar{c} - A^T \bar{y}^a$. Then, we have

$$\begin{aligned} e^T X^a \bar{s}^a &= e^T X^a (\bar{c} - A^T \bar{y}^a) = e^T X^a \bar{c} \\ &= c^T x^a - x_1^a(c_1 - a_1^T y^a) = n - 1. \end{aligned}$$

Thus,

$$\frac{\exp(\mathcal{B}(\Omega^+))}{\exp(\mathcal{B}(\Omega))} = \prod_{j=1}^{n} \frac{\bar{s}_j^a}{s_j^a}$$

2.2. ANALYTIC CENTER

$$\begin{aligned} &= \prod_{j=1}^{n} x_j^a \bar{s}_j^a \\ &\leq (\frac{1}{n} \sum_{j=1}^{n} x_j^a \bar{s}_j^a)^n \\ &= (\frac{n-1}{n})^n \leq \exp(-1). \end{aligned}$$

□

Now suppose we translate the first hyperplane by a β, $0 \leq \beta \leq 1$, the fractional distance to the analytic center, i.e.,

$$\Omega^+ := \{y : (1-\beta)c_1 + \beta a_1^T y^a - a_1^T y \geq 0,\ c_j - a_j^T y \geq 0,\ j = 2, ..., n\}.$$

If $\beta = 0$, then there is no translation; if $\beta = 1$, then the hyperplane is translated through the analytic center as in the above theorem. Regarding $\mathcal{B}(\Omega)$ and $\mathcal{B}(\Omega^+)$, we have the following inequality:

Corollary 2.8

$$\mathcal{B}(\Omega^+) \leq \mathcal{B}(\Omega) - \beta.$$

Now suppose we translate $k(<n)$ hyperplanes, say $1, 2, ..., k$, cutting through Ω; i.e., use multiple cuts passing through Ω, and let

$$\Omega^+ := \{y : (1-\beta_j)c_j + \beta_j a_j^T y^a - a_j^T y \geq 0,\ j = 1, ..., k,$$

$$c_j - a_j^T y \geq 0,\ j = k+1, ..., n\},$$

where $0 \leq \beta_j \leq 1$, $j = 1, 2, ..., k$. Then, we have the following corollary:

Corollary 2.9

$$\mathcal{B}(\Omega^+) \leq \mathcal{B}(\Omega) - \sum_{j=1}^{k} \beta_j.$$

This corollary will play an important role in establishing the current best complexity result for linear inequality and linear programming problems.

These corollaries show the shrinking nature of the coordinate-aligned ellipsoids after a cut is translated. They enable us to develop an algorithm that resembles the central-section method. Recall that the volume of the maximal coordinate-aligned ellipsoid contained in the polytope $\mathcal{S}_\Omega$ is

$$V(E(\Omega)) := V(E_{s^a}) = \pi_n \cdot \exp(\mathcal{B}(\Omega)).$$

If k hyperplanes are translated with $0 < \beta_j \le 1$, $j = 1, \ldots, k$,

$$V(E(\Omega^+)) \le \exp(-\sum_{j=1}^{k} \beta_j) \cdot V(E(\Omega)).$$

Since $\exp(-\sum_{j=1}^{k} \beta_j) < 1$, the volume shrinks geometrically, establishing the linear convergence of central-section methods.

Again, if a lower bound on the max-potential of the solution set is $n \log r$ and the max-potential of the initial containing set is $n \log R$, then a solution must be found in $O(n \log(R/r))$ central-section steps. Moreover, if we can translate multiple inequalities and the max-potential is reduced by $\theta(\sqrt{n})$ at each step, a solution must be found in $O(\sqrt{n} \log(R/r))$ central-section steps.

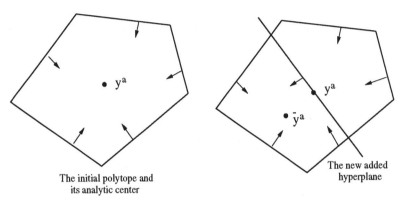

The initial polytope and its analytic center

The new added hyperplane

Figure 2.7. Addition of a new hyperplane.

In the following, we study the problem when placing additional hyperplanes through the analytic center of Ω (Figure 2.7). When a hyperplane is added, the new convex body is represented by

$$\Omega^+ = \{y : c - A^T y \ge 0, \quad a_{n+1}^T y^a - a_{n+1}^T y \ge 0\}.$$

Analytically, c is replaced by $(c; a_{n+1}^T y^a)$ and A is replaced by (A, a_{n+1}).

Again, the question is how the max-potential of the new polytope changes compared to that of Ω. Let

$$r(\Omega)_1 = \sqrt{a_{n+1}^T (A(X^a)^2 A^T)^{-1} a_{n+1}} = \sqrt{a_{n+1}^T (A(S^a)^{-2} A^T)^{-1} a_{n+1}}.$$

Then, we have an inequality as follows:

2.2. ANALYTIC CENTER

Theorem 2.10

$$\mathcal{B}(\Omega^+)) \leq \mathcal{B}(\Omega) + \log(r(\Omega)_1) + 2\log 2 - 1.5.$$

Proof. Again, x^a and (y^a, s^a) satisfy condition (2.3). Let

$$\bar{s}^a = c - A^T \bar{y}^a \quad \text{and} \quad \bar{s}^a_{n+1} = c_{n+1} - a^T_{n+1} \bar{y}^a. \tag{2.4}$$

Then,

$$\begin{aligned}
\bar{s}^a_{n+1} &= a^T_{n+1}(y^a - \bar{y}^a) \\
&= a^T_{n+1}(A(X^a)^2 A^T)^{-1}(A(X^a)^2 A^T)(y^a - \bar{y}^a) \\
&= a^T_{n+1}(A(X^a)^2 A^T)^{-1} A(X^a)^2 (A^T y^a - A^T \bar{y}^a) \\
&= a^T_{n+1}(A(X^a)^2 A^T)^{-1} A(X^a)^2 (-c + A^T y^a + c - A^T \bar{y}^a) \\
&= a^T_{n+1}(A(X^a)^2 A^T)^{-1} A(X^a)^2 (\bar{s}^a - (X^a)^{-1} e) \\
&= a^T_{n+1}(A(X^a)^2 A^T)^{-1} A X^a (X^a \bar{s}^a - e).
\end{aligned} \tag{2.5}$$

Note that we have

$$e^T X^a \bar{s}^a = e^T X^a (c - A^T \bar{y}^a) = e^T X^a c = n. \tag{2.6}$$

Thus, from (2.5)

$$\begin{aligned}
\frac{\exp(\mathcal{B}(\Omega^+))}{\exp(\mathcal{B}(\Omega))r(\Omega)_1} &= \frac{\bar{s}^a_{n+1}}{r(\Omega)_1} \prod_{j=1}^n \frac{\bar{s}^a_j}{s^a_j} \\
&= \frac{a^T_{n+1}(A(X^a)^2 A^T)^{-1} A X^a (X^a \bar{s}^a - e)}{r(\Omega)_1} \prod_{j=1}^n (x^a_j \bar{s}^a_j) \\
&\leq \frac{\|a^T_{n+1}(A(X^a)^2 A^T)^{-1} A X^a\| \|X^a \bar{s}^a - e\|}{r(\Omega)_1} \prod_{j=1}^n (x^a_j \bar{s}^a_j) \\
&= \|X^a \bar{s}^a - e\| \prod_{j=1}^n (x^a_j \bar{s}^a_j).
\end{aligned} \tag{2.7}$$

Let $\alpha = X^a \bar{s}^a$. Then, to evaluate (2.7) together with (2.6) we face a maximum problem

$$\begin{aligned}
\text{maximize} \quad & f(\alpha) = \|\alpha - e\| \prod_{j=1}^n \alpha_j \\
\text{s.t.} \quad & e^T \alpha = n, \; \alpha > 0.
\end{aligned}$$

This maximum is achieved, without loss of generality, at $\alpha_1 = \beta > 1$ and $\alpha_2 = \ldots = \alpha_n = (n-\beta)/(n-1) > 0$ (see Exercise 2.4). Hence,

$$\begin{aligned} f(\alpha) &\leq (\beta-1)\sqrt{\frac{n}{n-1}}\beta\left(\frac{n-\beta}{n-1}\right)^{n-1} \\ &\leq 4\sqrt{\frac{n}{n-1}}\frac{\beta-1}{2}\frac{\beta}{2}\left(\frac{n-\beta}{n-1}\right)^{n-1} \\ &\leq 4\sqrt{\frac{n}{n-1}}\left(\frac{n-.5}{n+1}\right)^{n+1} \\ &\leq \frac{4}{e^{1.5}}. \end{aligned}$$

To derive the last inequality we have verified that $\sqrt{\frac{n}{n-1}}(\frac{n-.5}{n+1})^{n+1}$ is an increasing function of n for $n \geq 2$ and its limit is $\exp(-1.5)$. This completes the proof.

□

Note that $r(\Omega)_1 = \sqrt{a_{n+1}^T(A(S^a)^{-2}A^T)^{-1}a_{n+1}}$ is the square-root of the maximal objective value of the problem:

$$\max \ a_{n+1}^T(y-y^a) \quad s.t. \quad y \in \{y : \|(S^a)^{-1}A^T(y-y^a)\| \leq 1\}.$$

This value can be viewed as the distance from the center to the tangent plane a_{n+1} of the ellipsoid constraint; see Figure 2.8.

A polytope can always be scaled to fit in a unit ball so that $r(\Omega)_1 \leq 1$ and $\log r(\Omega)_1 + 2\log 2 - 1.5 < 0$. Thus, the max-potential of the containing polytope tends linearly to $-\infty$ when appropriate hyperplanes are added.

In general, if k additional hyperplanes cut through the analytic center, i.e.,

$$\Omega^+ = \{y : c - A^T y \geq 0, \ a_{n+1}^T y^a - a_{n+1}^T y \geq 0, \cdots, a_{n+k}^T y^a - a_{n+k}^T y \geq 0\},$$

we have Corollary 2.11.

Corollary 2.11

$$\mathcal{B}(\Omega^+) \leq \mathcal{B}(\Omega) + \sum_{i=1}^{k}\log(r(\Omega)_i) + (k+1)\log(k+1) - (k + \frac{k}{k+1}),$$

where

$$r(\Omega)_i = \sqrt{a_{n+i}^T(A(S^a)^{-2}A^T)^{-1}a_{n+i}}, \quad i = 1,...,k.$$

2.3. PRIMAL AND PRIMAL-DUAL POTENTIAL FUNCTIONS

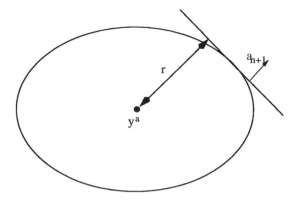

Figure 2.8. Illustration of $r(\Omega)_1$: the distance from y^a to the tangent plane a_{n+1}.

2.3 Primal and Primal-Dual Potential Functions

The analytic center defined in Section 2.2 can be represented in two other equivalent ways. One uses a (primal) setting defined below, while the other uses both the primal and dual settings.

From the duality theorem, we have a (homogeneous) linear programming problem related to finding a point in $\Omega = \{y : c - A^T y \geq 0\}$. We call it the *primal problem*:

$$\begin{aligned} \text{minimize} \quad & c^T x \\ \text{s.t.} \quad & Ax = 0, \ x \geq 0. \end{aligned}$$

Then the problem of finding a point in Ω is the dual.

If Ω is nonempty, then the minimal value of the primal problem is 0; if Ω is bounded and has an interior, then the interior of $\mathcal{X}_\Omega := \{x \in \mathcal{R}^n : Ax = 0, \ x \geq 0\}$ is nonempty and $x = 0$ is the unique primal minimal solution.

2.3.1 Primal potential function

One can define a potential function for $\mathcal{X}_\Omega$ as

$$\mathcal{P}(x, \Omega) = n \log(c^T x) - \sum_{j=1}^{n} \log x_j, \quad x \in \overset{\circ}{\mathcal{X}}_\Omega . \tag{2.8}$$

This is the so-called Karmarkar potential function. We now show that this quantity represents the logarithmic volume of a coordinate-aligned ellipsoid whose intersection with $\mathcal{A}_\Omega$ contains $\mathcal{S}_\Omega = \mathcal{R}^n_+ \cap \mathcal{A}_\Omega$.

Recall that we are interested in finding a point $s \in \mathcal{S}_\Omega$, which is equivalent to finding a point in Ω. Let $x \in \overset{\circ}{\mathcal{X}}_\Omega$. Then, for all $s \in \mathcal{S}_\Omega$ we must have
$$\|Xs\|^2 \leq (x^T s)^2 = (c^T x)^2. \tag{2.9}$$
The equality is true since $x^T s = x^T(c - A^T y) = c^T x - y^T Ax = c^T x$.

Let $\bar{s}$ be the point that minimizes $\|Xs\|$ subject to $s \in \mathcal{S}_\Omega$. Then,
$$\bar{s} = s(x) := c - A^T y(x) \quad \text{where} \quad \bar{y} = y(x) := (AX^2 A^T)^{-1} AX^2 c.$$
Since $\bar{s}^T X^2 (s - \bar{s}) = 0$ for any $s \in \mathcal{S}_\Omega$, we have, for any $s \in \mathcal{S}_\Omega$,
$$\|Xs\|^2 = \|X(s - \bar{s}) + X\bar{s}\|^2 = \|X(s - \bar{s})\|^2 + \|X\bar{s}\|^2,$$
or
$$\|X(s - \bar{s})\|^2 = \|Xs\|^2 - \|X\bar{s}\|^2 \leq \|Xs\|^2 \leq (c^T x)^2.$$
Thus, let E_x be the coordinate-aligned ellipsoid
$$E_x := \{s \in \mathcal{R}^n : \|X(s - \bar{s})\| \leq c^T x\}$$
that is centered at $\bar{s}$ (Figure 2.9). Then, we must have
$$\mathcal{S}_\Omega \subset (E_x \cap \mathcal{A}_\Omega).$$
Furthermore, the volume of E_x is
$$V(E_x) = \frac{\pi_n (c^T x)^n}{\det(X)} = \frac{\pi_n (c^T x)^n}{\prod_{j=1}^n x_j}, \tag{2.10}$$
where π_n is the volume of the unit ball in $\mathcal{R}^n$. Thus,
$$\mathcal{P}(x, \Omega) = \log V(E_x) - \log \pi_n.$$

In the next chapter, we will show that Karmarkar's algorithm actually generates sequences $\{0 < x^k \in \overset{\circ}{\mathcal{X}}_\Omega\}$ such that
$$\mathcal{P}(x^{k+1}, \Omega) \leq \mathcal{P}(x^k, \Omega) - .2$$
for $k = 0, 1, 2, ...$, as long as the ellipsoid is still too "fat". In other "words",
$$\frac{V(E_{x^{k+1}})}{V(E_{x^k})} \leq \exp(-.2).$$

2.3. PRIMAL AND PRIMAL-DUAL POTENTIAL FUNCTIONS 61

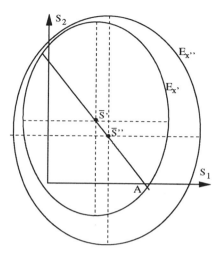

Figure 2.9. Coordinate-aligned (primal) ellipsoids centered at points $\bar{s}$'s on an affine set A; they also contain the intersection of A and the positive orthant.

That is, the volume of the containing ellipsoids shrinks at a constant rate. Note that E_{x^k} contains the solution set $\mathcal{S}_\Omega$. Therefore, Karmarkar's algorithm conceptually resembles the ellipsoid method.

Since the primal potential function represents the volume of a coordinate-aligned ellipsoid containing $\mathcal{S}_\Omega$, let us minimize it over all $x \in \overset{\circ}{\mathcal{X}}_\Omega$. Note that the primal potential function is homogeneous of degree 0 and there is an $0 < s^a \in \mathcal{S}_\Omega$, so that we can fix $c^T x = (s^a)^T x = n$. Then the problem becomes

$$\text{minimize} \quad \mathcal{P}(x, \Omega) = n \log n - \sum_{j=1}^{n} \log x_j$$
$$\text{s.t.} \quad c^T x = n, \ x \in \overset{\circ}{\mathcal{X}}_\Omega .$$

The problem minimizes a strictly convex function over a bounded feasible set ($c^T x = (s^a)^T x = n$ and $x > 0$ imply that x is bounded). Thus, the minimizer is unique and the optimality conditions are

$$x \in \mathcal{X}_\Omega \text{ and } Xs = X(c - A^T y) = e \text{ for some } s \in \mathcal{S}_\Omega. \tag{2.11}$$

One can see that s (or y) in (2.11) is in fact the analytic center of $\mathcal{S}_\Omega$ (or of Ω), since this condition is identical to condition (2.2) for defining the analytic center of Ω. Let x^a, called the *primal analytic center*, satisfy these conditions. Then, we have Theorem 2.12.

Theorem 2.12 *There is a unique minimal-volume coordinate-aligned ellipsoid E_{x^a}, where $x^a \in \mathring{\mathcal{X}}_\Omega$, whose intersection with $\mathcal{A}_\Omega$ contains polytope $\mathcal{S}_\Omega$. Moreover, polytope $\mathcal{S}_\Omega$ contains the intersection of $\mathcal{A}_\Omega$ and the ellipsoid obtained from E_{x^a} by shrinking it from its center by a factor of n. In fact, the two ellipsoids E_{x^a} and E_{s^a} (in Theorem 2.6) are concentric, and they can be obtained from each other by enlarging or shrinking with a factor n (Figures 2.5 and 2.9).*

Proof. The uniqueness of E_{x^a} is already demonstrated earlier.

To prove E_{x^a} and E_{s^a} are concentric, we only need to prove that the center of E_{x^a} is the analytic center of $\mathcal{S}_\Omega$. Recall that $\bar{s}$ is the center of E_{x^a} with $\bar{s} = c - A^T \bar{y}$, where

$$\bar{y} = (A(X^a)^2 A^T)^{-1} A(X^a)^2 c.$$

On the other hand, from (2.11) we have

$$X^a(c - A^T y^a) = e \quad \text{or} \quad X^a c = e + X^a A^T y^a,$$

where y^a is the analytic center of Ω. Thus

$$\begin{aligned} &X^a(c - A^T \bar{y}) \\ &= X^a c - X^a A^T (A(X^a)^2 A^T)^{-1} A(X^a)^2 c \\ &= e + X^a A^T y^a - X^a A^T (A(X^a)^2 A^T)^{-1} A X^a (e + X^a A^T y^a) \\ &= e - X^a A^T (A(X^a)^2 A^T)^{-1} A X^a e = e. \end{aligned}$$

Thus, $\bar{y} = y^a$ and $\bar{s} = s^a$ since $\bar{y}$ also satisfies the optimality condition of (2.2) and such a point is unique.

Since $X^a s^a = e$, $c^T x^a = n$ and $\bar{s} = s^a$, if we shrink E_{x^a} by a factor of n, the smaller ellipsoid is

$$\frac{1}{n} E_{x^a} = \{t \in \mathcal{R}^n : \|(S^a)^{-1}(t - s^a)\| \leq 1\},$$

which is exact E_{s^a}.

$\square$

2.3.2 Primal-dual potential function

For $x \in \mathring{\mathcal{X}}_\Omega$ and $s \in \mathring{\mathcal{S}}_\Omega$ consider a primal-dual potential function, which has the form

$$\begin{aligned} \psi_n(x,s) &:= n \log(x^T s) - \sum \log(x_j s_j) \\ &= n \log(c^T x) - \sum \log x_j - \sum \log s_j \\ &= \mathcal{P}(x, \Omega) - \mathcal{B}(s, \Omega). \end{aligned} \quad (2.12)$$

2.4. POTENTIAL FUNCTIONS FOR LP, LCP, AND PSP

This is the logarithmic ratio of the volume of E_x over the volume of E_s. We also have, from the arithmetic-geometric mean inequality

$$\psi_n(x,s) = n\log(x^T s) - \sum \log(x_j s_j) \geq n\log n,$$

and from $X^a s^a = e$

$$\psi_n(x^a, s^a) = n\log n.$$

Thus, x^a and s^a minimizes the primal dual potential function. Furthermore, $n \log n$ is the precise logarithmic ratio of the volumes of two concentric ellipsoids whose radii are differentiated by a factor n (Figures 2.5 and 2.9). This fact further confirms our results in Theorem 2.12.

2.4 Potential Functions for LP, LCP, and PSP

The potential functions in Sections 2.2 and 2.3 are used to find the analytic center of a polytope. In this section, we show how potential functions can be used to solve linear programming problems, linear complementarity, and positive semi-definite programming.

We assume that for a given LP data set (A, b, c), both the primal and dual have interior feasible point. We also let z^* be the optimal value of the standard form (LP) and (LD). Denote the feasible sets of (LP) and (LD) by $\mathcal{F}_p$ and $\mathcal{F}_d$, respectively. Denote by $\mathcal{F} = \mathcal{F}_p \times \mathcal{F}_d$, and the interior of $\mathcal{F}$ by $\overset{\circ}{\mathcal{F}}$.

2.4.1 Primal potential function for LP

Consider the level set

$$\Omega(z) = \{y \in \mathcal{R}^m : c - A^T y \geq 0, -z + b^T y \geq 0\}, \qquad (2.13)$$

where $z < z^*$. Since both (LP) and (LD) have interior feasible point for given (A, b, c), $\Omega(z)$ is bounded and has an interior for any finite z, even $\Omega := \mathcal{F}_d$ is unbounded (Exercise 1.26). Clearly, $\Omega(z) \subset \Omega$, and if $z_2 \geq z_1$, $\Omega(z_2) \subset \Omega(z_1)$ and the inequality $-z + b^T y$ is translated from $z = z_1$ to $z = z_2$.

From the duality theorem again, finding a point in $\Omega(z)$ has a homogeneous primal problem

$$\begin{aligned}
\text{minimize} \quad & c^T x' - z x'_0 \\
\text{s.t.} \quad & Ax' - b x'_0 = 0, (x', x'_0) \geq 0.
\end{aligned}$$

For (x', x'_0) satisfying
$$Ax' - bx'_0 = 0, \ (x', x'_0) > 0,$$
let $x := x'/x'_0 \in \overset{\circ}{\mathcal{F}}_p$, i.e.,
$$Ax = b, \ x > 0.$$
Then, the primal potential function for $\Omega(z)$ (Figure 2.10), as described in the preceding section, is
$$\mathcal{P}(x', \Omega(z)) = (n+1)\log(c^T x' - zx'_0) - \sum_{j=0}^{n} \log x'_j$$
$$= (n+1)\log(c^T x - z) - \sum_{j=1}^{n} \log x_j =: \mathcal{P}_{n+1}(x, z).$$

The latter, $\mathcal{P}_{n+1}(x, z)$, is the Karmarkar potential function in the standard LP form with a lower bound z for z^*. Based on our discussion in Section 2.3.1, we see that it represents the volume of a coordinate-aligned ellipsoid whose intersection with $\mathcal{A}_{\Omega(z)}$ contains $\mathcal{S}_{\Omega(z)}$.

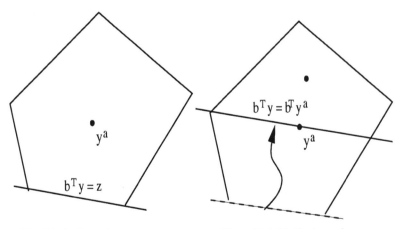

Figure 2.10. Intersections of a dual feasible region and the objective hyperplane; $b^T y \geq z$ on the left and $b^T y \geq b^T y^a$ on the right.

One algorithm for solving (LD) is suggested in Figure 2.10. If the objective hyperplane is repeatedly translated to the analytic center, the sequence

2.4. POTENTIAL FUNCTIONS FOR LP, LCP, AND PSP

of new analytic centers will converge to an optimal solution and the potentials of the new polytopes will decrease to $-\infty$. This will be discussed in more detail in Chapter 4.

As we illustrated before, one can represent $\Omega(z)$ differently:

$$\Omega(z) = \{y: \ c - A^T y \geq 0, \overbrace{-z + b^T y \geq 0, \cdots, -z + b^T y \geq 0}^{\rho \text{ times}}\}, \qquad (2.14)$$

i.e., "$-z + b^T y \geq 0$" is copied ρ times. Geometrically, this representation does not change $\Omega(z)$, but it changes the location of its analytic center. Since the last ρ inequalities in $\Omega(z)$ are identical, they must share the same slack value and the same corresponding primal variable. Let (x', x'_0) be the primal variables. Then the primal problem can be written as

$$\begin{aligned}
\text{minimize} \quad & c^T x' \overbrace{- zx'_0 - \cdots - zx'_0}^{\rho \text{ times}} \\
\text{s.t.} \quad & Ax' \overbrace{- bx'_0 - \cdots - bx'_0}^{\rho \text{ times}} = 0, \ (x', x'_0) \geq 0.
\end{aligned}$$

Let $x = x'/(\rho x'_0) \in \overset{\circ}{\mathcal{F}}_p$. Then, the primal potential function for the new $\Omega(z)$ given by (2.14) is

$$\begin{aligned}
\mathcal{P}(x, \Omega(z)) &= (n + \rho) \log(c^T x' - z(\rho x'_0)) - \sum_{j=1}^{n} \log x'_j - \rho \log x'_0 \\
&= (n + \rho) \log(c^T x - z) - \sum_{j=1}^{n} \log x_j + \rho \log \rho \\
&=: \mathcal{P}_{n+\rho}(x, z) + \rho \log \rho.
\end{aligned}$$

The function

$$\mathcal{P}_{n+\rho}(x, z) = (n + \rho) \log(c^T x - z) - \sum_{j=1}^{n} \log x_j \qquad (2.15)$$

is an extension of the Karmarkar potential function in the standard LP form with a lower bound z for z^*. It represents the volume of a coordinate-aligned ellipsoid whose intersection with $\mathcal{A}_{\Omega(z)}$ contains $\mathcal{S}_{\Omega(z)}$, where "$-z + b^T y \geq 0$" is duplicated ρ times.

2.4.2 Dual potential function for LP

We can also develop a dual potential function, symmetric to the primal, for $(y,s) \in \overset{\circ}{\mathcal{F}}_d$

$$\mathcal{B}_{n+\rho}(y,s,z) = (n+\rho)\log(z - b^T y) - \sum_{j=1}^{n} \log s_j, \qquad (2.16)$$

where z is a upper bound of z^*. One can show that it represents the volume of a coordinate-aligned ellipsoid whose intersection with the affine set $\{x : Ax = b\}$ contains the primal level set

$$\{x \in \mathcal{F}_p : \overbrace{c^T x - z \le 0, \cdots, c^T x - z \le 0}^{\rho \text{ times}}\},$$

where "$c^T x - z \le 0$" is copied ρ times (Exercise 2.9). For symmetry, we may write $\mathcal{B}_{n+\rho}(y,s,z)$ simply by $\mathcal{B}_{n+\rho}(s,z)$, since we can always recover y from s using equation $A^T y = c - s$.

2.4.3 Primal-dual potential function for LP

A primal-dual potential function for linear programming will be used later. For $x \in \overset{\circ}{\mathcal{F}}_p$ and $(y,s) \in \overset{\circ}{\mathcal{F}}_d$ it is defined by

$$\psi_{n+\rho}(x,s) := (n+\rho)\log(x^T s) - \sum_{j=1}^{n} \log(x_j s_j), \qquad (2.17)$$

where $\rho \ge 0$.

We have the relation:

$$\begin{aligned}
\psi_{n+\rho}(x,s) &= (n+\rho)\log(c^T x - b^T y) - \sum_{j=1}^{n} \log x_j - \sum_{j=1}^{n} \log s_j \\
&= \mathcal{P}_{n+\rho}(x, b^T y) - \sum_{j=1}^{n} \log s_j \\
&= \mathcal{B}_{n+\rho}(s, c^T x) - \sum_{j=1}^{n} \log x_j.
\end{aligned}$$

Since

$$\psi_{n+\rho}(x,s) = \rho \log(x^T s) + \psi_n(x,s) \ge \rho \log(x^T s) + n \log n,$$

2.4. POTENTIAL FUNCTIONS FOR LP, LCP, AND PSP

then, for $\rho > 0$, $\psi_{n+\rho}(x,s) \to -\infty$ implies that $x^T s \to 0$. More precisely, we have
$$x^T s \leq \exp(\frac{\psi_{n+\rho}(x,s) - n\log n}{\rho}).$$

We have the following theorem:

Theorem 2.13 *Define the level set*
$$\Psi(\delta) := \{(x,y,s) \in \overset{\circ}{\mathcal{F}}: \ \psi_{n+\rho}(x,s) \leq \delta\}.$$

i)
$$\Psi(\delta^1) \subset \Psi(\delta^2) \quad if \quad \delta^1 \leq \delta^2.$$

ii)
$$\overset{\circ}{\Psi}(\delta) = \{(x,y,s) \in \mathcal{F}: \ \psi_{n+\rho}(x,s) < \delta\}.$$

iii) *For every δ, $\Psi(\delta)$ is bounded and its closure $\hat{\Psi}(\delta)$ has non-empty intersection with the solution set.*

Later we will show that a potential reduction algorithm generates sequences $\{x^k, y^k, s^k\} \in \overset{\circ}{\mathcal{F}}$ such that
$$\psi_{n+\sqrt{n}}(x^{k+1}, y^{k+1}, s^{k+1}) \leq \psi_{n+\sqrt{n}}(x^k, y^k, s^k) - .05$$

for $k = 0, 1, 2, \ldots$. This indicates that the level sets shrink at least a constant rate independently of m or n.

2.4.4 Potential function for LCP

A potential function can be defined for the linear complementarity problem introduced in Section 1.3.7. Let
$$\mathcal{F} := \{(x,s): \ s = Mx + q, \quad (x,s) \geq 0\}$$

have an interior feasible point $(x,s) > 0$. Then, for $(x,s) \in \overset{\circ}{\mathcal{F}}$ consider the primal-dual potential function
$$\psi_{n+\rho}(x,s) := (n+\rho)\log(x^T s) - \sum_{j=1}^{n} \log(x_j s_j),$$

where $\rho \geq 0$. Again,

$$\psi_{n+\rho}(x,s) = \rho \log(x^T s) + \psi_n(x,s) \geq \rho \log(x^T s) + n \log n,$$

then, for $\rho > 0$, $\psi_{n+\rho}(x,s) \to -\infty$ implies that the complementarity gap $x^T s \to 0$. More precisely, we have

$$x^T s \leq \exp(\frac{\psi_{n+\rho}(x,s) - n \log n}{\rho}).$$

We have the following corollary:

Corollary 2.14 *Let the LCP be monotone and have non-empty interior, and define the level set*

$$\Psi(\delta) := \{(x,y,s) \in \overset{\circ}{\mathcal{F}}: \psi_{n+\rho}(x,s) \leq \delta\}.$$

i)
$$\Psi(\delta^1) \subset \Psi(\delta^2) \quad \text{if} \quad \delta^1 \leq \delta^2.$$

ii)
$$\overset{\circ}{\Psi}(\delta) = \{(x,y,s) \in \mathcal{F} : \psi_{n+\rho}(x,s) < \delta\}.$$

iii) *For every δ, $\Psi(\delta)$ is bounded and its closure $\hat{\Psi}(\delta)$ has non-empty intersection with the solution set.*

We will also show that a potential reduction algorithm generates sequences $\{x^k, s^k\} \in \overset{\circ}{\mathcal{F}}$ for the monotone LCP such that

$$\psi_{n+\sqrt{n}}(x^{k+1}, y^{k+1}, s^{k+1}) \leq \psi_{n+\sqrt{n}}(x^k, y^k, s^k) - .05$$

for $k = 0, 1, 2, \ldots$.

The potential function definition also applies to the nonlinear complementarity problem described in Section 1.3.10, where $s = Mx+q$ is replaced by $s = f(x)$.

2.4.5 Potential function for PSP

The potential functions for PSP of Section 1.3.8 are analogous to those for LP. For given data, we assume that both (PSP) and (PSD) have interior

2.4. POTENTIAL FUNCTIONS FOR LP, LCP, AND PSP

feasible points. Then, for any $X \in \overset{\circ}{\mathcal{F}}_p$ and $(y, S) \in \overset{\circ}{\mathcal{F}}_d$, the primal potential function is defined by

$$\mathcal{P}_{n+\rho}(X, z) := (n + \rho) \log(C \bullet X - z) - \log \det(X), \quad z \leq z^*;$$

the dual potential function is defined by

$$\mathcal{B}_{n+\rho}(y, S, z) := (n + \rho) \log(z - b^T y) - \log \det(S), \quad z \geq z^*,$$

where $\rho \geq 0$ and z^* designates the optimal objective value.

For $X \in \overset{\circ}{\mathcal{F}}_p$ and $(y, S) \in \overset{\circ}{\mathcal{F}}_d$ the primal-dual potential function for PSP is defined by

$$\begin{aligned}
\psi_{n+\rho}(X, S) &:= (n + \rho) \log(X \bullet S) - \log(\det(X) \cdot \det(S)) \\
&= (n + \rho) \log(C \bullet X - b^T y) - \log \det(X) - \log \det(S) \\
&= \mathcal{P}_{n+\rho}(X, b^T y) - \log \det(S) \\
&= \mathcal{B}_{n+\rho}(S, C \bullet X) - \log \det(X),
\end{aligned}$$

where $\rho \geq 0$. Note that if X and S are diagonal matrices, these definitions reduce to those for LP.

Note that we still have (Exercise 2.10)

$$\psi_{n+\rho}(X, S) = \rho \log(X \bullet S) + \psi_n(X, S) \geq \rho \log(X \bullet S) + n \log n.$$

Then, for $\rho > 0$, $\psi_{n+\rho}(X, S) \to -\infty$ implies that $X \bullet S \to 0$. More precisely, we have

$$X \bullet S \leq \exp(\frac{\psi_{n+\rho}(X, S) - n \log n}{\rho}).$$

We also have the following corollary:

Corollary 2.15 *Let (PSP) and (PSD) have non-empty interior and define the level set*

$$\Psi(\delta) := \{(X, y, S) \in \overset{\circ}{\mathcal{F}}: \ \psi_{n+\rho}(X, S) \leq \delta\}.$$

i)

$$\Psi(\delta^1) \subset \Psi(\delta^2) \quad \text{if} \quad \delta^1 \leq \delta^2.$$

ii)

$$\overset{\circ}{\Psi}(\delta) = \{(X, y, S) \in \mathcal{F}: \ \psi_{n+\rho}(X, S) < \delta\}.$$

iii) *For every δ, $\Psi(\delta)$ is bounded and its closure $\hat{\Psi}(\delta)$ has non-empty intersection with the solution set.*

2.5 Central Paths of LP, LCP, and PSP

Many interior-point algorithms find a sequence of feasible points along a "central" path that connects the analytic center and the solution set. We now present this one of the most important foundations for the development of interior-point algorithms.

2.5.1 Central path for LP

Consider a linear program in the standard form (LP) and (LD). Assume that $\mathring{\mathcal{F}} \neq \emptyset$, i.e., both $\mathring{\mathcal{F}}_p \neq \emptyset$ and $\mathring{\mathcal{F}}_d \neq \emptyset$, and denote z^* the optimal objective value.

The central path can be expressed as

$$\mathcal{C} = \left\{ (x, y, s) \in \mathring{\mathcal{F}} \colon Xs = \frac{x^T s}{n} e \right\}$$

in the primal-dual form. We also see

$$\mathcal{C} = \left\{ (x, y, s) \in \mathring{\mathcal{F}} \colon \psi_n(x, s) = n \log n \right\}.$$

For any $\mu > 0$ one can derive the central path simply by minimizing the primal LP with a logarithmic barrier function:

$$(P) \quad \begin{array}{ll} \text{minimize} & c^T x - \mu \sum_{j=1}^n \log x_j \\ \text{s.t.} & Ax = b, \ x \geq 0. \end{array}$$

Let $x(\mu) \in \mathring{\mathcal{F}}_p$ be the (unique) minimizer of (P). Then, for some $y \in \mathcal{R}^m$ it satisfies the optimality conditions

$$\begin{array}{rcl} Xs & = & \mu e \\ Ax & = & b \\ -A^T y - s & = & -c. \end{array} \quad (2.18)$$

Consider minimizing the dual LP with the barrier function:

$$(D) \quad \begin{array}{ll} \text{maximize} & b^T y + \mu \sum_{j=1}^n \log s_j \\ \text{s.t.} & A^T y + s = c, \ s \geq 0. \end{array}$$

Let $(y(\mu), s(\mu)) \in \mathring{\mathcal{F}}_d$ be the (unique) minimizer of (D). Then, for some $x \in \mathcal{R}^n$ it satisfies the optimality conditions (2.18) as well. Thus, both minimizers $x(\mu)$ and $(y(\mu), s(\mu))$ are on the central path with $x(\mu)^T s(\mu) = n\mu$.

2.5. CENTRAL PATHS OF LP, LCP, AND PSP

Another way to derive the central path is to consider again the dual level set $\Omega(z)$ of (2.13) for any $z < z^*$ (Figure 2.11).

Then, the analytic center $(y(z), s(z))$ of $\Omega(z)$ and a unique point $(x'(z), x'_0(z))$ satisfies

$$Ax'(z) - bx'_0(z) = 0, \ X'(z)s = e, \ s = c - A^T y, \text{ and } x'_0(z)(b^T y - z) = 1.$$

Let $x(z) = x'(z)/x'_0(z)$, then we have

$$Ax(z) = b, \ X(z)s(z) = e/x'_0(z) = (b^T y(z) - z)e.$$

Thus, the point $(x(z), y(z), s(z))$ is on the central path with $\mu = b^T y(z) - z$ and $c^T x(z) - b^T y(z) = x(z)^T s(z) = n(b^T y(z) - z) = n\mu$. As we proved earlier in Section 2.4, $(x(z), y(z), s(z))$ exists and is uniquely defined, which imply the following theorem:

Theorem 2.16 *Let both (LP) and (LD) have interior feasible points for the given data set (A, b, c). Then for any $0 < \mu < \infty$, the central path point $(x(\mu), y(\mu), s(\mu))$ exists and is unique.*

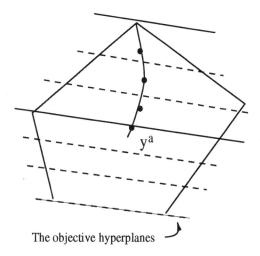

Figure 2.11. The central path of $y(z)$ in a dual feasible region.

The following theorem further characterizes the central path and utilizes it to solving linear programs.

Theorem 2.17 *Let $(x(\mu), y(\mu), s(\mu))$ be on the central path.*

i) *The central path point $(x(\mu), s(\mu))$ is bounded for $0 < \mu \leq \mu^0$ and any given $0 < \mu^0 < \infty$.*

ii) *For $0 < \mu' < \mu$,*
$$c^T x(\mu') < c^T x(\mu) \quad \text{and} \quad b^T y(\mu') > b^T y(\mu).$$

iii) *$(x(\mu), s(\mu))$ converges to an optimal solution pair for (LP) and (LD). Moreover, the limit point $x(0)_{P^*}$ is the analytic center on the primal optimal face, and the limit point $s(0)_{Z^*}$ is the analytic center on the dual optimal face, where (P^*, Z^*) is the strict complementarity partition of the index set $\{1, 2, ..., n\}$.*

Proof. Note that
$$(x(\mu^0) - x(\mu))^T (s(\mu^0) - s(\mu)) = 0,$$
since $(x(\mu^0) - x(\mu)) \in \mathcal{N}(A)$ and $(s(\mu^0) - s(\mu)) \in \mathcal{R}(A^T)$. This can be rewritten as
$$\sum_j^n \left(s(\mu^0)_j x(\mu)_j + x(\mu^0)_j s(\mu)_j \right) = n(\mu^0 + \mu) \leq 2n\mu^0,$$
or
$$\sum_j^n \left(\frac{x(\mu)_j}{x(\mu^0)_j} + \frac{s(\mu)_j}{s(\mu^0)_j} \right) \leq 2n.$$

Thus, $x(\mu)$ and $s(\mu)$ are bounded, which proves (i).

We leave the proof of (ii) as an exercise.

Since $x(\mu)$ and $s(\mu)$ are both bounded, they have at least one limit point which we denote by $x(0)$ and $s(0)$. Let $x^*_{P^*}$ ($x^*_{Z^*} = 0$) and $s^*_{Z^*}$ ($s^*_{P^*} = 0$), respectively, be the unique analytic centers on the primal and dual optimal faces: $\{x_{P^*} : A_{P^*} x_{P^*} = b, \ x_{P^*} \geq 0\}$ and $\{s_{Z^*} : s_{Z^*} = c_{Z^*} - A_{Z^*}^T y \geq 0, \ c_{P^*} - A_{P^*}^T y = 0\}$. Again, we have
$$\sum_j^n \left(s^*_j x(\mu)_j + x^*_j s(\mu)_j \right) = n\mu,$$
or
$$\sum_{j \in P^*} \left(\frac{x^*_j}{x(\mu)_j} \right) + \sum_{j \in Z^*} \left(\frac{s^*_j}{s(\mu)_j} \right) = n.$$

2.5. CENTRAL PATHS OF LP, LCP, AND PSP

Thus, we have
$$x(\mu)_j \geq x_j^*/n > 0, \ j \in P^*$$
and
$$s(\mu)_j \geq s_j^*/n > 0, \ j \in Z^*.$$
This implies that
$$x(\mu)_j \to 0, \ j \in Z^*$$
and
$$s(\mu)_j \to 0, \ j \in P^*.$$
Furthermore,
$$\left(\prod_{j \in P^*} \frac{x_j^*}{x(\mu)_j}\right)\left(\prod_{j \in Z^*} \frac{s_j^*}{s(\mu)_j}\right) \leq 1$$
or
$$\left(\prod_{j \in P^*} x_j^*\right)\left(\prod_{j \in Z^*} s_j^*\right) \leq \left(\prod_{j \in P^*} x(\mu)_j\right)\left(\prod_{j \in Z^*} s(\mu)_j\right).$$

However, $(\prod_{j \in P^*} x_j^*)(\prod_{j \in Z^*} s_j^*)$ is the maximal value of the potential function over all interior point pairs on the optimal faces, and $x(0)_{P^*}$ and $s(0)_{Z^*}$ is one interior point pair on the optimal face. Thus, we must have

$$\left(\prod_{j \in P^*} x_j^*\right)\left(\prod_{j \in Z^*} s_j^*\right) = \left(\prod_{j \in P^*} x(0)_j\right)\left(\prod_{j \in Z^*} s(0)_j\right).$$

Therefore,
$$x(0)_{P^*} = x_{P^*}^* \quad \text{and} \quad s(0)_{Z^*} = s_{Z^*}^*,$$
since $x_{P^*}^*$ and $s_{Z^*}^*$ are the unique maximizer pair of the potential function. This also implies that the limit point of the central path is unique.

$\square$

We usually define a neighborhood of the central path as
$$\mathcal{N}(\eta) = \left\{(x, y, s) \in \overset{\circ}{\mathcal{F}} : \ \|Xs - \mu e\| \leq \eta \mu \quad \text{and} \quad \mu = \frac{x^T s}{n}\right\},$$
where $\|.\|$ can be any norm, or even a one-sided "norm" as
$$\|x\|_{-\infty} = |\min(0, \min(x))|.$$
We have the following theorem:

Theorem 2.18 *Let $(x, y, s) \in \mathcal{N}(\eta)$ for constant $0 < \eta < 1$.*

i) *The $\mathcal{N}(\eta) \cap \{(x, y, s) : x^T s \leq n\mu^0\}$ is bounded for any given $\mu^0 < \infty$.*

ii) *Any limit point of $\mathcal{N}(\eta)$ as $\mu \to 0$ is an optimal solution pair for (LP) and (LD). Moreover, for any $j \in P^*$*

$$x_j \geq \frac{(1-\eta)x_j^*}{n},$$

where x^* is any optimal primal solution; for any $j \in Z^*$

$$s_j \geq \frac{(1-\eta)s_j^*}{n},$$

where s^* is any optimal dual solution.

2.5.2 Central path for LCP

Consider the monotone LCP in Section 1.3.7 where $\overset{\circ}{\mathcal{F}} \neq \emptyset$. The central path can be expressed as

$$\mathcal{C} = \left\{ (x, s) \in \overset{\circ}{\mathcal{F}} : Xs = \mu e, \ 0 < \mu < \infty \right\}.$$

We have the following corollary:

Corollary 2.19 *Let the monotone LCP have interior feasible points. Then for any $0 < \mu < \infty$, the central path point $(x(\mu), s(\mu))$ exists and is unique. Moreover,*

i) *The central pair $(x(\mu), s(\mu))$ is bounded where $0 < \mu \leq \mu^0$ for any given $0 < \mu^0 < \infty$.*

ii) *$(x(\mu), s(\mu))$ converges to an optimal solution of (LCP), and the limit point is a maximal complementarity solution.*

The central path definition and the corollary also apply to the monotone complementarity problem described in Section 1.3.10, where $s = Mx + q$ is replaced by $s = f(x)$ and $f(\cdot)$ is a monotone function.

2.5.3 Central path for PSP

Consider a PSP problem in Section 1.3.8 and Assume that $\overset{\circ}{\mathcal{F}} \neq \emptyset$, i.e., both $\overset{\circ}{\mathcal{F}}_p \neq \emptyset$ and $\overset{\circ}{\mathcal{F}}_d \neq \emptyset$. The central path can be expressed as

$$\mathcal{C} = \left\{ (X, y, S) \in \overset{\circ}{\mathcal{F}} : XS = \mu I, \ 0 < \mu < \infty \right\},$$

or a symmetric form

$$\mathcal{C} = \left\{ (X, y, S) \in \overset{\circ}{\mathcal{F}} \colon \ X^{.5} S X^{.5} = \mu I, \ 0 < \mu < \infty \right\},$$

where $X^{.5} \in \mathcal{M}_+^n$ is the "square root" matrix of $X \in \mathcal{M}_+^n$, i.e., $X^{.5} X^{.5} = X$. We also see

$$\mathcal{C} = \left\{ (X, y, S) \in \overset{\circ}{\mathcal{F}} \colon \ \psi_n(X, S) = n \log n \right\}.$$

When X and S are diagonal matrices, this definition is identical to LP.
We also have the following corollary:

Corollary 2.20 *Let both (PSP) and (PSD) have interior feasible points. Then for any $0 < \mu < \infty$, the central path point $(X(\mu), y(\mu), S(\mu))$ exists and is unique. Moreover,*

i) *the central path point $(X(\mu), S(\mu))$ is bounded where $0 < \mu \le \mu^0$ for any given $0 < \mu^0 < \infty$.*

ii) *For $0 < \mu' < \mu$,*

$$C \bullet X(\mu') < C \bullet X(\mu) \quad \text{and} \quad b^T y(\mu') > b^T y(\mu).$$

iii) *$(X(\mu), S(\mu))$ converges to an optimal solution pair for (PSP) and (PSD), and the rank of the limit of $X(\mu)$ is maximal among all optimal solutions of (PSP) and the rank of the limit $S(\mu)$ is maximal among all optimal solutions of (PSD).*

2.6 Notes

General convex problems, such as membership, separation, validity, and optimization, can be solved by the central-section method; see Grötschel, Lovász and Schrijver [170].

Levin [244] and Newman [318] considered the center of gravity of a convex body; Elzinga and Moore [110] considered the center of the max-volume sphere contained in a convex body. A number of Russian mathematicians (for example, Tarasov, Khachiyan and Érlikh [403]) considered the center of the max-volume ellipsoid inscribing the body; Huard and Liêu [190, 191] considered a generic center in the body that maximizes a distance function; and Vaidya [438] considered the volumetric center, the maximizer of the determinant of the Hessian matrix of the logarithmic barrier function. See Kaiser, Morin and Trafalis [210] for a complete survey.

Grünbaum [172] first proved Theorem 2.1, with a more geometric proof given by Mityagin [286]. Dyer and Frieze [104] proved that computing the volume of a convex polytope, either given as a list of facets or vertices, is itself #P-Hard. Furthermore, Elekes [109] has shown that no polynomial time algorithm can compute the volume of a convex body with less than exponential relative error. Bárány and Fürendi [42] further showed that for $\Omega \in \mathcal{R}^m$, any polynomial time algorithm that gives an upper and lower bound on the volume of Ω, represented as $\overline{V}(\Omega)$ and $\underline{V}(\Omega)$, respectively, necessarily has an exponential gap between them. They showed

$$\overline{V}(\Omega)/\underline{V}(\Omega) \geq (cm/\log m)^m,$$

where c is a constant independent of m. In other words, there is no polynomial time algorithm that would compute $\overline{V}(\Omega)$ and $\underline{V}(\Omega)$ such that

$$\overline{V}(\Omega)/\underline{V}(\Omega) < (cm/\log m)^m.$$

Recently, Dyer, Frieze and Kannan [105] developed a random polynomial time algorithm that can, with high probability, find a good estimate for the volume of Ω.

Apparently, the result that every convex body contains a unique ellipsoid of maximal volume and is contained in a unique ellipsoid of minimal volume, was discovered independently by several mathematicians—see, for example, Danzer, Grünbaum and Klee [94]. These authors attributed the first proof to K. Löwner. John [208] later proved Theorem 2.2.

Tarasov, Khachiyan, and Érlikh [403] proved the central-section Theorem 2.3. Khachiyan and Todd [222] established a polynomial complexity bound for computing an approximate point of the center of the maximal inscribing ellipsoid if the convex body is represented by linear inequalities, Theorem 2.5 was proved by Shor [380] and Nemirovskii and Yudin [321].

The "analytic center" for a convex polyhedron given by linear inequalities was introduced by Huard [190], and later by Sonnevend [383]. The function $d(y, \Omega)$ is very similar to Huard's generic distance function, with one exception, where property (3) there was stated as "If $\Omega' \subset \Omega$, then $d(y, \Omega') \leq d(y, \Omega)$." The reason for the difference is that the distance function may return different values even if we have the same polytope but two different representations. The negative logarithmic function $d(y, \Omega)$, called the *barrier function*, was introduced by Frisch [126]. Theorem 2.6 was first proved by Sonnevend [383], also see Karmarkar [217] for a canonical form.

Todd [405] and Ye [465] showed that Karmarkar's potential function represents the logarithmic volume of a coordinate-aligned ellipsoid who contains the feasible region. The Karmarkar potential function in the standard form (LP) with a lower bound z for z^* was seen in Todd and Burrell [413], Anstreicher [24], Gay [133], and Ye and Kojima [477]. The primal potential function with $\rho > 1$ was proposed by Gonzaga [160], Freund [123], and Ye

[466, 468]. The primal-dual potential function was proposed by Tanabe [400], and Todd and Ye [415]. Potential functions for LCP and PSP were studied by Kojima et al. [230, 228], Alizadeh [9], and Nesterov and Nemirovskii [327]. Noma [333] proved Theorem 2.13 for the monotone LCP including LP.

McLinden [267] earlier, then Bayer and Lagarias [45, 46], Megiddo [271], and Sonnevend [383], analyzed the central path for linear programming and convex optimization. Megiddo [271] derived the central path simply minimizing the primal with a logarithmic barrier function as in Fiacco and McCormick [116]. The central path for LCP, with more general matrix M, was given by Kojima et al. [227] and Güler [174]; the central path theory for PSP was first developed by Nesterov and Nemirovskii [327]. McLinden [267] proved Theorem 2.17 for the monotone LCP, which includes LP.

2.7 Exercises

2.1 *Find the min-volume ellipsoid containing a half of the unit ball $\{x \in \mathcal{R}^n : \|x\| \leq 1\}$.*

2.2 *Verify Examples 2.1, 2.2, and 2.3.*

2.3 *Compare and contrast the center of gravity of a polytope and its analytic center.*

2.4 *Consider the maximum problem*

$$\begin{aligned} \text{maximize} \quad & f(x) = \|x - e\| \prod_{j=1}^{n} x_j \\ \text{s.t.} \quad & e^T x = n, \ x > 0 \in \mathcal{R}^n. \end{aligned}$$

Prove that its maximizer is achieved at $x_1 = \beta$ and $x_2 = \ldots = x_n = (n-\beta)/(n-1) > 0$ for some $1 < \beta < n$.

2.5 *Let $\overset{\circ}{\Omega} = \{y \in \mathcal{R}^m : c - A^T y > 0\} \neq \emptyset$, $\overset{\circ}{\Omega}' = \{y \in \mathcal{R}^m : c' - A^T y > 0\} \neq \emptyset$, and $c' \leq c$. Prove $\mathcal{B}(\Omega') \leq \mathcal{B}(\Omega)$.*

2.6 *Prove Corollaries 2.8 and 2.9.*

2.7 *Prove Corollary 2.11.*

2.8 *If $\Omega = \{y : c - A^T y \geq 0\}$ is nonempty, prove the minimal value of the primal problem described at the beginning of Section 2.3 is 0; if Ω is bounded and has an interior, prove the interior of $\mathcal{X}_\Omega := \{x \in \mathcal{R}^n : Ax = 0, \ x \geq 0\}$ is nonempty and $x = 0$ is the unique primal solution.*

2.9 Let (LP) and (LD) have interior. Prove the dual potential function $\mathcal{B}_{n+1}(y, s, z)$, where z is a upper bound of z^*, represents the volume of a coordinate-aligned ellipsoid whose intersection with the affine set $\{x : Ax = b\}$ contains the primal level set $\{x \in \mathcal{F}_p : c^T x \leq z\}$.

2.10 Let $X, S \in \mathcal{M}^n$ be both positive definite. Then prove

$$\psi_n(X, S) = n \log(X \bullet S) - \log(\det(X) \cdot \det(S)) \geq n \log n.$$

2.11 Consider linear programming and the level set

$$\Psi(\delta) := \{(x, y, s) \in \mathcal{\overset{\circ}{F}} : \psi_{n+\rho}(x, s) \leq \delta\}.$$

Prove that

$$\Psi(\delta^1) \subset \Psi(\delta^2) \quad \text{if} \quad \delta^1 \leq \delta^2,$$

and for every δ $\Psi(\delta)$ is bounded and its closure $\hat{\Psi}(\delta)$ has non-empty intersection with the solution set.

2.12 Consider the linear program

$$\max \ b^T y \quad \text{s.t.} \quad 0 \leq y_1 \leq 1, \ 0 \leq y_2 \leq 1.$$

Draw the feasible region the the (dual) potential level sets

$$\{y : \mathcal{B}_5(y, s, 2) \leq 0\} \quad \text{and} \quad \{y : \mathcal{B}_5(y, s, 2) \leq -10\},$$

respectively, for

1. $b = (1; 0)$;
2. $b = (1; 1)/2$;
3. $b = (2; 1)/3$.

2.13 Consider the polytope

$$\{y \in \mathcal{R}^2 : 0 \leq y_1, \ 0 \leq y_2 \leq 1, \ y_1 + y_2 \leq z\}.$$

Describe how the analytic center changes as z decreases from 10 to 1.

2.14 Consider the linear program

$$\max \ b^T y \quad \text{s.t.} \quad 0 \leq y_1 \leq 1, \ 0 \leq y_2 \leq 1.$$

Draw the feasible region, central path, and solution set for

2.7. EXERCISES

1. $b = (1;0)$;
2. $b = (1;1)/2$;
3. $b = (2;1)/3$.

Finally, sketch a neighborhood for the third central path.

2.15 *Prove (ii) of Theorem 2.17.*

2.16 *Prove Theorem 2.18.*

Chapter 3

Computation of Analytic Center

As we mentioned in the preceding chapter, a favorable property of the analytic center is that it is relatively easy to compute. In this chapter, we discuss how to compute the analytic center using the dual, primal, and primal-dual algorithms in three situations: 1) from an approximate analytic center, 2) from an interior-point, and 3) from an exterior point.

3.1 Proximity to Analytic Center

Before we introduce numerical procedures to compute it, we need to discuss how to measure proximity to the analytic center. Recall that Ω (or $\mathcal{S}_\Omega$) is a bounded polytope in $\mathcal{R}^m$ (or $\mathcal{R}^n$) defined by n ($> m$) linear inequalities, i.e.,

$$\Omega = \{y \in \mathcal{R}^m : s = c - A^T y \geq 0\}$$

or

$$\mathcal{S}_\Omega = \{s \in \mathcal{R}^n : s = c - A^T y \geq 0 \text{ for some } y\}.$$

For a point $y \in \overset{\circ}{\Omega}$, the potential function of Ω is given by (2.1), simply written by $\mathcal{B}(y)$ in this section. Ideally, a measure of closeness of $y \in \Omega$ to the analytic center y^a would be

$$\mathcal{B}(y^a) - \mathcal{B}(y) = \left(\max_{y \in \Omega} \mathcal{B}(y)\right) - \mathcal{B}(y).$$

The problem is that we have no knowledge of y^a or $\mathcal{B}(y^a)$.

82   CHAPTER 3. COMPUTATION OF ANALYTIC CENTER

Since y^a is the maximizer of $\mathcal{B}(y)$, a measure would be the residual of the optimality condition (2.2) at y, which can be rewritten as

$$\nabla \mathcal{B}(y) = AS^{-1}e = 0.$$

Note that the negative Hessian

$$-\nabla^2 \mathcal{B}(y) = AS^{-2}A^T,$$

is a positive definite matrix for any $s > 0$. Therefore, for one measure of the proximity, we consider the $(-\nabla^2 \mathcal{B}(y)^{-1})$-norm of the gradient vector $\nabla \mathcal{B}(y)$

$$\begin{align} \eta_d(s)^2 &:= \|\nabla \mathcal{B}(y)\|^2_{-\nabla^2 \mathcal{B}(y)^{-1}} \\ &= \|AS^{-1}e\|^2_{(AS^{-2}A^T)^{-1}} \\ &= e^T S^{-1} A^T (AS^{-2}A^T)^{-1} AS^{-1} e \\ &= \|p(s)\|^2, \end{align} \tag{3.1}$$

where

$$\begin{align} p(s) &:= S^{-1} A^T (AS^{-2}A^T)^{-1} \nabla \mathcal{B}(y) \\ &= -S^{-1} A^T (AS^{-2}A^T)^{-1} AS^{-1} e. \end{align} \tag{3.2}$$

The vector $p(s)$ can be viewed the normalized gradient vector of $\mathcal{B}$ at $s = c - A^T y$. It can also be viewed the scaled Newton direction; see Section 3.2.1 below.

Setting

$$x(s) = S^{-1}(I - S^{-1} A^T (AS^{-2} A^T)^{-1} AS^{-1})e, \tag{3.3}$$

we have

$$Ax(s) = 0,$$

$$p(s) = Sx(s) - e \quad \text{and} \quad \eta_d(s) = \|Sx(s) - e\|.$$

If $\eta_d(s) = 0$, then $\nabla \mathcal{B}(y) = 0$, $y = y^a$, $s = s^a$, and also $x(s) = x^a$ which minimizes the primal (homogeneous) potential function $\mathcal{P}(x, \Omega)$ of (2.8) for $x \in \overset{\circ}{\mathcal{X}}_\Omega$.

Denote $\mathcal{P}(x, \Omega)$ simply by $\mathcal{P}(x)$ in this section, i.e.,

$$\mathcal{P}(x) = n \log(c^T x) - \sum_{j=1}^{n} \log x_j.$$

As we discussed earlier, this quantity represents the logarithmic volume of a coordinate-aligned ellipsoid that contains $\mathcal{S}_\Omega$, and the minimization of the potential function results in the analytic centers of Ω or $\mathcal{S}_\Omega$.

3.1. PROXIMITY TO ANALYTIC CENTER

Another measure is to use the norm of a scaled gradient projection of the primal potential function. Note that $\mathcal{P}(x)$ is homogeneous of degree 0 so that we may fix $c^T x = n$. Then

$$\nabla \mathcal{P}(x) = \frac{n}{c^T x} c - X^{-1} e = c - X^{-1} e.$$

The scaled gradient projection of $X(c - X^{-1}e)$ onto the null space of AX becomes

$$\begin{aligned} p(x) &:= (I - XA^T(AX^2A^T)^{-1}AX)X(c - X^{-1}e) \\ &= (I - XA^T(AX^2A^T)^{-1}AX)(Xc - e). \end{aligned}$$

In general (including the case $c^T x \ne n$), we let

$$p(x) := (I - XA^T(AX^2A^T)^{-1}AX)(Xc - e) \tag{3.4}$$

and

$$\begin{aligned} \eta_p(x)^2 &:= \|p(x)\|^2 \\ &= (Xc - e)^T (I - XA^T(AX^2A^T)^{-1}AX)(Xc - e). \end{aligned} \tag{3.5}$$

$p(x)$ can be viewed the scaled Newton direction as well; see Section 3.3.1 below.

Let

$$\begin{aligned} y(x) &= (AX^2A^T)^{-1}AX(Xc - e) = (AX^2A^T)^{-1}AX^2c \\ s(x) &= c - A^T y(x). \end{aligned} \tag{3.6}$$

(Recall that $\bar{s} = s(x)$ is the center of the containing ellipsoid E_x discussed in Section 2.3.1.) Then, we have

$$p(x) = Xs(x) - e \quad \text{and} \quad \eta_p(x) = \|Xs(x) - e\|.$$

Clearly, if $\eta_p(x) = 0$, $x = x^a$ and $y(x) = y^a$, and $s(x) = s^a$.

The third measure is to use both the primal and dual. For an $x \in \mathcal{X}_\Omega$ and a $y \in \Omega$ or $s = c - A^T y \in \mathcal{S}_\Omega$, the measure would be defined as

$$\eta(x, s) := \|Xs - e\| = \|Xs - e\|. \tag{3.7}$$

Again, if $\eta(x, s) = 0$, $x = x^a$, $y = y^a$, $s = s^a$, and $\psi_n(x, s) = n \log n$, where recall the primal-dual potential function

$$\psi_n(x, s) = n \log(x^T s) - \sum_{j=1}^n \log(x_j s_j).$$

Theorem 3.2 below proves the "equivalence" of these measures and others. The upshot is that if one measure of proximity is near 0, they are all near 0 and an approximate analytic center can be found using any of them. Therefore, we may use whichever measure suits our needs.

We first present a lemma whose proof is Exercise 3.1.

Lemma 3.1 *If $d \in \mathcal{R}^n$ such that $\|d\|_\infty < 1$ then*

$$e^T d \geq \sum_{i=1}^{n} \log(1 + d_i) \geq e^T d - \frac{\|d\|^2}{2(1 - \|d\|_\infty)}.$$

We present the following theorem to equalize these measures.

Theorem 3.2 *Let (y, s) be an interior point and (y^a, s^a) be the analytic center of Ω or $\mathcal{S}_\Omega$, and let x be an interior point of $\mathcal{X}_\Omega$ and x^a be the primal potential minimizer with $c^T x^a = n$.*

i)

$$\eta_d(s) \leq \eta(x, s)$$

and

$$\eta_p(x) \leq \eta(x, s).$$

Conversely, if $\eta_d(s) < 1$ then there is an $x(s) \in \overset{\circ}{\mathcal{X}}_\Omega$ such that

$$\eta(x(s), s) \leq \eta_d(s),$$

and if $\eta_p(x) < 1$ then there is a $s(x) \in \overset{\circ}{\mathcal{S}}_\Omega$ such that

$$\eta(x, s(x)) \leq \eta_p(x).$$

ii) *If $\eta(x, s) < 1$, then there is a $\hat{x} \geq 0$ with $A\hat{x} = 0$ and $c^T \hat{x} = s^T \hat{x} = n$ such that*

$$\eta(\hat{x}, s) \leq \frac{\eta(x, s)}{\sqrt{1 - \eta(x, s)^2/n}}.$$

iii) *If $\eta(x, s) < 1$ with $c^T x = n$, then*

$$\psi_n(x, s) - \psi_n(x^a, s^a) \leq \frac{\eta(x, s)^2}{2(1 - \eta(x, s))},$$

$$\mathcal{P}(x) - \mathcal{P}(x^a) \leq \frac{\eta(x, s)^2}{2(1 - \eta(x, s))},$$

and

$$\mathcal{B}(y^a) - \mathcal{B}(y) \leq \frac{\eta(x, s)^2}{2(1 - \eta(x, s))}.$$

3.1. PROXIMITY TO ANALYTIC CENTER

iv) *If $\eta(x,s) < 1$, then*

$$\|S^{-1}s^a - e\| \leq \frac{\eta(x,s)}{1-\eta(x,s)}$$

and

$$\|X^{-1}x^a - e\| \leq \frac{\eta(x,s)}{1-\eta(x,s)}.$$

Proof. i) Given $s > 0$ we can verify that $\eta_d(s)$ is the minimal value and $x(s)$ is the minimizer of the least-squares problem

$$\begin{array}{ll} \text{minimize} & \|Sx - e\| \\ \text{s.t.} & Ax = 0. \end{array}$$

Since x in $\eta(x,s)$ is a feasible point for this problem, we must have

$$\eta_d(s) = \eta(x(s), s) \leq \eta(x,s).$$

Conversely, setting $x = x(s)$ will do it.

Similarly, given $x > 0$ we can verify that $\eta_p(x)$ is the minimal value and $(y(x), s(x))$ is the minimizer of the least-squares problem

$$\begin{array}{ll} \text{minimize} & \|Xs - e\| \\ \text{s.t.} & s = c - A^T y. \end{array}$$

Since (y,s) in $\eta(x,s)$ is any point for this problem, we must have

$$\eta_p(x) = \eta(x, s(x)) \leq \eta(x,s).$$

Conversely, setting $y = y(x)$ and $s = s(x)$ will do it.

ii) Let $\hat{x} = (n/x(s)^T s)x(s)$ and $\eta(\hat{x}, s) = \|S\hat{x} - e\|$. Then

$$A\hat{x} = 0 \quad \text{and} \quad c^T \hat{x} = s^T \hat{x} = n.$$

Furthermore,

$$\begin{aligned} \eta_d(s)^2 &= \|Sx(s) - e\|^2 \\ &= \|Sx(s) - (x(s)^T s/n)e + (x(s)^T s/n)e - e\|^2 \\ &= \|Sx(s) - (x(s)^T s/n)e\|^2 + \|(x(s)^T s/n)e - e\|^2 \\ &= \|S\hat{x} - e\|^2 (x(s)^T s/n)^2 + (1 - x(s)^T s/n)^2 n \\ &= \eta(\hat{x}, s)^2 (x(s)^T s/n)^2 + (1 - x(s)^T s/n)^2 n. \end{aligned}$$

Thus, we have

$$(x(s)^T s/n)^2 (\eta(\hat{x}, s)^2 + n) - 2(x(s)^T s/n)n + n - \eta_d(s)^2 = 0.$$

Consider this relation as a quadratic equation with variable $x(s)^T s/n$. Since it has a real root, we have
$$4n^2 - 4(\eta(\hat{x}, s)^2 + n)(n - \eta_d(s)^2) \geq 0$$
or
$$\eta(\hat{x}, s)^2 \leq \frac{n\eta_d(s)^2}{n - \eta_d(s)^2},$$
which gives the desired result.

iii) Let $\eta = \eta(x, s) < 1$. From Lemma 3.1 and $c^T x = s^T x = n$,
$$\sum_{j=1}^n \log(x_j s_j) \geq e^T X s - n - \frac{\eta^2}{2(1-\eta)} = -\frac{\eta^2}{2(1-\eta)}.$$

Denote by x^a and y^a ($s^a = c - A^T y^a$) the center pair of Ω. Noting that $X^a s^a = e$, we have
$$\psi_n(x, s) - \psi_n(x^a, s^a) = \sum_{j=1}^n \log(x_j^a s_j^a) - \sum_{j=1}^n \log(x_j s_j) \leq \frac{\eta^2}{2(1-\eta)}. \quad (3.8)$$

The left-hand side of (3.8) can be written as
$$\sum_{j=1}^n \log x_j^a - \sum_{j=1}^n \log x_j + \sum_{j=1}^n \log s_j^a - \sum_{j=1}^n \log s_j = \mathcal{P}(x) - \mathcal{P}(x^a) + \mathcal{B}(y^a) - \mathcal{B}(y).$$

Since y^a maximizes $\mathcal{B}(y)$ over the interior of Ω and x^a minimizes $\mathcal{P}(x)$ over the interior of $\mathcal{X}$, we have
$$\mathcal{B}(y^a) - \mathcal{B}(y) \geq 0$$
and
$$\mathcal{P}(x) - \mathcal{P}(x^a) \geq 0.$$
Thus, we have the desired result.

iv) From (i), $\eta_d(s) \leq \eta(x, s)$. Let $y^0 = y$, $s^0 = s$, and the sequence $\{y^k, s^k\}$ be generated by the dual Newton procedure (3.9) in the next section. Then, using the result of Theorem 3.3 we show by induction that
$$\|(S^0)^{-1} s^k - e\| \leq \sum_{j=1}^{2^k - 1} \eta_d(s)^j.$$

Obviously, this relation is true for $k = 1$ by the definition of $\eta_d(s^0) = \eta_d(s)$. Now assuming it is true for k. Since we have
$$s^{k+1} = s^k - S^k(S^k x(s^k) - e),$$

3.2. DUAL ALGORITHMS

$$\begin{aligned}
\|(S^0)^{-1}(s^{k+1} - s^k)\| &= \|(S^0)^{-1}S^k(S^k x(s^k) - e)\| \\
&\leq \|(S^0)^{-1}S^k\| \|(S^k x(s^k) - e)\| \\
&\leq \left(1 + \sum_{j=1}^{2^k-1} \eta_d(s)^j\right) \|(S^k x(s^k) - e)\| \\
&\leq \left(1 + \sum_{j=1}^{2^k-1} \eta_d(s)^j\right) \eta_d(s)^{2^k} \\
&= \left(\sum_{j=0}^{2^k-1} \eta_d(s)^j\right) \eta_d(s)^{2^k}.
\end{aligned}$$

Thus,

$$\begin{aligned}
\|(S^0)^{-1}s^{k+1} - e\| &\leq \|(S^0)^{-1}s^k - e\| + \|(S^0)^{-1}(s^{k+1} - s^k)\| \\
&\leq \left(\sum_{j=1}^{2^k-1} \eta_d(s)^j\right) + \left(\sum_{j=0}^{2^k-1} \eta_d(s)^j\right) \eta_d(s)^{2^k} \\
&= \sum_{j=1}^{2^{k+1}-1} \eta_d(s)^j.
\end{aligned}$$

Similarly, we can prove the primal result using the primal Newton procedure 3.13 and Theorem 3.8 below.

□

3.2 Dual Algorithms

In this section we consider dual algorithms in which we need only a dual initial point $y \in \mathcal{R}^m$. These algorithms will primarily update dual solutions that converge to the analytic center of Ω; primal solutions will come as by-products. Each solution is generated by solving a system of linear equations. Thus, the problem of computing the analytic center is solved by separating it into a sequence of simple problems.

3.2.1 Dual Newton procedure

Given $y \in \Omega$, we call it an *η-approximate (analytic) center* if $\eta_d(s) \leq \eta < 1$. The dual Newton procedure would start from such a (y, s) and any x with $Ax = 0$ (e.g., $x = 0$).

Rewrite the system of equations (2.2) as
$$\begin{aligned} x - S^{-1}e &= 0 \\ Ax &= 0 \\ -A^T y - s + c &= 0, \end{aligned}$$
and apply the Newton step:
$$\begin{aligned} d_x + S^{-2} d_s &= -x + S^{-1}e \\ A d_x &= 0 \\ -A^T d_y - d_s &= 0. \end{aligned}$$

Multiplying A to the top equation and noting $Ax = 0$ and $Ad_x = 0$, we have
$$A S^{-2} d_s = A S^{-1} e,$$
which together with the third equation give
$$d_y = -(A S^{-2} A^T)^{-1} A S^{-1} e \quad \text{and} \quad d_s = A^T (A S^{-2} A^T)^{-1} A S^{-1} e. \tag{3.9}$$
(Thus, to compute (d_y, d_s) we don't need any x.) Finally, we update (y, s) to
$$y^+ := y + d_y \quad \text{and} \quad s^+ := s + d_s.$$
Note that from (3.2) and (3.3)
$$d_s = -S p(s) = -S(Sx(s) - e).$$

We can estimate the effectiveness of the procedure via the following theorem.

Theorem 3.3 *Let (y, s) be an interior point and $\eta_d(s) < 1$. Then,*
$$s^+ > 0 \quad \text{and} \quad \eta_d(s^+) \leq \eta_d(s)^2.$$

Proof. Note that from the proof of (i) of Theorem 3.2
$$\eta_d(s^+) = \|S^+ x(s^+) - e\| \leq \|S^+ x(s) - e\|,$$
where $x(s) > 0$. But
$$\begin{aligned} \|S^+ x(s) - e\|^2 &= \|(2S - S^2 x(s)) x(s) - e\|^2 \\ &= \sum_{j=1}^n (s_j x(s)_j - 1)^4 \\ &\leq (\sum_{j=1}^n (s_j x(s)_j - 1)^2)^2 \\ &= \|Sx(s) - e\|^4 \\ &= \eta_d(s)^4 < 1. \end{aligned}$$
$\square$

3.2. DUAL ALGORITHMS

Therefore, the dual Newton procedure, once a (y, s) satisfies $\eta_d(s) \le \eta < 1$, will generate a sequence $\{y^k\}$ that converges to y^a quadratically. The next question is how to generate an η-approximate center with $\eta < 1$.

3.2.2 Dual potential algorithm

Let y be any interior point in Ω. Let us apply the Newton procedure with a controlled step-size:

$$y^+ = y - \frac{\alpha}{\eta_d(s)}(AS^{-2}A^T)^{-1}AS^{-1}e,$$

where constant $0 < \alpha < 1$. Observe that if $\alpha = \eta_d(s) = \|p(s)\|$ of (3.2), this reduces to the Newton procedure. In general, $\eta_d(s) \ge 1$, which implies the step size $\alpha/\eta_d(s) < 1$.

Let

$$d_y = -\frac{\alpha}{\eta_d(s)}(AS^{-2}A^T)^{-1}AS^{-1}e$$

and

$$d_s = -A^T d_y,$$

which is the maximizer of the ball-constrained linear problem with a radius α:

$$\begin{aligned}\text{maximize} \quad & \nabla \mathcal{B}(y)^T d_y \\ \text{s.t.} \quad & d_y^T(-\nabla^2 \mathcal{B}(y))d_y \le \alpha^2,\end{aligned}$$

or

$$\begin{aligned}\text{maximize} \quad & -e^T S^{-1} A^T d_y \\ \text{s.t.} \quad & d_y^T AS^{-2}A^T d_y \le \alpha^2.\end{aligned}$$

Note that

$$s^+ = s + d_s = s - A^T d_y = S(e - S^{-1}A^T d_y).$$

Thus we must have $s^+ > 0$, since $\|S^{-1}d_s\| = \|S^{-1}A^T d_y\| \le \alpha < 1$, i.e., (y^+, s^+) remains in the interior of Ω.

Recall from (3.2) and (3.1)

$$p(s) = -S^{-1}A^T(AS^{-2}A^T)^{-1}AS^{-1}e = Sx(s) - e.$$

Then,

$$d_y = -\frac{\alpha}{\|p(s)\|}(AS^{-2}A^T)^{-1}AS^{-1}e,$$

$$d_s = -A^T d_y = -\frac{\alpha}{\|p(s)\|}Sp(s),$$

and

$$e^T S^{-1} d_s = \alpha \|p(s)\|^2.$$

Furthermore, $s^+ = s + d_s = s - \alpha Sp(s)/\|p(s)\| = S(e - \alpha p(s)/\|p(s)\|)$, i.e., $\|S^{-1}s^+ - e\| = \alpha$. Thus, from Lemma 3.1 we have

$$\begin{aligned}\mathcal{B}(y^+) - \mathcal{B}(y) &\geq e^T S^{-1}(s^+ - s) - \frac{\|S^{-1}(s^+ - s)\|^2}{2(1 - \|S^{-1}(s^+ - s)\|_\infty)} \\ &\geq \alpha\|p(s)\| - \frac{\alpha^2}{2(1-\alpha)}.\end{aligned} \quad (3.10)$$

Hence as long as $\|p(s)\| \geq 3/4$ we have

$$\mathcal{B}(y^+) - \mathcal{B}(y) \geq \delta,$$

where constant

$$\delta = 3\alpha/4 - \frac{\alpha^2}{2(1-\alpha)} > 0.$$

Note that if $\alpha = 1/3$ then $\delta \geq 1/6$. In other words, the potential function is increased by a constant. Note that the potential function is bounded above by the max-potential $\mathcal{B}(y^a)$. Thus, in a finite time we must have $\eta_d(s) = \|p(s)\| < 3/4 < 1$, which implies that the quadratic-convergence condition is satisfied or y is an 3/4-approximate center.

Theorem 3.4 *The number of total iterations of the dual potential algorithm to generate an approximate analytic center of y^a of Ω, starting from $y \in \overset{\circ}{\Omega}$, is bounded by $O(\mathcal{B}(y^a) - \mathcal{B}(y))$.*

Recall that $\mathcal{B}(y)$ represents the logarithmic volume of the coordinate-aligned ellipsoid centered at $s \in \mathcal{S}_\Omega$ and inscribing $\mathcal{R}_+^n$. Thus, the dual potential algorithm generates a sequence of ellipsoids whose volumes increases at a constant rate; see Figure 3.1.

3.2.3 Central-section algorithm

The next question is how to compute the analytic center if an interior point y is not known. This can be done with a central-section method. Consider the following set:

$$\Omega(\hat{c}) = \{y \in \mathcal{R}^m : A^T y \leq \hat{c}\}.$$

Obviously, $\Omega(c) = \Omega$, whose interior is assumed nonempty and bounded. Then, for any given $\hat{c} \geq c$, $\overset{\circ}{\Omega}(\hat{c})$ is also nonempty and bounded. Moreover, $y = 0$ is an interior point in $\Omega(\hat{c})$ if $\hat{c} > 0$. Let us choose c^0 such that $c^0 \geq e$ and $c^0 \geq c$. Then from the result in the preceding section we will generate an approximate center for $\Omega^0 := \Omega(c^0)$ in $O(\mathcal{B}(\Omega^0) - \mathcal{B}(0, \Omega^0))$ iterations.

3.2. DUAL ALGORITHMS

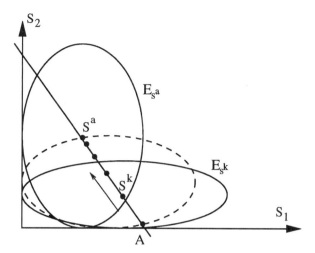

Figure 3.1. Illustration of the dual potential algorithm; it generates a sequence of contained coordinate-aligned ellipsoids whose volumes increase.

Let y^0 be an approximate center for Ω^0. From now on, we will generate a sequence of $\{c^k\}$ and $\{y^k\}$, where $c \leq c^{k+1} \leq c^k$ and y^k is an approximate center for $\Omega^k := \Omega(c^k)$. Moreover,

$$\mathcal{B}(\Omega) \leq \mathcal{B}(\Omega^{k+1}) \leq \mathcal{B}(\Omega^k) - \delta,$$

where δ is a positive constant, until $c^{k+1} = c$. This process terminates with y^{k+1} as an approximate center for $\Omega = \Omega(c)$.

We first describe a conceptual central-section algorithm.

Algorithm 3.1 (Conceptual Algorithm) Let (y^0, s^0) be the analytic center of $\Omega^0 = \Omega(c^0)$ and let β be a constant in $(0,1)$. Set $k := 0$.
While $c^k \neq c$ **do**:

1. *Translating Inequality:* Find i such that $c_i^k > c_i$ and update

$$c_i^{k+1} = \max\{c_i, \beta(c_i^k - a_i^T y^k) + a_i^T y^k\} = \max\{c_i, \beta s_i^k + a_i^T y^k\},$$

$$c_j^{k+1} = c_j^k \quad for \quad j \neq i.$$

Then, from the central-section theorem in the preceding chapter, we have either

$$\mathcal{B}(\Omega^{k+1}) \leq \mathcal{B}(\Omega^k) - (1-\beta) \quad if \quad c_i^{k+1} > c_i; \qquad (3.11)$$

or
$$\mathcal{B}(\Omega^{k+1}) < \mathcal{B}(\Omega^k) \quad \text{if} \quad c_i^{k+1} = c_i. \tag{3.12}$$

Note that the latter case can happen only n times.

2. *Updating Center:* Compute the center y^{k+1} of Ω^{k+1}, using Newton's method from y^k which is an approximate center of Ω^{k+1}.

3. Let $k := k+1$ and return to Step 1.

Clearly, the central-section algorithm will stop after $O(\mathcal{B}(\Omega^0) - \mathcal{B}(\Omega)) + n$ iterations. If
$$\mathcal{B}(\Omega^0) \leq n \log R \quad \text{and} \quad \mathcal{B}(\Omega) \geq n \log r,$$
then, $O(n \log(R/r)) + n$ iterations suffice.

Numerically, we will never be able to compute the exact analytic center. We must use approximate centers instead of perfect centers in the central-section algorithm. We discuss this issue now.

Algorithm 3.2 (Using Approximate Centers) *Let (y^0, s^0) be an approximate analytic center of $\Omega^0 = \Omega(c^0)$, with $\eta_d(s^0) \leq \eta < 1$, and let β be a constant in $(0, 1)$ such that $\eta + (1 - \beta)(1 + \eta) < 1$. Set $k := 0$.*
While $c^k \neq c$ **do:**

1. *Translating Inequality:* Find i such that $c_i^k > c_i$ and update
$$c_i^{k+1} = \max\{c_i, \beta s_i^k + a_i^T y^k\},$$
$$c_j^{k+1} = c_j^k \quad \text{for} \quad j \neq i.$$

2. *Updating Approximate Center:* Compute an approximate analytic center y^{k+1} of Ω^{k+1} so that $\eta_d(s^{k+1}) \leq \eta$, using one of the Newton procedures in Theorem 3.2 starting from y^k, which is an approximate center of Ω^{k+1}.

3. Let $k := k + 1$ and return to Step 1.

To show that y^k is an approximate center of Ω^{k+1}, we prove the following Lemma:

Lemma 3.5 *There exists a point $x^+ > 0$ such that*
$$Ax^+ = 0 \quad \text{and} \quad \|X^+ s^+ - e\| \leq \eta + (1 - \beta)(1 + \eta) < 1,$$
where $s^+ = c^{k+1} - A^T y^k$.

3.2. DUAL ALGORITHMS

Proof. Let $x^+ = x(s^k) > 0$ with $s^k = c^k - A^T y^k$. Then,

$$Ax^+ = Ax(s^k) = 0$$

and

$$\|x^+ s^k - e\| = \eta_d(s^k) \leq \eta.$$

Note that $s_j^+ = s_j^k$ for $j \neq i$ and $s_i^k \geq s_i^+ \geq \beta s_i^k$. Thus,

$$\begin{aligned}
\|X^+ s^+ - e\| &= \|X^+ s^k - e + X^+ s^+ - X^+ s^k\| \\
&\leq \|X^+ s^k - e\| + \|X^+ s^+ - X^+ s^k\| \\
&\leq \|X^+ s^k - e\| + |x_i^+ s_i^k(\beta - 1)| \\
&\leq \eta + (1 - \beta)(1 + \eta).
\end{aligned}$$

□

Lemma 3.5 shows that, after an inequality is translated, y^k is still in the "quadratic convergence" region of the center of Ω^{k+1}, because we choose $\eta + (1 - \beta)(1 + \eta) < 1$. Thus, a closer approximate center, y^{k+1} with $\eta_d(s^{k+1}) \leq \eta$ for Ω^{k+1}, can be updated from y^k in a constant number of Newton's steps. We now verify that the potential function is still reduced by a constant for a small η after a translation.

Lemma 3.6 *Let (y^k, s^k) be an approximate center for Ω^k with $\eta_d(s^k) \leq \eta$ and let Ω^{k+1} be defined above. Then, if $c_i^{k+1} > c_i$ in the update*

$$\mathcal{B}(\Omega^{k+1}) \leq \mathcal{B}(\Omega^k) - \delta \quad \text{for a constant} \quad \delta > 0;$$

otherwise $c_i^{k+1} = c_i$ and

$$\mathcal{B}(\Omega^{k+1}) \leq \mathcal{B}(\Omega^k).$$

The latter case can happen only n times.

Proof. The proof of the first case is similar to Theorem 2.7 of Chapter 2. Let (y^a, s^a) and (y_+^a, s_+^a) be the centers of Ω^k and Ω^{k+1}, respectively. Note that $A(S^a)^{-1} e = 0$, $(c^k)^T (S^a)^{-1} e = (s^a)^T (S^a)^{-1} e = n$, and

$$s_+^a = c^{k+1} - A^T y_+^a,$$

where

$$c_j^{k+1} = c_j^k, \; j \neq i \quad \text{and} \quad c_i^{k+1} = c_i^k - (1 - \beta)s_i^k.$$

Note that we still have

$$e^T (S^a)^{-1} s_+^a = e^T (S^a)^{-1} c^{k+1} = n - (1 - \beta)(s_i^k / s_i^a) \leq n - (1 - \beta)(1 - \eta).$$

The last inequality is due to (v) of Theorem 3.2. Therefore,

$$\frac{\exp(\mathcal{B}(\Omega^{k+1}))}{\exp(\mathcal{B}(\Omega^k))} = \prod_{j=1}^{n}\frac{(s_+^a)_j}{s_j^a}$$
$$\leq (\frac{n-(1-\beta)(1-\eta)}{n})^n$$
$$\leq \exp(-(1-\beta)(1-\eta)).$$

The proof of the latter case is straightforward.

□

From the lemma, we can conclude that

Theorem 3.7 *In $O(\mathcal{B}(\Omega^0) - \mathcal{B}(\Omega)) + n$ central-section steps, the algorithm will generate an approximate analytic center for Ω.*

3.3 Primal Algorithms

In this section we consider primal algorithms, in which we need only a primal initial point $x \in \mathcal{X}_\Omega$. These algorithms will primarily update primal solutions that converge to x^a; dual solutions are generated as by-products. Like the dual algorithms, each solution is found by solving a system of linear equations. Primal algorithms have several advantages, especially when additional inequalities are added to Ω. We will discuss it in detail in Chapter 8.

3.3.1 Primal Newton procedure

Given $x \in \mathcal{X}_\Omega$, we call it an η-approximate (analytic) center if $\eta_p(x) \leq \eta < 1$. The primal Newton procedure will start from such an x and any (y, s) with $s = c - A^T y$ (e.g., $y = 0$ and $s = c$).

Rewrite the system of equations (2.2) as

$$\begin{aligned}-X^{-1}e + s &= 0 \\ Ax &= 0 \\ -A^T y - s + c &= 0,\end{aligned}$$

and apply the Newton step:

$$\begin{aligned}X^{-2}d_x + d_s &= X^{-1}e - s \\ Ad_x &= 0 \\ -A^T d_y - d_s &= 0.\end{aligned}$$

3.3. PRIMAL ALGORITHMS

Multiplying AX^2 to the top equation and noting $Ad_x = 0$, we have

$$AX^2 d_s = -AX(Xs - e),$$

which together with the third equation give

$$d_y = (AX^2 A^T)^{-1} AX(Xs - e) \text{ and } d_s = -A^T (AX^{-2} A^T)^{-1} AX(Xs - e),$$

and

$$\begin{aligned} d_x &= -X(I - XA^T(AX^2A^T)^{-1}AX)(Xs - e) \\ &= -X(I - XA^T(AX^2A^T)^{-1}AX)(Xc - e). \end{aligned} \quad (3.13)$$

The last equality is due to $s = c - A^T y$ and $(I - XA^T(AX^2A^T)^{-1}AX)XA^T = 0$. (Thus, to compute d_x we don't need any (y, s).) Finally, we update x to

$$x^+ := x + d_x.$$

Note that from (3.4) and (3.6)

$$d_x = -Xp(x) = -X(Xs(x) - e).$$

We can estimate the effectiveness of the procedure via the following corollary:

Corollary 3.8 *Let x be an interior point of $\mathcal{X}_\Omega$ and $\eta_p(x) < 1$. Then,*

$$x^+ > 0, \quad Ax^+ = 0, \quad \text{and} \quad \eta_p(x^+) \leq \eta_p(x)^2.$$

We leave the proof of the corollary as an Exercise.

We see the primal Newton procedure, once an $x \in \mathcal{X}_\Omega$ satisfies $\eta_p(x) \leq \eta < 1$, will generate a sequence $\{x^k\}$ that converges to x^a quadratically. The next question is how to generate an η-approximate center with $\eta < 1$.

3.3.2 Primal potential algorithm

Consider the primal potential function $\mathcal{P}(x)$ for $x \in \overset{\circ}{\mathcal{X}}_\Omega = \{x : Ax = 0, x > 0\}$. Again, this quantity represents the logarithmic volume of a coordinate-aligned ellipsoid that contains $\mathcal{S}_\Omega$.

The analysis of the remaining two algorithms in this section requires a simple fact about the primal potential function. Given $x \in \overset{\circ}{\mathcal{R}}^n_+$ and $d_x \in \mathcal{R}^n$, let $x^+ = x + d_x$ and $\|X^{-1}d_x\| < 1$. Then, from the concavity of log function we have

$$n \log(c^T x^+) - n \log(c^T x) \leq \frac{n}{c^T x} c^T (x^+ - x) = \frac{n}{c^T x} c^T d_x,$$

and from Lemma 3.1 we have

$$-\sum \log x_j^+ + \sum \log x_j \leq -e^T X^{-1} d_x + \frac{\|X^{-1} d_x\|^2}{2(1 - \|X^{-1} d_x\|_\infty)}.$$

Thus,

$$\begin{aligned} \mathcal{P}(x^+) - \mathcal{P}(x) &\leq \frac{n}{c^T x} c^T d_x - e^T X^{-1} d_x + \frac{\|X^{-1} d_x\|^2}{2(1 - \|X^{-1} d_x\|_\infty)} \\ &= \nabla \mathcal{P}(x)^T d_x + \frac{\|X^{-1} d_x\|^2}{2(1 - \|X^{-1} d_x\|_\infty)}. \end{aligned} \qquad (3.14)$$

Karmarkar's algorithm

We now describe Karmarkar's algorithm to reduce the potential function. Since the primal potential function is homogeneous of degree 0, we can normalize $e^T x = n$ and work in the region

$$\mathcal{K}_p = \{x : \ Ax = 0, \ e^T x = n, \ x \geq 0\}.$$

This is the so-called Karmarkar canonical form. Its related LP canonical problem is given as

$$\begin{aligned} \text{minimize} \quad & c^T x \\ \text{s.t.} \quad & x \in \mathcal{K}_p. \end{aligned}$$

Starting from any x^0 in $\overset{\circ}{\mathcal{K}}_x$ we generate a sequence $\{x^k\}$ such that

$$\mathcal{P}(x^{k+1}) \leq \mathcal{P}(x^k) - 1/6$$

for $k = 0, 1, 2, \ldots$ until an approximate center of Ω is generated.

One observation regarding $\mathcal{K}_p$ is that if $Ae = 0$, e is the analytic center of $\mathcal{K}_p$. Unfortunately, in general we may not have $Ae = 0$. However, with a given $x^k > 0$ and $Ax^k = 0$, we may transform the LP problem into

$$\begin{aligned} (LP') \quad \text{minimize} \quad & (c^k)^T x' \\ \text{s.t.} \quad & x' \in \mathcal{K}'_p := \{x' : \ A^k x' = 0, \ e^T x' = n, \ x' \geq 0\}. \end{aligned}$$

where

$$c^k = X^k c \quad \text{and} \quad A^k = A X^k.$$

Note that if a pure affine scaling transformation,

$$x' = (X^k)^{-1} x,$$

3.3. PRIMAL ALGORITHMS

had been used, the last inequality constraint would become $e^T X^k x' = n$. But as we discussed before, the potential function is homogeneous, so that we can use a projective transformation,

$$x' = T(x) = \frac{n(X^k)^{-1} x}{e^T (X^k)^{-1} x} \quad \text{for} \quad x \in \mathcal{K}_p,$$

whose inverse transformation is

$$x = T^{-1}(x') = \frac{n X^k x'}{e^T X^k x'} \quad \text{for} \quad x' \in \mathcal{K}'_p.$$

Obviously, $T(x^k) = e$ is the analytic center for $\mathcal{K}'_p$. In other words, Karmarkar transforms x^k into the analytic center of $\mathcal{K}'_p$ in (LP').

Note that the potential function for (LP') is

$$\mathcal{P}'(x') = n \log((c^k)^T x') - \sum_{j=1}^{n} \log x'_j.$$

The difference of the potential function values at two points of $\overset{\circ}{\mathcal{K}}_p$ is invariant under this projective transformation, i.e.,

$$\mathcal{P}'(T(x^2)) - \mathcal{P}'(T(x^1)) = \mathcal{P}(x^2) - \mathcal{P}(x^1).$$

To reduce $\mathcal{P}'(x')$, one may reduce its linearized function

$$\nabla \mathcal{P}'(e)^T x' = (\frac{n}{(c^k)^T e} c^k - e)^T x' = (\frac{n}{c^T x^k} c^k - e)^T x' = \frac{n}{c^T x^k} (c^k)^T x' - n.$$

Since $c^T x^k$ and n are fixed, we simply solve the following ball-constrained linear problem:

$$\begin{aligned} \text{minimize} \quad & (c^k)^T (x' - e) \\ \text{s.t.} \quad & A^k(x' - e) = 0, \ e^T(x' - e) = 0, \ \|x' - e\| \le \alpha. \end{aligned}$$

According to the discussion in Chapter 1, the solution of the problem is

$$x' - e = -\alpha \frac{p^k}{\|p^k\|},$$

where

$$\begin{aligned} p^k &= \left(I - \begin{pmatrix} A^k \\ e \end{pmatrix}^T \begin{pmatrix} A^k(A^k)^T & 0 \\ 0 & n \end{pmatrix}^{-1} \begin{pmatrix} A^k \\ e \end{pmatrix} \right) c^k \\ &= c^k - (A^k)^T y^k - \lambda^k e, \end{aligned}$$

where $\lambda^k = (c^k)^T e/n = c^T x^k/n$, and
$$y^k = (A^k(A^k)^T)^{-1} A^k c^k.$$

Thus
$$p^k = X^k(c - A^T y^k) - (c^T x^k/n)e.$$

Using relation (3.14) we consider the difference of the potential values
$$\mathcal{P}'(x') - \mathcal{P}'(e) \leq \nabla \mathcal{P}'(e)(x' - e) + \frac{\alpha^2}{2(1-\alpha)} = -\alpha \frac{n\|p^k\|}{c^T x^k} + \frac{\alpha^2}{2(1-\alpha)}.$$

Thus, as long as $\|p^k\| \geq \frac{3}{4} \frac{c^T x^k}{n} > 0$, we have
$$\mathcal{P}'(x') - \mathcal{P}'(e) \leq -\delta,$$
for a positive constant $\delta = 3\alpha/4 - \alpha^2/2(1-\alpha)$. Again we have $\delta = 1/6$ if $\alpha = 1/3$. Let
$$x^{k+1} = T^{-1}(x').$$

Then,
$$\mathcal{P}(x^{k+1}) - \mathcal{P}(x^k) \leq -\delta.$$

Thus, in $O(\mathcal{P}(x^0) - \mathcal{P}(x^a))$ iterations, we shall generate a pair (x^k, y^k) such that
$$\|X^k(c - A^T y^k) - (c^T x^k/n)e\| \leq \frac{3}{4} \frac{c^T x^k}{n},$$
or
$$\|\frac{n}{c^T x^k} X^k(c - A^T y^k) - e\| < 3/4 < 1,$$
which indicates that y^k is an 3/4-approximate analytic center of Ω.

Affine potential algorithm

In this section, we use a simple affine scaling transformation to achieve a reduction in $\mathcal{P}$. Given $x \in \overset{\circ}{\mathcal{X}}_\Omega$ with $c^T x = n$, let us apply the Newton procedure with a controlled step-size:
$$x^+ = x - \frac{\alpha}{\eta_p(x)} X(I - XA^T(AX^2A^T)^{-1}AX)(Xc - e),$$

where constant $0 < \alpha < 1$. Observe that If $\alpha = \eta_p(x) = \|p(x)\|$ of (3.4), this reduces to the Newton procedure. In general, $\eta_p(x) \geq 1$, which implies the step size $\alpha/\eta_p(x) < 1$.

Let
$$d_x = -\frac{\alpha}{\eta_p(x)} X(I - XA^T(AX^2A^T)^{-1}AX)(Xc - e)$$

3.3. PRIMAL ALGORITHMS

and $d'_x = X^{-1}d_x$, which is an affine transformation. Then, d'_x is the minimizer of the ball-constrained linear problem with a radius α:

$$\begin{aligned}\text{minimize} \quad & \nabla \mathcal{P}(x)^T X d'_x \\ \text{s.t.} \quad & AX d'_x = 0, \ \|d'_x\| \leq \alpha,\end{aligned}$$

where the gradient vector of the potential function at x with $c^T x = n$ is

$$\nabla \mathcal{P}(x) = \frac{n}{c^T x} c - X^{-1} e = c - X^{-1} e.$$

Note that

$$x^+ = x + d_x = X(e + X^{-1} d_x) = X(e + d'_x).$$

Thus we must have $x^+ > 0$, since $\|d'_x\| = \alpha < 1$, i.e., $x^+ = x + d_x$ remains in the interior of $\mathcal{X}_\Omega$.

From (3.4) and (3.5) we have

$$d_x = -\frac{\alpha}{\|p(x)\|} X p(x),$$

and

$$\nabla \mathcal{P}(x)^T d_x = -\alpha \|p(x)\|^2.$$

Furthermore, $x^+ = X(e - \alpha p(x)/\|p(x)\|)$ and $\|X^{-1} x^+ - e\| = \alpha$. Thus, from (3.14) we have

$$\begin{aligned}\mathcal{P}(x^+) - \mathcal{P}(x) &\leq \nabla \mathcal{P}(x)^T (x^+ - x) + \frac{\|X^{-1}(x^+ - x)\|^2}{2(1 - \|X^{-1}(x^+ - x)\|_\infty)} \\ &\geq -\alpha \|p(x)\| + \frac{\alpha^2}{2(1 - \alpha)} \ .\end{aligned}$$

Hence as long as $\|p(x)\| \geq 3/4$ we have

$$\mathcal{P}(x^+) - \mathcal{P}(x) \leq -\delta,$$

where constant

$$\delta = 3\alpha/4 - \frac{\alpha^2}{2(1 - \alpha)} > 0.$$

In other words, the potential function is increased by a constant. Note that the potential function is bounded below by the min-potential $\mathcal{P}(x^a)$. Thus, in a finite time we must have $\eta_p(x) = \|p(x)\| < 3/4 < 1$, which implies that the quadratic-convergence condition is satisfied or x is an 3/4-approximate center of $\mathcal{X}_\Omega$.

Theorem 3.9 *The number of total iterations of the primal potential algorithm to generate an approximate analytic center of x^a, starting from $x \in \overset{\circ}{\mathcal{X}}_\Omega$, is bounded by $O(\mathcal{P}(x) - \mathcal{P}(x^a))$.*

Since $\mathcal{P}(x)$ represents the logarithmic volume of the coordinate-aligned ellipsoid centered at $s(x)$ and containing $\mathcal{S}_\Omega$, the primal potential algorithm generates a sequence of ellipsoids whose volumes decreases at a constant rate, see Figure 3.2.

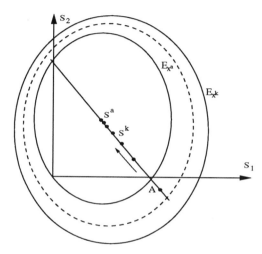

Figure 3.2. Illustration of the primal potential algorithm; it generates a sequence of containing coordinate-aligned ellipsoids whose volumes decrease.

3.3.3 Affine scaling algorithm

In the affine potential algorithm above, since $AXe = 0$ we have

$$p(x) = Xs(x) - e,$$

where again

$$y(x) = (AX^2A^T)^{-1}AX^2c \quad \text{and} \quad s(x) = c - A^Ty.$$

Here, $(y(x), s(x))$ is usually called the *dual affine scaling estimate* at x.

Because the problem is homogeneous in x, we have assumed that $c^Tx = n$. Then, for some step-size $\theta = \frac{\alpha}{\eta_p(x)} > 0$, the affine potential algorithm

3.4. PRIMAL-DUAL (SYMMETRIC) ALGORITHMS

produces
$$\begin{aligned} x^+ &= x - \theta X p(x) \\ &= (1+\theta)x - \theta X^2 s(x) \\ &= (1+\theta)X(e - \frac{\theta}{1+\theta} X s(x)). \end{aligned}$$

Again since scaling is immaterial, we simply let

$$x^+ := X(e - \frac{\theta}{1+\theta} X s(x)).$$

This update is called the *affine scaling algorithm step* and the vector $X^2 s(x) = X(I - XA^T(AX^2A^T)^{-1}AX)Xc$ is called the *affine scaling direction*.

Note that $e^T p(x) = c^T x - n = 0$, we must have

$$\max\{Xs(x)\} = \max\{p(x)\} + 1 \geq 1.$$

The following theorem can be established using Exercise 3.1 and similar analyses in the above section. Note that the step-size here is generally larger than the one previously used.

Theorem 3.10 *Choose*

$$\frac{\theta}{1+\theta} = \frac{\lambda}{\max\{Xs(x)\}}$$

for constant $0 < \lambda \leq 2/3$. Then,

$$\mathcal{P}(x^+) - \mathcal{P}(x) \leq -\frac{\lambda(2-3\lambda)}{2\sqrt{2}(1-\lambda)} \quad \text{if} \quad \lambda < 2/3$$

or

$$\mathcal{P}(x^+) - \mathcal{P}(x) \leq -\frac{\sqrt{2}}{3n} \quad \text{if} \quad \lambda = 2/3.$$

3.4 Primal-Dual (Symmetric) Algorithms

In this section we consider symmetric primal-dual algorithms in which both primal and dual variables are treated equally. These algorithms are most popular and effective in practice.

3.4.1 Primal-dual Newton procedure

Let (y, s) be an interior point of Ω or $\mathcal{S}_\Omega$, and let x be an interior point of $\mathcal{X}_\Omega$. The primal-dual Newton step for system (2.2) is:

$$\begin{aligned} Sd_x + Xd_s &= e - Xs, \\ Ad_x &= 0, \\ -A^T d_y - d_s &= 0. \end{aligned} \quad (3.15)$$

Multiplying AS^{-1} to the top equation and noting $Ax = 0$ and $Ad_x = 0$, we have

$$AXS^{-1}d_s = -AS^{-1}(Xs - e) = AS^{-1}e,$$

which together with the third equation give

$$d_y = -(AXS^{-1}A^T)^{-1}AS^{-1}e \quad \text{and} \quad d_s = A^T(AXS^{-1}A^T)^{-1}AS^{-1}e,$$

and

$$d_x = -S^{-1}(I - XA^T(AXS^{-1}A^T)^{-1}AS^{-1})(Xs - e).$$

Note that we need both primal and dual interior-points to start the procedure. We update (x, y, s) to

$$x^+ = x + d_x, \quad y^+ = y + d_y, \quad s^+ = s + d_s.$$

We can estimate the effectiveness of the procedure via the following theorem:

Theorem 3.11 *If the starting point of the Newton procedure satisfies $\eta(x, s) < 1$, then*

$$x^+ > 0, \quad Ax^+ = 0, \quad s^+ > 0$$

and

$$\eta(x^+, s^+) \leq \frac{\sqrt{2}\eta(x, s)^2}{4(1 - \eta(x, s))}.$$

Proof. To prove the result we first see that

$$\|X^+ s^+ - e\| = \|D_x d_s\|.$$

Multiplying the both sides of the first equation of (3.15) by $(XS)^{-1/2}$, we see

$$Dd_x + D^{-1}d_s = r := (XS)^{-1/2}(e - Xs),$$

3.4. PRIMAL-DUAL (SYMMETRIC) ALGORITHMS

where $D = S^{1/2}X^{-1/2}$. Let $p = Dd_x$ and $q = D^{-1}d_s$. Note that $p^T q = d_x^T d_s \geq 0$ (For LP $d_x^T d_s = 0$, but here we relax this property to include cases $d_x^T d_s > 0$ for solving other broader problems). Then,

$$
\begin{aligned}
\|D_x d_s\|^2 &= \|Pq\|^2 \\
&= \sum_{j=1}^{n}(p_j q_j)^2 \\
&\leq \left(\sum_{p_j q_j > 0} p_j q_j\right)^2 + \left(\sum_{p_j q_j < 0} p_j q_j\right)^2 \\
&\leq 2\left(\sum_{p_j q_j > 0}^{n} p_j q_j\right)^2 \\
&\leq 2\left(\sum_{p_j q_j > 0}^{n} (p_j + q_j)^2/4\right)^2 \\
&\leq 2\left(\|r\|^2/4\right)^2.
\end{aligned}
$$

Furthermore,

$$\|r\|^2 \leq \|(XS)^{-1/2}\|^2 \|e - Xs\|^2 \leq \frac{\eta^2(x,s)}{1 - \eta(x,s)},$$

which gives the desired result. We leave the proof of $x^+, s^+ > 0$ as an Exercise.

□

Clearly, the primal-dual Newton procedure, once an approximate center pair pair $x \in \mathcal{X}_\Omega$ and $s \in \mathcal{S}_\Omega$ satisfies $\eta(x,s) \leq \eta < 2/3$, will generate a sequence $\{x^k, y^k, s^k\}$ that converges to (x^a, y^a, s^a) quadratically.

The next question is how to generate such a pair with

$$\eta(x,s) = \|Xs - e\| \leq \eta < 2/3, \tag{3.16}$$

3.4.2 Primal-dual potential algorithm

Given any $x \in \overset{\circ}{\mathcal{X}}_\Omega$ and $s \in \overset{\circ}{\mathcal{S}}_\Omega$ with $s^T x = n$, we show how to use the primal-dual algorithm to generate an approximate analytic center pair using the primal-dual potential function. We use the same strategy: apply the Newton procedure with a controlled step-size.

Lemma 3.12 *Let the directions (d_x, d_y, d_s) be generated by equation (3.15), and let*

$$\theta = \frac{\alpha\sqrt{\min(Xs)}}{\|(XS)^{-1/2}(e-Xs)\|}, \qquad (3.17)$$

where α is a positive constant less than 1 and $\min(v \in \mathcal{R}^n) = \min_j \{v_j|\ j = 1,...,n\}$. Then, we have

$$\psi_n(x + \theta d_x, s + \theta d_s) - \psi_n(x,s)$$
$$\leq -\alpha\sqrt{\min(Xs)}\|(XS)^{-1/2}(e-Xs)\| + \frac{\alpha^2}{2(1-\alpha)}.$$

Proof. It can be verified from the proof of the primal-dual Newton procedure that

$$\tau := \max(\|\theta S^{-1} d_s\|_\infty, \|\theta X^{-1} d_x\|_\infty) < 1.$$

This implies that

$$x^+ := x + \theta d_x > 0 \quad \text{and} \quad s^+ := s + \theta d_s > 0.$$

Then, from Lemma 3.1 and (3.14) we derive

$$\psi_n(x^+, s^+) - \psi_n(x,s)$$
$$\leq \theta e^T(Xd_s + Sd_x) - \theta e^T(S^{-1}d_s + X^{-1}d_x)$$
$$+ \frac{\|\theta S^{-1} d_s\|^2 + \|\theta X^{-1} d_x\|^2}{2(1-\tau)}.$$

The choice of θ in (3.17) implies that

$$\|\theta S^{-1} d_s\|^2 + \|\theta X^{-1} d_x\|^2 \leq \alpha^2.$$

Hence, we have $\tau \leq \alpha$ and

$$\frac{\|\theta S^{-1} d_s\|^2 + \|\theta X^{-1} d_x\|^2}{2(1-\tau)} \leq \frac{\alpha^2}{2(1-\alpha)}. \qquad (3.18)$$

Moreover,

$$\theta e^T(Xd_s + Sd_x) - \theta e^T(S^{-1}d_s + X^{-1}d_x)$$
$$= \theta\left(e^T(Xd_s + Sd_x) - e^T(S^{-1}d_s + X^{-1}d_x)\right)$$
$$= \theta\left(e^T(Xd_s + Sd_x) - e^T(XS)^{-1}(Xd_s + Sd_x)\right)$$
$$= \theta(e - (XS)^{-1}e)^T(Xd_s + Sd_x)$$
$$= \theta(e - (XS)^{-1}e)^T(e - Xs) \quad \text{(from (3.15))}$$
$$= -\theta(e - XS)^T(XS)^{-1}(e - XS)$$
$$= -\theta\|(XS)^{-1/2}(e-Xs)\|^2$$
$$= -\alpha\sqrt{\min(Xs)}\|(XS)^{-1/2}(e-Xs)\|. \qquad (3.19)$$

3.4. PRIMAL-DUAL (SYMMETRIC) ALGORITHMS

Therefore, we have the desired result combining (3.18) and (3.19). □

Theorem 3.13 *Let x^+ and s^+ be defined in Lemma 3.12. Then, if $\eta(x,s) \geq \eta$ for a positive constant $\eta < 1$, we can choose α such that*

$$\psi_n(x + \theta d_x, s + \theta d_s) - \psi_n(x, s) \leq -\delta$$

for a positive constant δ.

Proof. Note that

$$\alpha\sqrt{\min(Xs)}\|(XS)^{-1/2}(e - Xs)\| = \alpha\sqrt{\min(Xs)}\|(XS)^{-1/2}e - (XS)^{1/2}e\|.$$

Let $z = Xs$ and $z_1 = \min(z) \leq 1$ (since $e^T z = n$). If $z_1 \leq 1/2$, then

$$\alpha\sqrt{\min(z)}\|Z^{-1/2}e - Z^{1/2}e\| \geq \alpha|1 - z_1| \geq \alpha/2.$$

Thus, we can select an α such that

$$\psi_n(x^+, s^+) - \psi_n(x, s) \leq -\frac{\alpha}{2} + \frac{\alpha^2}{2(1-\alpha)} \leq -\delta.$$

If $z_1 \geq 1/2$, then

$$\alpha\sqrt{\min(z)}\|Z^{-1/2}e - Z^{1/2}e\| \geq (\alpha/\sqrt{2})\|Z^{-1/2}e - Z^{1/2}e\|.$$

Let $z_n = \max(z) \geq 1$. If $z_n \geq 2$, then

$$(\alpha/\sqrt{2})\|Z^{-1/2}e - Z^{1/2}e\| \geq (\alpha/\sqrt{2})|\sqrt{z_n} - 1/\sqrt{z_n}| \geq \alpha/2.$$

Again, we have

$$\psi_n(x^+, s^+) - \psi_n(x, s) \leq -\frac{\alpha}{2} + \frac{\alpha^2}{2(1-\alpha)} \leq -\delta.$$

Finally, let $\min(z) \geq 1/2$ and $\max(z) \leq 2$. Then

$$\alpha\sqrt{\min(z)}\|Z^{-1/2}e - Z^{1/2}e\| = \alpha\sqrt{\min(z)}\|Z^{-1/2}(e - z)\| \geq (\alpha/2)\|e - z\|.$$

Thus, if $\|e - z\| = \|e - Xs\| \geq \eta$, we can select an α such that

$$\psi_n(x^+, s^+) - \psi_n(x, s) \leq -\delta.$$

□

Thus, until $\|Xs - e\| < \eta < 1$, the primal-dual potential function at (x^+, s^+) will be reduced by a constant for some α. Therefore, in $O(\psi_n(x^0, s^0) - n \log n)$ iterations, we shall generate a pair (x, y), such that

$$\eta(x, s) = \|Xs - e\| = \|X(c - A^T y) - e\| < \eta,$$

which indicates that x and y are an approximate analytic center pair for Ω. Note that this complexity bound depends only on the initial point (x^0, s^0).

Recall that the primal-dual potential function represents the logarithmic ratio of the volume of the containing ellipsoid E_x over the volume of the contained ellipsoid E_s. The progress of the algorithm can be illustrated in Figure 3.3.

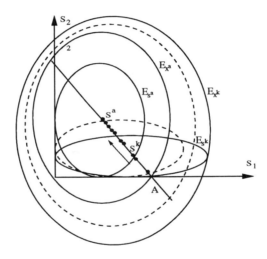

Figure 3.3. Illustration of the primal-dual potential algorithm; it generates a sequence of containing and contained coordinate-aligned ellipsoids whose logarithmic volume-ratio reduces to $n \log n$.

3.5 Notes

The proof of Lemma 3.1 is due to Karmarkar [217]; that of Theorem 3.2 is mostly due to Gonzaga [161, 162], Renegar and Shub [358, 360], Roos and Vial [367], Tseng [422], Vaidya [436], and Goffin et al. [144].

The dual algorithm with a starting interior point, described in this chapter, is similar to the one of Vaidya [436].

3.5. NOTES

The primal or dual affine potential algorithm was proposed by Gonzaga [160], Freund [123], and Ye [466, 468]. The affine scaling algorithm was originally developed by Dikin in 1967 [98, 99], and was rediscovered by Barnes [43], Cavalier and Soyster [79], Kortanek and Shi [235], Sherali, Skarpness and Kim [378], Vanderbei and Lagarias [448], and Vanderbei, Meketon and Freedman [449].

The primal-dual algorithm described in this chapter is adapted from the one in Kojima, Mizuno and Yoshise [231, 230, 232] and Monteiro and Adler [298, 299]. The primal-dual procedure result in (iv) of Theorem 3.2 was proved by Mizuno [288].

Finally, we remark the relation among potential reduction algorithms. We provide a simple argument that the primal-dual potential function is also reduced in either the primal or the dual potential reduction algorithm described earlier.

Given x and (y, s) in the interior of $\mathcal{X}_\Omega$ and $\mathcal{S}_\Omega$, respectively. We have shown in the preceding chapter that

$$\psi_n(x,s) = n\log(s^T x) - \sum_{j=1}^n \log(x_j s_j) = \mathcal{P}(x) - \mathcal{B}(y).$$

Thus, if we update the dual (y, s) to (y^+, s^+) such that

$$\mathcal{B}(y^+) \geq \mathcal{B}(y) + \delta,$$

then we must also have

$$\psi_n(x, s^+) \leq \psi_n(x, s) - \delta;$$

or if we update the primal x to x^+ such that

$$\mathcal{P}(x^+) \leq \mathcal{P}(x) - \delta,$$

then we must also have

$$\psi_n(x^+, s) \leq \psi_n(x, s) - \delta.$$

Thus, to make either the primal or dual potential reduction leads to the same reduction in the primal-dual potential function. Therefore, all these algorithms must stop in $O(\psi_n(x^0, s^0) - n\log n)$ iterations. Again, this complexity bound depends only on the initial point (x^0, s^0). Moreover, the primal algorithm does not need knowledge of s^0 and the dual algorithm does not need knowledge of x^0, while the primal-dual algorithm uses both x^0 and s^0.

3.6 Exercises

3.1 *Prove a slightly stronger variant of Lemma 3.1: If $d \in \mathcal{R}^n$ such that $0 \leq \max\{d\} < 1$ then*

$$-e^T d \geq \sum_{i=1}^{n} \log(1 - d_i) \geq -e^T d - \frac{\|d\|^2}{2(1 - \max\{d\})}.$$

3.2 *Given $s > 0$ verify that $x(s)$ is the minimizer of the least-squares problem*

$$\text{minimize} \quad \|Sx - e\|$$
$$\text{s.t.} \quad Ax = 0.$$

Given $x > 0$ verify that $y(x)$ is the minimizer of the least-squares problem

$$\text{minimize} \quad \|Xs - e\|$$
$$\text{s.t.} \quad s = c - A^T y.$$

3.3 *Let x be an interior point of $\mathcal{X}_\Omega$ and $\eta_p(x) < 1$. Prove that in the primal Newton procedure (3.13),*

$$x^+ > 0, \quad Ax^+ = 0, \quad \text{and} \quad \eta_p(x^+) \leq \eta_p(x)^2.$$

3.4 *Prove the primal inequality in (iv) of Theorem 3.2.*

3.5 *Let $e \in \mathcal{K}_p = \{x : Ax = 0, \ e^T x = n, \ x \geq 0\}$. Then, prove that*

$$\mathcal{K}_p \subset \{x : \|x - e\| \leq \sqrt{n(n-1)}\}.$$

3.6 *Consider the projective transformation and Karmarkar's potential function. Prove*

$$\mathcal{P}'(T(x^2)) - \mathcal{P}'(T(x^1)) = \mathcal{P}(x^2) - \mathcal{P}(x^1).$$

3.7 *Using Exercise 3.1 to show Theorem 3.10.*

3.8 *If the starting point of the primal-dual Newton procedure satisfies $\eta(x, s) < 1$, prove the update*

$$x^+ > 0, \quad Ax^+ = 0, \quad s^+ > 0.$$

3.9 *In each Newton procedure of Chapter 3, a system of linear equations needs to be solved. Prove that the solution $d = (d_x, d_y, d_s)$ to each system is unique.*

3.10 *Do some algorithm appear to use less information than others? Do some algorithm appear more efficient than others? Why?*

3.11 *Given $x \in \overset{\circ}{\mathcal{F}}_p$ and $(y, s) \in \overset{\circ}{\mathcal{F}}_d$ for (LP), design primal, dual, and primal-dual algorithms for finding a point in the neighborhood of the central path and determine its iteration bound.*

Chapter 4

Linear Programming Algorithms

In the preceding chapter we have used several interior algorithms to compute an approximate analytic center of a polytope specified by inequalities. The goal of this chapter is to extend these algorithms to solving the standard linear programming problem described in Section 1.3.5.

We assume that both $\overset{\circ}{\mathcal{F}}_p$ and $\overset{\circ}{\mathcal{F}}_d$ are nonempty. Thus, the optimal faces for both (LP) and (LD) are bounded. Furthermore, we assume $b \neq 0$, since otherwise (LD) reduces to finding a feasible point in $\mathcal{F}_d$.

Let z^* denote the optimal value and $\mathcal{F} = \mathcal{F}_p \times \mathcal{F}_d$. In this chapter, we are interested in finding an ϵ-approximate solution for the LP problem:

$$c^T x - z^* \leq \epsilon \quad \text{and} \quad z^* - b^T y \leq \epsilon.$$

For simplicity, we assume that a central path pair (x^0, y^0, s^0) with $\mu^0 = (x^0)^T s^0 / n$ is known. We will use it as our initial point throughout this chapter.

4.1 Karmarkar's Algorithm

We first show how Karmarkar's algorithm may be adapted to solve (LP) and (LD). The basic idea is the following: Given a polytope $\Omega(z^k)$ of (2.13), i.e.,

$$\Omega(z^k) = \{y: \ c - A^T y \geq 0, -z^k + b^T y \geq 0\},$$

where $z^k < z^*$, and a $x^k \in \overset{\circ}{\mathcal{F}}_p$, we step toward its analytic center using Karmarkar's algorithm presented in the previous chapter. Each step results in a reduction of the associated Karmarkar potential function $\mathcal{P}_{n+1}(x, z^k) = \mathcal{P}(x, \Omega(z^k))$, which represents the logarithmic volume of a coordinate-aligned ellipsoid containing $\Omega(z^k)$. When an approximate center y^{k+1} is generated, it is used to translate the hyperplane $b^T y \geq z^k$ to $b^T y \geq z^{k+1} := b^T y^{k+1}$, i.e., through y^{k+1}. Since $z^{k+1} > z^k$ and $\Omega(z^{k+1})$ has shrunk, both $\mathcal{P}_{n+1}(x, z^{k+1})$ and the max-potential $\mathcal{B}(\Omega(z^{k+1}))$, which represents the logarithmic volume of the max-volume coordinate-aligned ellipsoid contained in $\Omega(z)$, are further reduced. The algorithm terminates when the potential function has been reduced enough to guarantee an approximate solution for (LP) and (LD).

More specifically, starting from (x^0, y^0, s^0) and letting $z^0 = b^T y^0 < z^*$, the algorithm, generates sequences $\{x^k \in \overset{\circ}{\mathcal{F}}_p\}$, $\{y^k \in \mathcal{F}_d\}$, and $\{z^k \leq b^T y^k\}$, such that

$$\mathcal{P}_{n+1}(x^{k+1}, z^{k+1}) \leq \mathcal{P}_{n+1}(x^k, z^k) - \delta \quad \text{for} \quad k = 0, 1, \ldots.$$

where constant $\delta \geq .2$. Then when

$$\mathcal{P}_{n+1}(x^k, z^k) - \mathcal{P}_{n+1}(x^0, z^0) \leq (n+1) \log \frac{\epsilon}{c^T x^0 - z^0} - n \log 2,$$

from Proposition 4.2 which will be proved at the end of the section, we must have

$$z^* - b^T y^k \leq z^* - z^k \leq \epsilon.$$

That is, we have an ϵ-approximation solution for (LD). Moreover, there is a subsequence of $\{(x^k, y^k, s^k)\}$ whose points are approximate analytic centers of $\Omega(z^k)$. Thus, along this subsequence $c^T x^k - b^T y^k \to 0$.

Here is how the algorithm works. Given $x^k \in \overset{\circ}{\mathcal{F}}_p$ and $z^k \leq b^T y^k$, where $y^k \in \mathcal{F}_d$, we again transform the (LP) problem into Karmarkar's canonical form:

$$(LP') \quad \text{minimize} \quad (c^k)^T x'$$
$$\text{s.t.} \quad x' \in \mathcal{K}'_p.$$

where

$$\mathcal{K}'_p := \{x' \in \mathcal{R}^{n+1} : A^k x' = 0, \ e^T x' = n+1, \ x' \geq 0\},$$

$$c^k = \begin{pmatrix} X^k c \\ -z^k \end{pmatrix}, \quad \text{and} \quad A^k = (\ AX^k, \ -b\ ).$$

This is accomplished via an extended Karmarkar's projective transformation

$$x' = T(x) = \begin{pmatrix} \frac{(n+1)(X^k)^{-1} x}{1 + e^T (X^k)^{-1} x} \\ \frac{(n+1)}{1 + e^T (X^k)^{-1} x} \end{pmatrix} \quad \text{from} \quad x \in \mathcal{F}_p.$$

4.1. KARMARKAR'S ALGORITHM

Obviously, $T(x^k) = e$ is the analytic center for $\mathcal{K}'_p$. In other words, the projective transformation maps x^k to the analytic center of $\mathcal{K}'_p$ in (LP'). Each feasible point $x \in \mathcal{F}_p$ is also mapped to a feasible point $x' \in \mathcal{K}'_p$. Conversely, each feasible point $x' \in \mathcal{K}'_p$ can be transformed back to an $x \in \mathcal{F}_p$ via the inverse transformation, T^{-1}, given by

$$x = T^{-1}(x') = \frac{X^k x'[n]}{x'_{n+1}} \quad \text{from} \quad x' \in \mathcal{K}'_p,$$

where $x'[n]$ denotes the vector of the first n components of $x' \in \mathcal{R}^{n+1}$.

The projective transformation T also induces the potential function $\mathcal{P}'_{n+1}$ associated with (LP'):

$$\mathcal{P}'_{n+1}(x', z^k) = (n+1)\log((c^k)^T x') - \sum_{i=1}^{n+1} \log(x'_i), \quad x' \in \overset{\circ}{\mathcal{K}'_p}.$$

Again, the difference of the potential values at two points x^1, $x^2 \in \overset{\circ}{\mathcal{F}_p}$ is invariant under the projective transformation, i.e.,

$$\mathcal{P}'_{n+1}(T(x^2), z^k) - \mathcal{P}'_{n+1}(T(x^1), z^k) = \mathcal{P}_{n+1}(x^2, z^k) - \mathcal{P}_{n+1}(x^1, z^k).$$

Therefore, a reduction of the potential $\mathcal{P}'_{n+1}$ for the transformed problem induces the same amount of reduction of the potential $\mathcal{P}_{n+1}$ for the original problem.

To reduce $\mathcal{P}'_{n+1}(x', z^k)$, we again solve the following ball-constrained problem

$$\begin{array}{ll} \text{minimize} & (c^k)^T(x' - e) \\ \text{s.t.} & A^k(x' - e) = 0, \ e^T(x' - e) = 0, \ \|x' - e\| \leq \alpha. \end{array}$$

According to the discussion in Section 3.3.2, the solution of the problem is

$$x' - e = -\alpha \frac{p^k}{\|p^k\|},$$

where

$$\begin{aligned} p^k &= \left(I - \begin{pmatrix} A^k \\ e \end{pmatrix}^T \begin{pmatrix} A^k(A^k)^T & 0 \\ 0 & n \end{pmatrix}^{-1} \begin{pmatrix} A^k \\ e \end{pmatrix} \right) c^k \\ &= c^k - (A^k)^T y^k - \frac{c^T x^k - z^k}{n+1} e, \end{aligned}$$

and
$$\begin{aligned} y^k &= y(z^k) := y_2 + z^k y_1, \\ y_1 &= (A^k(A^k)^T)^{-1} b, \\ y_2 &= (A^k(A^k)^T)^{-1} A(X^k)^2 c. \end{aligned} \quad (4.1)$$

Again the difference of the potential values
$$\mathcal{P}'_{n+1}(x', z^k) - \mathcal{P}'_{n+1}(e, z^k) \leq -\alpha \frac{(n+1)\|p^k\|}{c^T x^k - z^k} + \frac{\alpha^2}{2(1-\alpha)}.$$

Thus, as long as $\|p^k\| \geq \eta \frac{c^T x^k - z^k}{n+1}$ for a constant $0 < \eta < 1$ we can choose an appropriate α such that
$$\mathcal{P}'_{n+1}(x', z^k) - \mathcal{P}'_{n+1}(e, z^k) \leq -\delta,$$
for a positive constant $\delta = \alpha\eta - \alpha^2/2(1-\alpha)$. (If η is slightly less than 1, α can be chosen such that $\delta \geq .2$.) Let
$$x^{k+1} = T^{-1}(x').$$

Then,
$$\mathcal{P}_{n+1}(x^{k+1}, z^k) - \mathcal{P}_{n+1}(x^k, z^k) \leq -\delta.$$

In this case, we simply let $z^{k+1} = z^k$.

However, when $\|p^k\| < \eta \frac{c^T x^k - z^k}{n+1}$, we can still reduce the potential function by increasing z^k to $z^{k+1} > z^k$. Note that p^k can be decomposed as
$$\begin{aligned} p^k = p(z^k) &:= \begin{pmatrix} X^k c \\ -z^k \end{pmatrix} - \begin{pmatrix} X^k A^T \\ -b^T \end{pmatrix} y(z^k) - \frac{c^T x^k - z^k}{n+1} e \\ &= \begin{pmatrix} X^k(c - A^T y(z^k)) \\ b^T y(z^k) - z^k \end{pmatrix} - \frac{c^T x^k - z^k}{n+1} e. \end{aligned} \quad (4.2)$$

Thus, $\|p^k\| < \eta \frac{c^T x^k - z^k}{n+1}$ implies that $(x^k, y(z^k))$ is an η-approximate center pair for $\Omega(z^k)$. Consequently,
$$\begin{pmatrix} X^k(c - A^T y(z^k)) \\ b^T y(z^k) - z^k \end{pmatrix} > 0,$$

that is
$$A^T y(z^k) < c \quad \text{and} \quad z^k < b^T y(z^k).$$

By the duality theorem, $z^{k+1} = b^T y(z^k) > z^k$ is a new lower bound for z^*, which makes $\mathcal{P}_{n+1}(x^k, z^{k+1}) < \mathcal{P}_{n+1}(x^k, z^k)$. We now update $\Omega(z^k)$ to $\Omega(z^{k+1})$, that is, we translate the inequality $b^T y \geq z^k$ through $y(z^k)$ to cut $\Omega(z^k)$.

4.1. KARMARKAR'S ALGORITHM

We may place a deeper cut and try to update z^k in each step of the algorithm. We can do a simple ratio test in all steps to obtain

$$\bar{z} = \arg\max_z \{b^T y(z) : A^T y(z) \leq c\}. \tag{4.3}$$

(This test returns value $\bar{z} = -\infty$ if there is no z such that $A^T y(z) \leq c$.) This leads us to a possible new dual feasible point and a new lower bound $b^T y(\bar{z}) \geq b^T y(z^k) > z^k$, and guarantees

$$\|p(b^T y(\bar{z}))\| \geq \frac{c^T x^k - b^T y(\bar{z})}{n+1},$$

to ensure a potential reduction of at least .2. We describe the algorithm as follows:

Algorithm 4.1 *Given a central path point $(x^0, y^0, s^0) \in \overset{\circ}{\mathcal{F}}$. Let $z^0 = b^T y^0$. Set $k := 0$.*
    While $(c^T x^k - z^k) > \epsilon$ **do:**

1. *Compute y_1 and y_2 from (4.1).*

2. *Compute $\bar{z}$ from (4.3). If $\bar{z} > z^k$ then set $y^{k+1} = y(\bar{z})$ and $z^{k+1} = b^T y^{k+1}$, otherwise set $y^{k+1} = y^k$ and $z^{k+1} = z^k$.*

3. *Let*
$$x' = e - \frac{\alpha}{\|p(z^{k+1})\|} p(z^{k+1})$$
   and
$$x^{k+1} = T^{-1}(x').$$

4. *Let $k := k+1$ and return to Step 1.*

Since $z^{k+1} \geq z^k$ and $\|p(z^{k+1})\| \geq \frac{c^T x^k - z^{k+1}}{n+1}$ (Exercise 4.1), we have

Lemma 4.1 *For $k = 0, 1, 2, \ldots$*

$$\mathcal{P}_{n+1}(x^{k+1}, z^{k+1}) \leq \mathcal{P}_{n+1}(x^k, z^k) - \delta,$$

where $\delta \geq .2$.

We now estimate how much we need to reduce the potential function in order to have $z^k = b^T y^k$ close to z^*. We establish a lower bound for the potential function value.

Proposition 4.2 *Given a central path point (x^0, y^0, s^0) with $\mu^0 = (x^0)^T s^0/n$, and consider the dual level set $\Omega(z)$ where $z^0 := b^T y^0 - \mu^0 \leq z < z^*$. There exists an $x(z) \in \mathcal{F}_p$ such that the max-potential of $\Omega(z)$*

$$\mathcal{B}(\Omega(z)) \geq \mathcal{B}(\Omega(z^0)) + (n+1)\log \frac{c^T x(z) - z}{c^T x^0 - z^0} - n\log 2, \quad (4.4)$$

and for all $x \in \overset{\circ}{\mathcal{F}}_p$ the primal potential

$$\mathcal{P}_{n+1}(x, z) \geq \mathcal{P}_{n+1}(x^0, z^0) + (n+1)\log \frac{c^T x(z) - z}{c^T x^0 - z^0} - n\log 2. \quad (4.5)$$

Proof. First, recall that the max-potential of $\Omega(z)$ is

$$\mathcal{B}(\Omega(z)) = \max_y \left\{ \sum_{j=1}^n \log(c_j - a_j^T y) + \log(b^T y - z) \right\},$$

so that we have

$$\mathcal{P}_{n+1}(x, z) - \mathcal{B}(\Omega(z)) \geq (n+1)\log(n+1), \quad \forall\, x \in \overset{\circ}{\mathcal{F}}_p \quad \forall\, z < z^*. \quad (4.6)$$

Let $y(z)$ be the analytic center of $\Omega(z)$ and let $s(z) = c - A^T y(z)$. Then, from the central path theory in Section 2.5.1, there is a central path point $0 < x(z) \in \mathcal{R}^n$ such that

$$Ax(z) = b \text{ and } X(z)s(z) = \mu(z)e,$$

where

$$\mu(z) = \frac{x(z)^T s(z)}{n} = \frac{c^T x(z) - b^T y(z)}{n} = b^T y(z) - z.$$

Thus,

$$\begin{aligned}
\mathcal{B}(\Omega(z)) &= \sum_{j=1}^n \log s_j(z) + \log(b^T y(z) - z) \\
&= \sum_{j=1}^n \log(\mu(z)/x_j(z)) + \log \mu(z) \\
&= (n+1)\log \mu(z) - \sum_{j=1}^n \log x(z)_j \\
&= (n+1)\log((n+1)\mu(z)) - \sum_{j=1}^n \log x(z)_j - (n+1)\log(n+1)
\end{aligned}$$

4.1. KARMARKAR'S ALGORITHM

$$\begin{aligned} &= (n+1)\log(c^T x(z) - z) - \sum_{j=1}^{n} \log x(z)_j - (n+1)\log(n+1) \\ &= (n+1)\log\frac{c^T x(z) - z}{c^T x^0 - z^0} - \sum_{j=1}^{n} \log \frac{x(z)_j}{x_j^0} \\ &\quad + \mathcal{P}_{n+1}(x^0, z^0) - (n+1)\log(n+1). \end{aligned} \quad (4.7)$$

Since $\mu(z) \leq \mu^0$ (see Exercise 4.3), we have

$$(x(z) - x^0)^T (s(z) - s^0) = 0 \quad \text{or} \quad x(z)^T s^0 + s(z)^T x^0 = n(\mu^0 + \mu(z)),$$

which implies that

$$\sum_{j=1}^{n} \left(\frac{x(z)_j}{x_j^0} \mu^0 + \frac{x_j^0}{x(z)_j} \mu(z) \right) = n(\mu^0 + \mu(z)).$$

Therefore,

$$\sum_{j=1}^{n} \left(\frac{x(z)_j}{x_j^0} \right) \leq n \left(1 + \frac{\mu(z)}{\mu^0} \right) \leq 2n,$$

which, from the arithmetic-geometric mean inequality, further implies

$$\prod_{j=1}^{n} \left(\frac{x(z)_j}{x_j^0} \right) \leq 2^n. \quad (4.8)$$

Combining inequalities (4.6), (4.7) and (4.8), we have

$$\begin{aligned} \mathcal{B}(\Omega(z)) &\geq \mathcal{P}_{n+1}(x^0, z^0) - (n+1)\log(n+1) \\ &\quad + (n+1)\log\frac{c^T x(z) - z}{c^T x^0 - z^0} - n\log 2 \\ &\geq \mathcal{B}(\Omega(z^0)) + (n+1)\log\frac{c^T x(z) - z}{c^T x^0 - z^0} - n\log 2. \end{aligned}$$

Also, for all $x \in \overset{\circ}{\mathcal{F}}_p$ we have from (4.6)

$$\begin{aligned} \mathcal{P}_{n+1}(x, z) &\geq \mathcal{B}(\Omega(z)) + (n+1)\log(n+1) \\ &\geq \mathcal{P}_{n+1}(x^0, z^0) + (n+1)\log\frac{c^T x(z) - z}{c^T x^0 - z^0} - n\log 2. \end{aligned}$$

These inequalities lead to the desired result.

□

The proposition indicates that if the net reduction of max-potential or the primal potential of $\Omega(z)$, where $z = b^T y$ for some $y \in \mathcal{F}_d$, is greater than $(n+1)\log(c^T x^0 - z^0)/\epsilon + n\log 2$, then we have

$$(n+1)\log\frac{c^T x(z) - z}{c^T x^0 - z^0} - n\log 2 \leq (n+1)\log\frac{\epsilon}{c^T x^0 - z^0} - n\log 2,$$

which implies that

$$z^* - b^T y = z^* - z \leq c^T x(z) - z \leq \epsilon$$

and

$$c^T x(z) - z^* \leq c^T x(z) - z \leq \epsilon,$$

i.e., $x(z)$ and y are ϵ-approximate solutions for (LP) and (LD).

In the proposition, $x(z)$ is chosen as a central path point. This need not be the case. The proposition holds for any $\hat{x} \in \overset{\circ}{\mathcal{F}}_p$ that is in the neighborhood $\mathcal{N}(\beta)$ of the central path, i.e., for some y it satisfies

$$\left\| \begin{pmatrix} \hat{X}(c - A^T y) \\ b^T y - z \end{pmatrix} - \frac{c^T \hat{x} - z}{n} e \right\| \leq \beta \frac{c^T \hat{x} - z}{n} \quad (4.9)$$

for a constant $0 \leq \beta < 1$. More specifically, we have Corollary 4.3.

Corollary 4.3 *Given the interior feasible point* (x^0, y^0, s^0) *with* $\mu^0 = (x^0)^T s^0 / n$, *and consider the dual level set* $\Omega(z)$ *where* $z^0 := b^T y^0 - \mu^0 \leq z < z^*$. *Let* $\hat{x} \in \overset{\circ}{\mathcal{F}}_p$ *satisfy condition (4.9). Then*

$$\mathcal{B}(\Omega(z)) \geq \mathcal{B}(\Omega(z^0)) + (n+1)\log\frac{c^T \hat{x} - z}{c^T x^0 - z^0} - O(n),$$

and for all $x \in \overset{\circ}{\mathcal{F}}_p$ *the primal potential*

$$\mathcal{P}_{n+1}(x, z) \geq \mathcal{P}_{n+1}(x^0, z^0) + (n+1)\log\frac{c^T \hat{x} - z}{c^T x^0 - z^0} - O(n).$$

These lead to the following theorem:

Theorem 4.4 *In at most* $O(n\log(c^T x^0 - z^0)/\epsilon + n)$ *iterations, Algorithm 4.1 will generate* $x^k \in \overset{\circ}{\mathcal{F}}_p$ *and* $y^k \in \mathcal{F}_d$ *with* $z^k \leq b^T y^k < z^*$ *such that*

$$c^T x^k - z^k \leq \epsilon$$

and, thereby,

$$c^T x^k - b^T y^k \leq \epsilon$$

at termination.

4.2 Path-Following Algorithm

While Karmarkar's algorithm reduces the primal potential function, the path-following algorithm reduces the max-potential $\mathcal{B}(\Omega(z^k))$ by downsizing $\Omega(z^k)$. Beginning with (y^k, s^k), an approximate analytic center of $\Omega(z^k)$ with $\eta_d(s^k) \leq \eta < 1$, where

$$s^k = \begin{pmatrix} c - A^T y^k \\ b^T y^k - z^k \end{pmatrix} > 0.$$

Let β be a constant in $(0,1)$. Then, similar to what has been done in the central-section algorithm of Chapter 3, we update z^{k+1} from z^k at the kth iteration

$$z^{k+1} = b^T y^k - \beta s^k_{n+1},$$

and then find an η-approximate center of $\Omega(z^{k+1})$. Accordingly,

$$\mathcal{B}(\Omega(z^{k+1})) \leq \mathcal{B}(\Omega(z^k)) - \delta,$$

where δ is a positive constant. This process stops with y^k as an approximate center for $\Omega(z^k)$, where $z^k \geq z^* - \epsilon$. The total number of iterations is bounded by $O(n \log(c^T x^0 - z^0)/\epsilon + n \log 2)$ from Proposition 4.2, which is the same bound as in Karmarkar's algorithm.

In an additional effort, Renegar developed a new method to improve the iteration bound by a factor $\sqrt{n}$. His algorithm can be described as the following. For $z^0 \leq z < z^*$ consider $\Omega(z)$ of (2.14) where $\rho = n$, that is,

$$\Omega(z) = \{y: \ c - A^T y \geq 0, \overbrace{-z + b^T y \geq 0, \cdots, -z + b^T y \geq 0}^{n \text{ times}}\},$$

where "$-z + b^T y \geq 0$" is copied n times. The n copies of the objective hyperplane have the effect of "pushing" the analytic center toward the optimal solution set.

Note that the slack vector $s \in \mathcal{R}^{2n}$ and

$$s = \begin{pmatrix} c - A^T y \\ b^T y - z \\ \cdots \\ b^T y - z \end{pmatrix} > 0.$$

Thus, $s_{n+1} = \cdots = s_{2n}$. The primal potential function associated with $\Omega(z)$ is $\mathcal{P}_{2n}(x, z)$ where $x \in \mathcal{\mathring{F}}_p$. Following the proof of Proposition 4.2, there are points $x(z) \in \mathcal{\mathring{F}}_p$ such that

$$\mathcal{B}(\Omega(z)) - \mathcal{B}(\Omega(z^0)) \geq 2n \log \frac{c^T x(z) - z}{c^T x^0 - z^0} - O(n).$$

Algorithm 4.2 Given an approximate analytic center y^0 of $\Omega(z^0)$ with $\eta_d(s^0) \le \eta < 1$, set $k := 0$.
    While $(c^T x(z^k) - b^T y^k) > \epsilon$ **do:**

1. Update $z^{k+1} = b^T y^k - \beta s_{2n}^k$.

2. Via the dual Newton procedure compute an approximate analytic center y^{k+1} with $\eta_d(s^{k+1}) \le \eta$ for $\Omega(z^{k+1})$. Let $x' = x(s^{k+1}) \in \mathcal{R}^{2n}$ given by (3.3) for $\Omega(z^{k+1})$, which minimizes

$$\text{minimize} \quad \|S^{k+1}x - e\|$$
$$\text{s.t.} \quad (A, \overbrace{-b, \cdots, -b}^{n \text{ times}})x = 0, \ x \in \mathcal{R}^{2n},$$

and let $x(z^{k+1}) = x'[n]/(nx'_{2n})$ where $x'[n]$ denotes the vector of the first n components of x'.

3. Let $k := k + 1$ and return to Step 1.

The following Lemma provides a guide for choosing constants η and β to ensure y^k remains an approximate analytic center of $\Omega(z^{k+1})$.

Lemma 4.5 Let $\beta = 1 - \eta/\sqrt{n}$. Then, there exists a point $x^+ > 0$ such that

$$(A, -b, \cdots, -b)x^+ = 0 \quad \text{and} \quad \|X^+ s^+ - e\| \le \eta + \eta(1+\eta),$$

where

$$s^+ = \begin{pmatrix} c - A^T y^k \\ b^T y^k - z^{k+1} \\ \cdots \\ b^T y^k - z^{k+1} \end{pmatrix} > 0.$$

Proof. Let $x^+ = x(s^k) > 0$ of (3.3) for $\Omega(z^k)$, which minimizes

$$\text{minimize} \quad \|S^k x - e\|$$
$$\text{s.t.} \quad (A, -b, \cdots, -b)x = 0, \ x \in \mathcal{R}^{2n}.$$

Since $s_{n+1}^k = \cdots = s_{2n}^k$, we have $x_{n+1}^+ = \cdots = x_{2n}^+$,

$$(A, -b, \cdots, -b)x^+ = Ax(s^k) = 0,$$

and

$$\|S^k x^+ - e\| = \eta_d(s^k) \le \eta.$$

4.2. PATH-FOLLOWING ALGORITHM

Note that $s_j^+ = s_j^k$ for $j \leq n$ and $s_j^+ = \beta s_j^k$ for $j \geq n+1$. Thus, noting that s^+ and s^k share the same first n components and $s_{n+1}^+ - s_{n+1}^k = \cdots = s_{n+1}^+ - s_{n+1}^k = (\beta - 1)s_{2n}^k$, we have

$$\begin{aligned}
\|X^+ s^+ - e\| &= \|X^+ s^k - e + X^+ s^+ - X^+ s^k\| \\
&\leq \|X^+ s^k - e\| + \|X^+ s^+ - X^+ s^k\| \\
&\leq \|X^+ s^k - e\| + \sqrt{n}|x_{2n}^+ s_{2n}^k (\beta - 1)| \\
&\leq \eta + \sqrt{n}(1 - \beta)(1 + \eta) \\
&\leq \eta + \eta(1 + \eta).
\end{aligned}$$

□

Lemma 4.5 shows that, even though it is not perfectly centered, y^k is in the "quadratic convergence" region of the center of $\Omega(z^{k+1})$, if we choose $\eta + \eta(1+\eta) < 1$. Thus, an η-approximate center y^{k+1} with $\eta_d(s^{k+1}) \leq \eta$ for $\Omega(z^{k+1})$ can be updated from y^k in a constant number of Newton's steps. For example, if choose $\eta = 1/5$, then one Newton step suffices.

We now verify that the max-potential of $\Omega(z^{k+1})$ is reduced by $\sqrt{n}\delta$ for a constant δ.

Lemma 4.6 *Let (y^k, s^k) be an approximate center for Ω^k with $\eta_d(s^k) \leq \eta$ and let $\Omega(z^{k+1})$ be given as in Step 1 of Algorithm 4.2. Then,*

$$\mathcal{B}(\Omega(z^{k+1})) \leq \mathcal{B}(\Omega(z^k)) - \sqrt{n}\delta \quad \text{for a constant} \quad \delta > 0.$$

Proof. The proof is very similar to Theorem 2.7 in Chapter 2. Let (y^a, s^a) and (y_+^a, s_+^a) be the centers of $\Omega(z^k)$ and $\Omega(z^{k+1})$, respectively. Note we have

$$s_{n+1}^a = \cdots = s_{2n}^a = b^T y^a - z^k.$$

Also,

$$(A, -b, \cdots, -b)(S^a)^{-1} e = 0,$$

$$(c, -z^k, \cdots, -z^k)^T (S^a)^{-1} e = (s^a)^T (S^a)^{-1} e = 2n,$$

and

$$s_+^a = \begin{pmatrix} c - A^T y_+^a \\ b^T y_+^a - z^{k+1} \\ \cdots \\ b^T y_+^a - z^{k+1} \end{pmatrix} > 0,$$

where

$$z^{k+1} = z^k + (1 - \beta)s_{2n}^k.$$

Then we have

$$\begin{aligned}
e^T(S^a)^{-1}s_+^a &= e^T(S^a)^{-1}(c; -z^{k+1}; \cdots; -z^{k+1}) \\
&= e^T(S^a)^{-1}(c; -z^k; \cdots; -z^k) - n(1-\beta)(s_{2n}^k/s_{2n}^a) \\
&= 2n - n(1-\beta)(s_{2n}^k/s_{2n}^a) \\
&\leq 2n - n(1-\beta)(1-\eta).
\end{aligned}$$

The last inequality is due to (iv) of Theorem 3.2. Therefore,

$$\begin{aligned}
\frac{\exp \mathcal{B}(\Omega(z^{k+1}))}{\exp \mathcal{B}(\Omega(z^k))} &= \prod_{j=1}^{2n} \frac{(s_+^a)_j}{s_j^a} \\
&\leq \left(\frac{2n - n(1-\beta)(1-\eta)}{2n}\right)^{2n} \\
&\leq \exp(-n(1-\beta)(1-\eta)) \\
&= \exp(-\sqrt{n}\eta(1-\eta)).
\end{aligned}$$

Taking the logarithm of each side completes the proof.

$\square$

From the lemma, Proposition 4.2 can be used to conclude the following theorem:

Theorem 4.7 *In at most $O(\sqrt{n}\log(c^T x^0 - z^0)/\epsilon + \sqrt{n})$ iterations, Algorithm 4.2 will generate a $(x(z^k), y^k) \in \overset{\circ}{\mathcal{F}}$ such that it is an approximate center for $\Omega(z^k)$, where*

$$c^T x(z^k) - b^T y^k \leq c^T x(z^k) - z^k \leq \epsilon.$$

4.3 Potential Reduction Algorithm

At this point, we can see the difference between Karmarkar's and the path-following algorithms. The former, called *potential reduction algorithms*, are equipped with a primal potential functions, which are solely used to measure the solution's progress. There is no restriction on either step-size or z-update during the iterative process; the greater the reduction of the potential function, the faster the convergence of the algorithm. The *path-following algorithms* are equipped with the max-potential, so that the z-update needs to be carefully chosen and each step needs to stay close to the central path. Thus, from a practical point of view, Karmarkar's algorithm has an advantage.

4.3. POTENTIAL REDUCTION ALGORITHM

The next question is whether or not we can improve the complexity bound by the same factor for potential reduction algorithms. Let $(x, y, s) \in \overset{\circ}{\mathcal{F}}$. Then consider the primal-dual potential function in Section 2.4.3:

$$\psi_{n+\rho}(x, s) = (n + \rho) \log(x^T s) - \sum_{j=1}^{n} \log(x_j s_j),$$

where $\rho \geq 0$. Let $z = b^T y$, then $s^T x = c^T x - z$ and we have

$$\psi_{n+\rho}(x, s) = \mathcal{P}_{n+\rho}(x, z) - \sum_{j=1}^{n} \log s_j.$$

Recall from Chapter 2 that when $\rho = 0$, $\psi_{n+\rho}(x, s)$ is minimized along the central path. However, when $\rho > 0$, $\psi_{n+\rho}(x, s) \to -\infty$ means that x and s converge to the optimal face, and the descent gets steeper as ρ increases. In this section we choose $\rho = \sqrt{n}$, which is greater than $\rho = 1$ used in Karmarkar's algorithm.

Whereas Karmarkar's algorithm reduced the primal potential function $\mathcal{P}_{n+\rho}(x, z)$ when stepping toward the analytic center of the dual level set and increasing z, there was no estimate on how much the function was reduced. The potential reduction algorithm of this section will give such a reduction bound. The process calculates steps for x and s, which guarantee a constant reduction in the primal-dual potential function. As the potential function decreases, both x and s are forced to an optimal solution pair.

Consider a pair of $(x^k, y^k, s^k) \in \overset{\circ}{\mathcal{F}}$. Fix $z^k = b^T y^k$, then the gradient vector of the primal potential function at x^k is

$$\nabla \mathcal{P}_{n+\rho}(x^k, z^k) = \frac{(n+\rho)}{(s^k)^T x^k} c - (X^k)^{-1} e = \frac{(n+\rho)}{c^T x^k - z^k} c - (X^k)^{-1} e.$$

We directly solve the ball-constrained linear problem for direction d_x:

$$\begin{aligned} \text{minimize} \quad & \nabla \mathcal{P}_{n+\rho}(x^k, z^k)^T d_x \\ \text{s.t.} \quad & A d_x = 0, \ \|(X^k)^{-1} d_x\| \leq \alpha. \end{aligned}$$

Let the minimizer be d_x. Then

$$d_x = -\alpha \frac{X^k p^k}{\|p^k\|},$$

where

$$p^k = p(z^k) := = (I - X^k A^T (A(X^k)^2 A^T)^{-1} A X^k) X^k \nabla \mathcal{P}_{n+\rho}(x^k, z^k).$$

Update
$$x^{k+1} = x^k + d_x = x^k - \alpha \frac{X^k p^k}{\|p^k\|}, \qquad (4.10)$$

and, in view of Section 3.3.2,

$$\mathcal{P}_{n+\rho}(x^{k+1}, z^k) - \mathcal{P}_{n+\rho}(x^k, z^k) \le -\alpha \|p^k\| + \frac{\alpha^2}{2(1-\alpha)}.$$

Thus, as long as $\|p^k\| \ge \eta > 0$, we may choose an appropriate α such that

$$\mathcal{P}_{n+\rho}(x^{k+1}, z^k) - \mathcal{P}_{n+\rho}(x^k, z^k) \le -\delta$$

for some positive constant δ. By the relation between $\psi_{n+\rho}(x, s)$ and $\mathcal{P}_{n+\rho}(x, z)$, the primal-dual potential function is also reduced. That is,

$$\psi_{n+\rho}(x^{k+1}, s^k) - \psi_{n+\rho}(x^k, s^k) \le -\delta.$$

However, even if $\|p^k\|$ is small, we will show that the primal-dual potential function can be reduced by a constant δ by increasing z^k and updating (y^k, s^k).

We focus on the expression of p^k, which can be rewritten as

$$\begin{aligned} p^k &= (I - X^k A^T (A(X^k)^2 A^T)^{-1} A X^k)(\frac{(n+\rho)}{c^T x^k - z^k} X^k c - e) \\ &= \frac{(n+\rho)}{c^T x^k - z^k} X^k s(z^k) - e, \end{aligned} \qquad (4.11)$$

where

$$s(z^k) = c - A^T y(z^k) \qquad (4.12)$$

and

$$\begin{aligned} y(z^k) &= y_2 - \frac{c^T x^k - z^k}{(n+\rho)} y_1, \\ y_1 &= (A(X^k)^2 A^T)^{-1} b, \\ y_2 &= (A(X^k)^2 A^T)^{-1} A(X^k)^2 c. \end{aligned} \qquad (4.13)$$

Regarding $\|p^k\| = \|p(z^k)\|$, we have the following lemma:

Lemma 4.8 *Let*

$$\mu^k = \frac{(x^k)^T s^k}{n} = \frac{c^T x^k - z^k}{n} \quad \text{and} \quad \mu = \frac{(x^k)^T s(z^k)}{n}.$$

If

$$\|p(z^k)\| < \min\left(\eta \sqrt{\frac{n}{n+\eta^2}}, 1 - \eta\right), \qquad (4.14)$$

then the following three inequalities hold:

$$s(z^k) > 0, \quad \|X^k s(z^k) - \mu e\| < \eta\mu, \quad \text{and} \quad \mu < (1 - .5\eta/\sqrt{n})\mu^k. \qquad (4.15)$$

4.3. POTENTIAL REDUCTION ALGORITHM

Proof. The proof is by contradiction.

i) If the first inequality of (4.15) is not true, then $\exists\, j$ such that $s_j(z^k) \leq 0$ and
$$\|p(z^k)\| \geq 1 - \frac{(n+\rho)}{n\mu^k} x_j s_j(z^k) \geq 1.$$

ii) If the second inequality of (4.15) does not hold, then
$$\begin{aligned}
\|p(z^k)\|^2 &= \|\frac{(n+\rho)}{n\mu^k} X^k s(z^k) - \frac{(n+\rho)\mu}{n\mu^k} e + \frac{(n+\rho)\mu}{n\mu^k} e - e\|^2 \\
&= (\frac{(n+\rho)}{n\mu^k})^2 \|X^k s(z^k) - \mu e\|^2 + \|\frac{(n+\rho)\mu}{n\mu^k} e - e\|^2 \\
&\geq (\frac{(n+\rho)\mu}{n\mu^k})^2 \eta^2 + (\frac{(n+\rho)\mu}{n\mu^k} - 1)^2 n \qquad (4.16) \\
&\geq \eta^2 \frac{n}{n+\eta^2},
\end{aligned}$$

where the last relation prevails since the quadratic term yields the minimum at
$$\frac{(n+\rho)\mu}{n\mu^k} = \frac{n}{n+\eta^2}.$$

iii) If the third inequality of (4.15) is violated, then
$$\frac{(n+\rho)\mu}{n\mu^k} \geq (1 + \frac{1}{\sqrt{n}})(1 - \frac{.5\eta}{\sqrt{n}}) \geq 1,$$

which, in view of (4.16), leads to
$$\begin{aligned}
\|p(z^k)\|^2 &\geq (\frac{(n+\rho)\mu}{n\mu^k} - 1)^2 n \\
&\geq ((1 + \frac{1}{\sqrt{n}})(1 - \frac{.5\eta}{\sqrt{n}}) - 1)^2 n \\
&\geq (1 - \frac{\eta}{2} - \frac{\eta}{2\sqrt{n}})^2 \\
&\geq (1 - \eta)^2.
\end{aligned}$$

□

The lemma says that, when $\|p(z^k)\|$ is small, then $(x^k, y(z^k), s(z^k))$ is in the neighborhood of the central path and $b^T y(z^k) > z^k$. Thus, we can increase z^k to $b^T y(z^k)$ to cut the dual level set $\Omega(z^k)$. We have the following potential reduction theorem to evaluate the progress.

Theorem 4.9 *Given* $(x^k, y^k, s^k) \in \overset{\circ}{\mathcal{F}}$. *Let* $\rho = \sqrt{n}$, $z^k = b^T y^k$, x^{k+1} *be given by (4.10), and* $y^{k+1} = y(z^k)$ *in (4.13) and* $s^{k+1} = s(z^k)$ *in (4.12). Then, either*

$$\psi_{n+\rho}(x^{k+1}, s^k) \leq \psi_{n+\rho}(x^k, s^k) - \delta$$

or

$$\psi_{n+\rho}(x^k, s^{k+1}) \leq \psi_{n+\rho}(x^k, s^k) - \delta$$

where $\delta > 1/20$.

Proof. If (4.14) does not hold, i.e.,

$$\|p(z^k)\| \geq \min\left(\eta\sqrt{\frac{n}{n+\eta^2}}, 1 - \eta\right),$$

then

$$\mathcal{P}_{n+\rho}(x^{k+1}, z^k) - \mathcal{P}_{n+\rho}(x^k, z^k) \leq -\alpha \min\left(\eta\sqrt{\frac{n}{n+\eta^2}}, 1 - \eta\right) + \frac{\alpha^2}{2(1-\alpha)},$$

hence from the relation between $\mathcal{P}_{n+\rho}$ and $\psi_{n+\rho}$,

$$\psi_{n+\rho}(x^{k+1}, s^k) - \psi_{n+\rho}(x^k, s^k) \leq -\alpha \min\left(\eta\sqrt{\frac{n}{n+\eta^2}}, 1 - \eta\right) + \frac{\alpha^2}{2(1-\alpha)}.$$

Otherwise, from Lemma 4.8 the inequalities of (4.15) hold:

i) The first of (4.15) indicates that y^{k+1} and s^{k+1} are in $\overset{\circ}{\mathcal{F}}_d$.

ii) Using the second of (4.15) and applying Lemma 3.1 to vector $X^k s^{k+1}/\mu$, we have

$$n \log(x^k)^T s^{k+1} - \sum_{j=1}^n \log(x_j^k s_j^{k+1})$$

$$= n \log n - \sum_{j=1}^n \log(x_j^k s_j^{k+1}/\mu)$$

$$\leq n \log n + \frac{\|X^k s^{k+1}/\mu - e\|^2}{2(1 - \|X^k s^{k+1}/\mu - e\|_\infty)}$$

$$\leq n \log n + \frac{\eta^2}{2(1-\eta)}$$

$$\leq n \log(x^k)^T s^k - \sum_{j=1}^n \log(x_j^k s_j^k) + \frac{\eta^2}{2(1-\eta)}.$$

4.3. POTENTIAL REDUCTION ALGORITHM

iii) According to the third of (4.15), we have

$$\sqrt{n}(\log(x^k)^T s^{k+1} - \log(x^k)^T s^k) = \sqrt{n}\log\frac{\mu}{\mu^k} \leq -\frac{\eta}{2}.$$

Adding the two inequalities in ii) and iii), we have

$$\psi_{n+\rho}(x^k, s^{k+1}) \leq \psi_{n+\rho}(x^k, s^k) - \frac{\eta}{2} + \frac{\eta^2}{2(1-\eta)}.$$

Thus, by choosing $\eta = .43$ and $\alpha = .3$ we have the desired result.

□

Theorem 4.9 establishes an important fact: the *primal-dual* potential function can be reduced by a constant no matter where x^k and y^k are. In practice, one can perform the line search to minimize the primal-dual potential function. This results in the following primal-dual potential reduction algorithm.

Algorithm 4.3 *Given a central path point* $(x^0, y^0, s^0) \in \overset{\circ}{\mathcal{F}}$. *Let* $z^0 = b^T y^0$. *Set* $k := 0$.
 While $(s^k)^T x^k \geq \epsilon$ **do**

1. *Compute* y_1 *and* y_2 *from (4.13).*

2. *If there exists* z *such that* $s(z) > 0$, *compute*

$$\bar{z} = \arg\min_z \psi_{n+\rho}(x^k, s(z)),$$

and if $\psi_{n+\rho}(x^k, s(\bar{z})) < \psi_{n+\rho}(x^k, s^k)$ *then* $y^{k+1} = y(\bar{z})$, $s^{k+1} = s(\bar{z})$ *and* $z^{k+1} = b^T y^{k+1}$; *otherwise,* $y^{k+1} = y^k$, $s^{k+1} = s^k$ *and* $z^{k+1} = z^k$.

3. *Let* $x^{k+1} = x^k - \bar{\alpha} X^k p(z^{k+1})$ *with*

$$\bar{\alpha} = \arg\min_{\alpha \geq 0} \psi_{n+\rho}(x^k - \alpha X^k p(z^{k+1}), s^{k+1}).$$

4. *Let* $k := k + 1$ *and return to Step 1.*

The performance of the algorithm results from the following corollary:

Corollary 4.10 *Let* $\rho = \sqrt{n}$. *Then, Algorithm 4.3 terminates in at most* $O(\sqrt{n}\log(c^T x^0 - b^T y^0)/\epsilon)$ *iterations with*

$$c^T x^k - b^T y^k \leq \epsilon.$$

Proof. In $O(\sqrt{n}\log((x^0)^T s^0/\epsilon))$ iterations

$$\begin{aligned}-\sqrt{n}\log((x^0)^T s^0/\epsilon) &= \psi_{n+\rho}(x^k, s^k) - \psi_{n+\rho}(x^0, s^0)\\ &\geq \sqrt{n}\log(x^k)^T s^k + n\log n - \psi_{n+\rho}(x^0, s^0)\\ &= \sqrt{n}\log((x^k)^T s^k/(x^0)^T s^0).\end{aligned}$$

Thus,
$$\sqrt{n}\log(c^T x^k - b^T y^k) = \sqrt{n}\log(x^k)^T s^k \leq \sqrt{n}\log\epsilon,$$

i.e.,
$$c^T x^k - b^T y^k = (x^k)^T s^k \leq \epsilon.$$

$\square$

4.4 Primal-Dual (Symmetric) Algorithm

Another technique for solving linear programs is the symmetric primal-dual algorithm. Once we have a pair $(x, y, s) \in \overset{\circ}{\mathcal{F}}$ with $\mu = x^T s/n$, we can generate a new iterate x^+ and (y^+, s^+) by solving for d_x, d_y and d_s from the system of linear equations:

$$\begin{aligned}Sd_x + Xd_s &= \gamma\mu e - Xs,\\ Ad_x &= 0,\\ -A^T d_y - d_s &= 0.\end{aligned} \quad (4.17)$$

Let $d := (d_x, d_y, d_s)$. To show the dependence of d on the current pair (x, s) and the parameter γ, we write $d = d(x, s, \gamma)$. Note that $d_x^T d_s = -d_x^T A^T d_y = 0$ here.

The system (4.17) is the Newton step starting from (x, s) which helps to find the point on the central path with duality gap $\gamma n\mu$, see Section 2.5.1. If $\gamma = 0$, it steps toward the optimal solution characterized by the system of equations (1.2); if $\gamma = 1$, it steps toward the central path point $(x(\mu), y(\mu), s(\mu))$ characterized by the system of equations (2.18); if $0 < \gamma < 1$, it steps toward a central path point with a smaller complementarity gap. In the algorithm presented in this section, we choose $\gamma = n/(n+\rho) < 1$. Each iterate reduces the primal-dual potential function by at least a constant δ, as does the previous potential reduction algorithm.

To analyze this algorithm, we present the following lemma, whose proof is very similar to Lemma 3.12 and will be omitted.

4.4. PRIMAL-DUAL (SYMMETRIC) ALGORITHM

Lemma 4.11 *Let the direction $d = (d_x, d_y, d_s)$ be generated by equation (4.17) with $\gamma = n/(n+\rho)$, and let*

$$\theta = \frac{\alpha\sqrt{\min(Xs)}}{\|(XS)^{-1/2}(\frac{x^Ts}{(n+\rho)}e - Xs)\|}, \qquad (4.18)$$

where α is a positive constant less than 1. Let

$$x^+ = x + \theta d_x, \quad y^+ = y + \theta d_y, \quad \text{and} \quad s^+ = s + \theta d_s.$$

Then, we have $(x^+, y^+, s^+) \in \overset{\circ}{\mathcal{F}}$ and

$$\psi_{n+\rho}(x^+, s^+) - \psi_{n+\rho}(x, s)$$

$$\leq -\alpha\sqrt{\min(Xs)}\|(XS)^{-1/2}(e - \frac{(n+\rho)}{x^Ts}Xs)\| + \frac{\alpha^2}{2(1-\alpha)}.$$

Let $v = Xs$. Then, we can prove the following lemma (Exercise 4.8):

Lemma 4.12 *Let $v \in \mathcal{R}^n$ be a positive vector and $\rho \geq \sqrt{n}$. Then,*

$$\sqrt{\min(v)}\|V^{-1/2}(e - \frac{(n+\rho)}{e^Tv}v)\| \geq \sqrt{3/4}.$$

Combining these two lemmas we have

$$\psi_{n+\rho}(x^+, s^+) - \psi_{n+\rho}(x, s)$$

$$\leq -\alpha\sqrt{3/4} + \frac{\alpha^2}{2(1-\alpha)} = -\delta$$

for a constant δ. This result will provide a competitive theoretical iteration bound, but a faster algorithm may be again implemented by conducting a line search along direction d to achieve the greatest reduction in the primal-dual potential function. This leads to

Algorithm 4.4 *Given $(x^0, y^0, s^0) \in \overset{\circ}{\mathcal{F}}$. Set $\rho \geq \sqrt{n}$ and $k := 0$.*
 While $(s^k)^T x^k \geq \epsilon$ do

1. Set $(x, s) = (x^k, s^k)$ and $\gamma = n/(n+\rho)$ and compute (d_x, d_y, d_s) from (4.17).

2. Let $x^{k+1} = x^k + \bar{\alpha}d_x$, $y^{k+1} = y^k + \bar{\alpha}d_y$, and $s^{k+1} = s^k + \bar{\alpha}d_s$ where

$$\bar{\alpha} = \arg\min_{\alpha \geq 0} \psi_{n+\rho}(x^k + \alpha d_x, s^k + \alpha d_s).$$

3. Let $k := k+1$ and return to Step 1.

Theorem 4.13 *Let $\rho = O(\sqrt{n})$. Then, Algorithm 4.4 terminates in at most $O(\sqrt{n}\log((x^0)^Ts^0/\epsilon))$ iterations with*

$$c^Tx^k - b^Ty^k \leq \epsilon.$$

4.5 Adaptive Path-Following Algorithms

Here we describe and analyze several additional primal-dual interior-point algorithms for linear programming. In some sense these methods follow the central path

$$\mathcal{C} = \left\{(x,s) \in \overset{\circ}{\mathcal{F}} : Xs = \mu e \quad \text{where} \quad \mu = \frac{x^T s}{n}\right\}$$

in primal-dual form, but certain algorithms allow a very loose approximation to the path.

Suppose we have a pair $(x, s) \in \mathcal{N}$, a neighborhood of $\mathcal{C}$, where $\mathcal{C} \subset \mathcal{N} \subset \overset{\circ}{\mathcal{F}}$. Consider the neighborhood

$$\mathcal{N}_2(\eta) = \left\{(x,s) \in \overset{\circ}{\mathcal{F}} : \|Xs - \mu e\| \leq \eta\mu \quad \text{where} \quad \mu = \frac{x^T s}{n}\right\}$$

for some $\eta \in (0,1)$. We will first analyze an adaptive-step path-following algorithm that generates a sequence of iterates in $\mathcal{N}_2(1/4)$. Actually, the algorithm has a predictor-corrector form, so that it also generates intermediate iterates in $\mathcal{N}_2(1/2)$.

Next we consider adaptive-step algorithms generating sequences of iterates in either

$$\mathcal{N}_\infty(\eta) = \left\{(x,s) \in \overset{\circ}{\mathcal{F}} : \|Xs - \mu e\|_\infty \leq \eta\mu \quad \text{where} \quad \mu = \frac{x^T s}{n}\right\}$$

or

$$\mathcal{N}_\infty^-(\eta) = \left\{(x,s) \in \overset{\circ}{\mathcal{F}} : \|Xs - \mu e\|_\infty^- \leq \eta\mu \quad \text{where} \quad \mu = \frac{x^T s}{n}\right\},$$

for any $\eta \in (0,1)$. Here, for any $z \in R^n$,

$$\|z\|_\infty^- := \|z^-\|_\infty$$

and

$$\|z\|_\infty^+ := \|z^+\|_\infty,$$

where $(z^-)_j := \min\{z_j, 0\}$ and $(z^+)_j := \max\{z_j, 0\}$ and $\|\cdot\|_\infty$ is the usual ℓ_∞ norm. Note that $\|z\|_\infty = \max\{\|z\|_\infty^+, \|z\|_\infty^-\}$ and that neither $\|\cdot\|_\infty^-$ nor $\|\cdot\|_\infty^+$ is a norm, although they obey the triangle inequality.

We easily see that

$$\mathcal{C} \subset \mathcal{N}_2(\eta) \subset \mathcal{N}_\infty(\eta) \subset \mathcal{N}_\infty^-(\eta) \subset \overset{\circ}{\mathcal{F}} \quad \text{for each } \eta \in (0,1).$$

4.5. ADAPTIVE PATH-FOLLOWING ALGORITHMS

Our results indicate that when we use a wider neighborhood of the central path, the worst-case number of iterations grows, while the practical behavior might be expected to improve.

Given $(x, s) \in \mathcal{N}$, we again generate a search direction $d = (d_x, d_y, d_s)$ using the primal-dual Newton method by solving (4.17). Having obtained the search direction d, we let

$$\begin{aligned} x(\theta) &:= x + \theta d_x, \\ y(\theta) &:= y + \theta d_y, \\ s(\theta) &:= s + \theta d_s. \end{aligned} \quad (4.19)$$

We will frequently let the next iterate be $(x^+, s^+) = (x(\bar{\theta}), s(\bar{\theta}))$, where $\bar{\theta}$ is as large as possible so that $(x(\theta), s(\theta))$ remains in the neighborhood $\mathcal{N}$ for $\theta \in [0, \bar{\theta}]$.

Let $\mu(\theta) = x(\theta)^T s(\theta)/n$ and $X(\theta) = \text{diag}(x(\theta))$. In order to get bounds on $\bar{\theta}$, we first note that

$$\begin{aligned} \mu(\theta) &= (1 - \theta)\mu + \theta\gamma\mu, & (4.20) \\ X(\theta)s(\theta) - \mu(\theta)e &= (1 - \theta)(Xs - \mu e) + \theta^2 D_x d_s, & (4.21) \end{aligned}$$

where $D_x = \text{diag}(d_x)$. Thus $D_x d_s$ is the second-order term in Newton's method to compute a new point of $\mathcal{C}$. Hence we can usually choose a larger $\bar{\theta}$ (and get a larger decrease in the duality gap) if $D_x d_s$ is smaller. In this section we obtain several bounds on the size of $D_x d_s$.

First, it is helpful to re-express $D_x d_s$. Let

$$\begin{aligned} p &:= X^{-.5} S^{.5} d_x, \\ q &:= X^{.5} S^{-.5} d_s, \\ r &:= (XS)^{-.5}(\gamma\mu e - Xs), \end{aligned} \quad (4.22)$$

Note that $p + q = r$ and $p^T q = 0$ so that p and q represent an orthogonal decomposition of r.

Lemma 4.14 *With the notations above,*

i)

$$\|Pq\| \leq \frac{\sqrt{2}}{4}\|r\|^2;$$

ii)

$$-\frac{\|r\|^2}{4} \leq p_j q_j \leq \frac{r_j^2}{4} \quad \text{for each } j;$$

iii)

$$\|Pq\|_\infty^- \le \frac{\|r\|^2}{4} \le \frac{n\|r\|_\infty^2}{4},$$

$$\|Pq\|_\infty^+ \le \frac{\|r\|_\infty^2}{4}, \quad \text{and}$$

$$\|Pq\|_\infty \le \frac{\|r\|^2}{4} \le \frac{n\|r\|_\infty^2}{4}.$$

The bounds in Lemma 4.14 cannot improved by much in the worst case: consider the case where

$$\begin{aligned} r &= e = (1,1,\cdots,1)^T, \\ p &= (1/2, 1/2, \cdots, 1/2, (1+\sqrt{n})/2)^T, \quad \text{and} \\ q &= (1/2, 1/2, \cdots, 1/2, (1-\sqrt{n})/2)^T. \end{aligned}$$

Lemma 4.14 has been partially proved in Theorem 3.11. We leave the rest proof of the lemma to the reader.

To use Lemma 4.14 we also need to bound r. The following result is useful:

Lemma 4.15 *Let r be as above.*

i) *If $\gamma = 0$, then $\|r\|^2 = n\mu$.*

ii) *If $\eta \in (0,1)$, $\gamma = 1$ and $(x,s) \in \mathcal{N}_2(\eta)$, then $\|r\|^2 \le \eta^2 \mu/(1-\eta)$.*

iii) *If $\eta \in (0,1)$, $\gamma \in (0,1)$, $\gamma \le 2(1-\eta)$ and $(x,s) \in \mathcal{N}_\infty^-(\eta)$, then $\|r\|^2 \le n\mu$. Moreover, if $(x,s) \in \mathcal{N}_\infty(\eta)$ then*

$$\sqrt{1-\eta} \ge r_j/\sqrt{\mu} \ge -\sqrt{1+\eta},$$

so $\|r\|_\infty^2 \le (1+\eta)\mu$.

Proof. i) If $\gamma = 0$, $r = -(XS)^{-.5}Xs$, so $\|r\|^2 = x^T s = n\mu$.

ii) Now $r = (XS)^{-.5}(\mu e - Xs)$, so $\|r\| \le \frac{1}{\sqrt{(1-\eta)\mu}}\eta\mu$, which yields the desired result.

iii) In this case

$$\begin{aligned} \|r\|^2 &= \sum_{j=1}^n \frac{(\gamma\mu - x_j s_j)^2}{x_j s_j} \\ &= \sum_{j=1}^n \left(\frac{(\gamma\mu)^2}{x_j s_j} - 2\gamma\mu + x_j s_j\right) \\ &\le \frac{n(\gamma\mu)^2}{(1-\eta)\mu} - 2n\gamma\mu + n\mu \quad (\text{since } x_j s_j \ge (1-\eta)\mu) \\ &\le n\mu \quad (\text{since } \gamma \le 2(1-\eta)). \end{aligned}$$

4.5. ADAPTIVE PATH-FOLLOWING ALGORITHMS

Now suppose $(x,s) \in \mathcal{N}_\infty(\eta)$, so that $x_j s_j \in [(1-\eta)\mu, (1+\eta)\mu]$ for each j. Thus, for each j,

$$\frac{\gamma\mu}{\sqrt{1-\eta}\sqrt{\mu}} - \sqrt{1-\eta}\sqrt{\mu} \geq r_j = \frac{\gamma\mu}{\sqrt{x_j s_j}} - \sqrt{x_j s_j} \geq \frac{\gamma\mu}{\sqrt{1+\eta}\sqrt{\mu}} - \sqrt{1+\eta}\sqrt{\mu},$$

which yields the final result since $0 \leq \gamma \leq 2(1-\eta)$.

□

4.5.1 Predictor-corrector algorithm

In this section we describe and analyze an algorithm that takes a single "corrector" step to the central path after each "predictor" step to decrease μ. Although it is possible to use more general values of η, we will work with nearly-centered pairs in $\mathcal{N}_2(\eta)$ with $\eta = 1/4$ (iterates after the corrector step), and intermediate pairs in $\mathcal{N}_2(2\eta)$ (iterates after a predictor step).

Algorithm 4.5 Given $(x^0, s^0) \in \mathcal{N}_2(\eta)$ with $\eta = 1/4$. Set $k := 0$.
  While $(x^k)^T s^k > \epsilon$ do:

1. *Predictor step:* set $(x,s) = (x^k, s^k)$ and compute $d = d(x,s,0)$ from (4.17); compute the largest $\bar\theta$ so that

$$(x(\theta), s(\theta)) \in \mathcal{N}_2(2\eta) \quad \text{for} \quad \theta \in [0, \bar\theta].$$

2. *Corrector step:* set $(x', s') = (x(\bar\theta), s(\bar\theta))$ and compute $d' = d(x', s', 1)$ from (4.17); set $(x^{k+1}, s^{k+1}) = (x' + d'_x, s' + d'_s)$.

3. *Let* $k := k+1$ *and return to Step 1.*

To analyze this method, we start by showing

Lemma 4.16 For each k, $(x^k, s^k) \in \mathcal{N}_2(\eta)$.

Proof. The claim holds for $k=0$ by hypothesis. For $k > 0$, let (x', s') be the result of the predictor step at the kth iteration and let $d' = d(x', s', 1)$, as in the description of the algorithm. Let $x'(\theta)$ and $s'(\theta)$ be defined as in (4.19) and p', q' and r' as in (4.22) using x', s' and d'. Let $\mu'(\theta) := x'(\theta)^T s'(\theta)/n$ for all $\theta \in [0,1]$ with $\mu' := \mu'(0) = (x')^T s'/n$ and $\mu^{k+1} := \mu'(1) = (x^{k+1})^T s^{k+1}/n$.
  From (4.20),

$$\mu'(\theta) = \mu' \quad \text{for all } \theta, \tag{4.23}$$

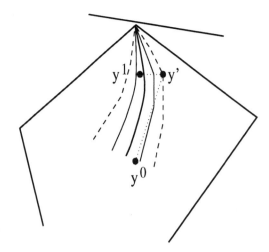

Figure 4.1. Illustration of the predictor-corrector algorithm; the predictor step moves y^0 in a narrower neighborhood of the central path to y' on the boundary of a wider neighborhood and the corrector step then moves y' to y^1 back in the narrower neighborhood.

and, in particular, $\mu^{k+1} = \mu'$. From (4.21),

$$\begin{aligned} X'(\theta)s'(\theta) - \mu'(\theta)e &= (1-\theta)(X's' - \mu'e) + \theta^2 D'_x d'_s \\ &= (1-\theta)(X's' - \mu'e) + \theta^2 P'q', \end{aligned} \quad (4.24)$$

where $X'(\theta) = \mathrm{diag}(x'(\theta))$, etc. But by Lemma 4.14(i), Lemma 4.15(ii) and $(x', s') \in \mathcal{N}(2\eta)$ with $\eta = 1/4$,

$$\|P'q'\| \leq \frac{\sqrt{2}}{4}\|r'\|^2 \leq \frac{\sqrt{2}}{4}\frac{(2\eta)^2}{1-2\eta}\mu' < \frac{1}{4}\mu'.$$

It follows that

$$\|X'(\theta)s'(\theta) - \mu'e\| \leq (1-\theta)\frac{\mu'}{2} + \theta^2\frac{\mu'}{4} \leq \frac{1}{2}\mu'. \quad (4.25)$$

Thus $X'(\theta)s'(\theta) \geq \frac{\mu'}{2}e > 0$ for all $\theta \in [0, 1]$, and this implies that $x'(\theta) > 0$, $s'(\theta) > 0$ for all such θ by continuity. In particular, $x^{k+1} > 0$, $s^{k+1} > 0$, and (4.25) gives $(x^{k+1}, s^{k+1}) \in \mathcal{N}_2(1/4)$ as desired when we set $\theta = 1$.

□

4.5. ADAPTIVE PATH-FOLLOWING ALGORITHMS

Now let $(x, s) = (x^k, s^k)$, $d = d(x, s, 0)$, $\mu = \mu^k = x^T s/n$, and p, q and r be as in (4.22); these quantities all refer to the predictor step at iteration k. By (4.20),

$$\mu' = (1 - \bar{\theta})\mu, \quad \text{or}$$
$$\mu^{k+1} = (1 - \bar{\theta})\mu^k. \tag{4.26}$$

Hence the improvement in the duality gap at the kth iteration depends on the size of $\bar{\theta}$.

Lemma 4.17 *With the notation above, the step-size in the predictor step satisfies*

$$\bar{\theta} \geq \frac{2}{1 + \sqrt{1 + 4\|Pq/\mu\|/\eta}}.$$

Proof. By (4.21) applied to the predictor step,

$$\begin{aligned}
\|X(\theta)s(\theta) - \mu(\theta)e\| &= \|(1-\theta)(Xs - \mu e) + \theta^2 Pq\| \\
&\leq (1-\theta)\|Xs - \mu e\| + \theta^2 \|Pq\| \\
&\leq (1-\theta)\eta\mu + \theta^2 \|Pq\|,
\end{aligned}$$

after using Lemma 4.16. We see that for

$$0 \leq \theta \leq \frac{2}{1 + \sqrt{1 + 4\|Pq/\mu\|/\eta}}$$

$$\begin{aligned}
\|X(\theta)s(\theta) - \mu(\theta)e\|/\mu &\leq (1-\theta)\eta + \theta^2 \|Pq/\mu\| \\
&\leq 2\eta(1-\theta).
\end{aligned}$$

This is because the quadratic term in θ:

$$\|Pq/\mu\|\theta^2 + \eta\theta - \eta \leq 0$$

for θ between zero and the root

$$\frac{-\eta + \sqrt{\eta^2 + 4\|Pq/\mu\|\eta}}{2\|Pq/\mu\|} = \frac{2}{1 + \sqrt{1 + 4\|Pq/\mu\|/\eta}}.$$

Thus,

$$\|X(\theta)s(\theta) - \mu(\theta)e\| \leq 2\eta(1-\theta)\mu = 2\eta\mu(\theta)$$

or $(x(\theta), s(\theta)) \in \mathcal{N}_2(2\eta)$ for

$$0 \leq \theta \leq \frac{2}{1 + \sqrt{1 + 4\|Pq/\mu\|/\eta}}.$$

□

We can now show

Theorem 4.18 *Let $\eta = 1/4$. Then Algorithm 4.5 will terminate in at most $O(\sqrt{n}\log((x^0)^T s^0/\epsilon))$ iterations with*
$$c^T x^k - b^T y^k \leq \epsilon.$$

Proof. Using Lemma 4.14(i) and Lemma 4.15(i), we have
$$\|Pq\| \leq \frac{\sqrt{2}}{4}\|r\|^2 = \frac{\sqrt{2}}{4}n\mu,$$

so that
$$\bar{\theta} \geq \frac{2}{1+\sqrt{1+\sqrt{2}n/\eta}} = \frac{2}{1+\sqrt{1+4\sqrt{2}n}}$$

at each iteration. Then (4.26) and Lemma 4.17 imply that
$$\mu^{k+1} \leq \left(1 - \frac{2}{1+\sqrt{1+4\sqrt{2}n}}\right)\mu^k$$

for each k. This yields the desired result.

$\square$

4.5.2 Wide-neighborhood algorithm

In this section we consider algorithms of the following form based on $\gamma \in (0,1)$ and $\mathcal{N}$, where $\mathcal{N}$ is a wide neighborhood of either $\mathcal{N}_\infty$ or $\mathcal{N}_\infty^-$.

Algorithm 4.6 *Let $\eta \in (0,1)$ and $\gamma \in (0,1)$ with $\gamma \leq 2(1-\eta)$. Given $(x^0, s^0) \in \mathcal{N}(\eta)$. Set $k := 0$.*
 While $(x^k)^T s^k > \epsilon$ do:

1. *Set $(x,s) = (x^k, s^k)$ and compute $d = d(x,s,\gamma)$ from (4.17).*

2. *Compute the largest $\bar{\theta}$ so that*
$$(x(\theta), s(\theta)) \in \mathcal{N} \quad \text{for} \quad \theta \in [0, \bar{\theta}];$$
 set $(x^{k+1}, s^{k+1}) = (x(\bar{\theta}), s(\bar{\theta}))$.

3. *Let $k := k+1$ and return to Step 1.*

4.5. ADAPTIVE PATH-FOLLOWING ALGORITHMS

Again the selection of $\bar\theta$ makes this an adaptive-step method. We will analyze this algorithm for $\mathcal{N} = \mathcal{N}_\infty(\eta)$ and $\mathcal{N}_\infty^-(\eta)$, where $\eta \in (0,1)$. In either case, computing $\bar\theta$ involves the solution of at most $2n$ single-variable quadratic equations.

Note that, if $\mu^k := (x^k)^T s^k / n$, (4.20) implies

$$\mu^{k+1} = (1 - \bar\theta(1-\gamma))\mu^k, \qquad (4.27)$$

so we wish to bound $\bar\theta$ from below.

Lemma 4.19 *Let $\eta \in (0,1)$, $\gamma \in (0,1)$, and $\mathcal{N} = \mathcal{N}_\infty(\eta)$ or $\mathcal{N}_\infty^-(\eta)$. Let x, s, d and $\bar\theta$ be as in the kth iteration of Algorithm 4.6, and define p, q and r by (4.22). Then*

$$\bar\theta \geq \theta_2 := \min\left\{1, \frac{\eta\gamma\mu^k}{\|Pq\|_\infty}\right\} \text{ if } \mathcal{N} = \mathcal{N}_\infty(\eta),$$

$$\bar\theta \geq \theta_2^- := \min\left\{1, \frac{\eta\gamma\mu^k}{\|Pq\|_\infty^-}\right\} \text{ if } \mathcal{N} = \mathcal{N}_\infty^-(\eta).$$

Proof. Suppose first $\mathcal{N} = \mathcal{N}_\infty^-(\eta)$. Then, for each $\theta \in [0, \theta_2^-]$, (4.20) and (4.21) imply

$$\begin{aligned} &X(\theta)s(\theta) - \mu(\theta)e \\ &= (1-\theta)(Xs - \mu e) + \theta^2 Pq, \\ &\geq -((1-\theta)\|Xs - \mu e\|_\infty^- + \theta^2 \|Pq\|_\infty^-)e \\ &\geq -((1-\theta)\eta\mu^k + \theta\eta\gamma\mu^k)e \\ &= -\eta\mu(\theta)e. \end{aligned}$$

Hence, as in the proof of Lemma 4.16, $(x(\theta), s(\theta)) \in \mathcal{N}_\infty^-(\eta)$ for $\theta \in [0, \theta_2^-]$, when $\bar\theta \geq \theta_2^-$. If $\mathcal{N} = \mathcal{N}_\infty(\eta)$, a similar proof gives

$$\eta\mu(\theta)e \geq X(\theta)s(\theta) - \mu(\theta)e \geq -\eta\mu(\theta)e$$

for $\theta \in [0, \theta_2]$, which again implies $\bar\theta \geq \theta_2$.

$\square$

We can now prove the following theorem:

Theorem 4.20 *Let $\eta \in (0,1)$ and $\gamma \in (0,1)$ be constants with $\gamma \leq 2(1-\eta)$. Then Algorithm 4.6, with $\mathcal{N} = \mathcal{N}_\infty(\eta)$ or $\mathcal{N}_\infty^-(\eta)$, will terminate in $O(n \log((x^0)^T s^0/\epsilon))$ iterations.*

Proof. In either case, each iterate lies in $\mathcal{N}_\infty^-(\eta)$, where

$$\|Pq\|_\infty^- \leq \|Pq\|_\infty \leq \|r\|^2/4 \leq n\mu^k/4,$$

using Lemma 4.14(iii) and Lemma 4.15(iii). Hence

$$\theta_2^- \geq \theta_2 \geq 4\eta\gamma/n.$$

Then Lemma 4.19 and (4.27) give

$$\mu^{k+1} \leq \left(1 - \frac{4\eta\gamma(1-\gamma)}{n}\right)\mu^k, \tag{4.28}$$

which yields the result.

$\square$

The algorithms for the neighborhoods $\mathcal{N}_\infty(\eta)$ and $\mathcal{N}_\infty^-(\eta)$ generate sequences of points lying in the boundaries of these sets. Since the results hold for arbitrary $\eta \in (0,1)$, the algorithms can generate sequences of points in a wide area of the feasible region. In particular,

$$\mathcal{N}_\infty^-(1) = \overset{\circ}{\mathcal{F}},$$

so when η is close to 1, the neighborhood $\mathcal{N}_\infty^-(\eta)$ spreads over almost all of the feasible region $\mathcal{F}$, and the points generated by the algorithm based on $\mathcal{N}_\infty^-(\eta)$ are close to the boundary rather than the central path.

4.6 Affine Scaling Algorithm

Soon after Karmarkar published his projective interior-point algorithm, many researchers rediscovered the affine scaling algorithm, which was originally developed by Dikin in 1967. The algorithm is one of the simplest and most efficient interior-point algorithms for LP.

Given an $x^k \in \overset{\circ}{\mathcal{F}}_p$, the affine scaling algorithm solves the following problem to generate the direction:

$$\begin{array}{ll} \text{minimize} & c^T(x - x^k) \\ \text{s.t.} & A(x - x^k) = 0, \ \|(X^k)^{-1}(x - x^k)\| \leq \alpha, \end{array}$$

where $\alpha \leq 1$ and $\{x : \|(X^k)^{-1}(x - x^k)\| \leq \alpha\}$ is called the *Dikin ellipsoid* with radius α—a coordinate-aligned ellipsoid contained in the interior of the positive orthant.

4.6. AFFINE SCALING ALGORITHM

Let its minimizer be x^{k+1}. Then

$$x^{k+1} - x^k = -\alpha \frac{X^k p^k}{\|p^k\|}, \tag{4.29}$$

and, in view of Section 1.5.4,

$$\begin{aligned} p^k &= (I - X^k A^T (A(X^k)^2 A^T)^{-1} A X^k) X^k c \\ &= X^k c - X^k A^T y^k \\ &= X^k (c - A^T y^k) \end{aligned}$$

and

$$\begin{aligned} y^k := y(x^k) &= (A(X^k)^2) A^T)^{-1} A(X^k)^2 c \\ s^k := s(x^k) &= c - A^T y^k. \end{aligned}$$

We can assume that $\|p^k\| > 0$, for if not, x^k is already an optimal solution. Thus,

$$c^T x^{k+1} - c^T x^k = -\alpha \|p^k\| < 0.$$

The update (4.29) is called *small step*. A *large step* update is given by

$$x^{k+1} - x^k = -\alpha \frac{X^k p^k}{\max(p^k)}. \tag{4.30}$$

We can safely assume that $\max(p^k) > 0$, since if not, the LP problem is either unbounded, or any $x \in \mathcal{F}_p$ is optimal if $p^k = 0$.

To exclude these trivial cases, we impose the following assumption for the affine scaling algorithm.

Assumption 4.1 *The objective function $c^T x$ is not constant over the feasible region of (LP) and it is bounded from below.*

Proposition 4.21 *The following statements hold for the affine scaling algorithm:*

i) $x^k \in \overset{\circ}{\mathcal{F}}_p$ *for all* $k \geq 0$;

ii) *The sequence of objective function values $\{c^T x^k\}$ strictly decreases and converges to a finite value;*

iii) $X^k s^k \longrightarrow 0$ *as* $k \longrightarrow \infty$.

Proof. Assume that $x^k \in \overset{\circ}{\mathcal{F}}_p$. Using (4.30), we obtain

$$(X^k)^{-1} x^{k+1} = e - \alpha \frac{p^k}{\max(p^k)} \geq e - \alpha e = (1-\alpha)e > 0,$$

138   CHAPTER 4. LINEAR PROGRAMMING ALGORITHMS

from which i) follows. Using (4.30) and noting
$$c^T X^k p^k = \|p^k\|^2,$$
we obtain
$$\begin{aligned} c^T x^{k+1} &= c^T x^k - \frac{\alpha}{\max(p^k)} c^T X^k p^k \\ &\leq c^T x^k - \frac{\alpha}{\|p^k\|} \|p^k\|^2 \\ &= c^T x^k - \alpha \|X^k s^k\|. \end{aligned} \quad (4.31)$$

Hence,
$$0 < \|X^k s^k\| \leq \alpha^{-1}(c^T x^k - c^T x^{k+1}), \quad \forall k \geq 0,$$
from which ii) and iii) obviously follows.

$\square$

Proposition 4.21 does not guarantee that the sequence $\{x^k\}$ converges and not even that $\{x^k\}$ is a bounded sequence. We now concentrate our efforts in showing that the sequence $\{x^k\}$ converges.

Theorem 4.22 *For the affine scaling algorithm, the following statements hold.*

i) *There exists a constant $M(A, c) > 0$ such that*
$$\|X^k p^k\| \leq M(A, c)\|p^k\|^2 = M(A, c) c^T X^k p^k, \quad \forall k \geq 0. \quad (4.32)$$

ii) *The sequence $\{x^k\}$ converges to a point $\bar{x} \in \mathcal{F}_p$.*

iii) *For all $k \geq 0$, we have $\|x^k - \bar{x}\| \leq M(A, c)(c^T x^k - c^T \bar{x})$.*

Proof. The proof of i) follows Exercise 4.13. Using (4.30), (4.31) and (4.32), we obtain
$$\|x^{k+1} - x^k\| \leq M(A, c)(c^T x^k - c^T x^{k+1}), \quad \forall k \geq 0. \quad (4.33)$$
Since $\{c^T x^k\}$ converges, (4.33) implies that
$$\sum_{k=0}^{\infty} \|x^{k+1} - x^k\| \leq \lim_{k \to \infty} M(A, c)(c^T x^0 - c^T x^k) < \infty.$$

4.6. AFFINE SCALING ALGORITHM

This implies that $\{x^k\}$ is a Cauchy sequence, and therefore a convergent sequence. Clearly, $\bar{x} := \lim_{k\to\infty} x^k \in \mathcal{F}_p$. We now show iii). From (4.33), we also obtain

$$\|x^k - x^l\| \leq M(A,c)(c^T x^k - c^T x^l), \quad \forall l > k \geq 0.$$

Letting $l \to \infty$, we obtain

$$\|x^k - \bar{x}\| \leq M(A,c)(c^T x^k - c^T \bar{x}), \quad \forall k \geq 0.$$

□

The question is whether or not the limit $\bar{x}$ is an optimal solution. In what follows, let $\bar{z} = c^T \bar{x}$, and let (P, Z) be the partition of $\bar{x}$ such that $\bar{x}_P > 0$ and $\bar{x}_Z = 0$. Since $\|X^k s^k\| \to 0$ (using Theorem 4.21(iii)) and $\bar{x}_P > 0$, we see that $s_P^k \to 0$. Furthermore, since $\{(y^k, s^k)\}$ is a bounded sequence from Exercise 4.14, it must have an accumulation point $(\bar{y}, \bar{s})$. Clearly, $A^T \bar{y} + \bar{s} = c$, $\bar{X}\bar{s} = 0$, and $\bar{s}_P = 0$.

Lemma 4.23 *The set $\{x \in \mathcal{F}_p : x_Z = 0\}$ is a face of $\mathcal{F}_p$ where the objective function is constant, and for an $x \in \mathcal{F}_p$, we have*

$$c^T(x - \bar{x}) = \bar{s}^T(x - \bar{x}) = \bar{s}_Z^T x_Z.$$

This lemma shows that $|Z| > 0$. We now show that $\bar{x}$ is an optimal solution of problem (LP) if the step-size $\alpha \in (0, 2/3]$. An important ingredient used in the proof of these results is the following (partial) potential function.

Definition 4.1 *The potential function with respect to Z is defined as*

$$\mathcal{P}_Z(x, \bar{z}) = |Z| \log(c^T x - \bar{z}) - \sum_{j \in Z} \log x_j \tag{4.34}$$

for every $x \in \overset{\circ}{\mathcal{F}}_p$, such that $c^T x > \bar{z}$.

The fact that the limit point $\bar{x}$ is an optimal solution of problem (LP) follows from two results stated below, namely Lemma 4.24 and Theorem 4.25. Lemma 4.24 follows as an immediate consequence of Theorem 4.22(iii). Theorem 4.26 combines these two results to obtain the conclusion that $\bar{x}$ is an optimal solution of (LP).

Lemma 4.24 *For the affine scaling algorithm,*

$$\frac{c^T x^k - \bar{z}}{\sum_{j \in Z} x_j^k} \geq \frac{1}{\sqrt{|Z|} M(A, c)}, \quad \forall k \geq 0. \tag{4.35}$$

Proof. By Theorem 4.22(iii), we know that for some $M(A,c) \geq 0$,
$$\|x^k - \bar{x}\| \leq M(A,c)(c^T x^k - \bar{z}), \quad \forall k \geq 0.$$
Since $\bar{x}_Z = 0$, this implies
$$\|x_Z^k\| \leq M(A,c)(c^T x^k - \bar{z}), \quad \forall k \geq 0.$$
Then, we obtain
$$\sum_{j \in Z} x_j^k \leq \sqrt{|Z|} \|x_Z^k\| \leq \sqrt{|Z|} M(A,c)(c^T x^k - \bar{z}), \quad \forall k \geq 0.$$
□

The proof of the next theorem will be omitted. It is similar to the affine scaling theorem described in Section 3.3.3.

Theorem 4.25 *For the affine scaling algorithm, assume that $\alpha \leq 2/3$. If $\bar{x}$ is not an optimal solution lying in the interior of the optimal face of problem (LP) then there exist a constant $\delta > 0$ and an integer K such that*
$$\mathcal{P}_Z(x^{k+1}, \bar{z}) - \mathcal{P}_Z(x^k, \bar{z}) < -\delta, \quad \forall k \geq K. \tag{4.36}$$

Combining Lemma 4.24 and Theorem 4.25, we can now show that $\bar{x}$ is an optimal solution of (LP).

Theorem 4.26 *For the affine scaling algorithm, assume that $\alpha \leq 2/3$. Then the limit $\bar{x}$ of the primal sequence $\{x^k\}$ is an optimal solution lying in the interior of the optimal face of (LP).*

Proof. Assume by contradiction that $\bar{x}$ is not an optimal solution lying in the relative interior of the optimal face of (LP). By Theorem 4.25, it follows that
$$\lim_{k \to \infty} \mathcal{P}_Z(x^k, \bar{z}) = -\infty. \tag{4.37}$$
Noting the fact that
$$|Z| \log(\sum_{j \in Z} x_j^k) - \sum_{j \in Z} \log x_j^k \geq |Z| \log |Z|,$$
we obtain
$$\begin{aligned}\mathcal{P}_Z(x^k, \bar{z}) &= |Z| \log \left(\frac{c^T x - \bar{z}}{\sum_{j \in Z} x_j^k} \right) + |Z| \log(\sum_{j \in Z} x_j^k) - \sum_{j \in Z} \log x_j^k \\ &\geq |Z| \log \frac{c^T x^k - \bar{z}}{\sum_{j \in Z} x_j^k} + |Z| \log |Z|.\end{aligned} \tag{4.38}$$

4.7. EXTENSIONS TO QP AND LCP

Relations (4.37) and (4.38) then imply

$$\lim_{k\to\infty} \frac{c^T x^k - \bar{z}}{\sum_{j\in Z} x_j^k} = 0,$$

which contradicts Lemma 4.24.

□

4.7 Extensions to QP and LCP

Many LP algorithms discussed in this chapter can be extended to solving convex quadratic and monotone linear complementarity problems. We present one extension, the primal-dual potential reduction algorithm, here. More results will be introduced in the subsequent chapters. Since solving convex QP problems reduces to solving monotone LCP problems, we present the algorithm in the LCP format of (1.5).

Once we have an interior feasible point (x, y, s) with $\mu = x^T s/n$, we can generate a new iterate x^+ and (y^+, s^+) by solving for d_x, d_y and d_s from the system of linear equations:

$$\begin{aligned} Sd_x + Xd_s &= \gamma\mu e - Xs, \\ M\begin{pmatrix} d_x \\ d_y \end{pmatrix} - \begin{pmatrix} d_s \\ 0 \end{pmatrix} &= 0, \end{aligned} \quad (4.39)$$

where we choose $\gamma = n/(n + \sqrt{n}) < 1$. Then, we assign $x^+ = x + \bar{\alpha}d_x$, $y^+ = y + \bar{\alpha}d_y$, and $s^+ = s + \bar{\alpha}d_s$ where

$$\bar{\alpha} = \arg\min_{\alpha \geq 0} \psi_{n+\rho}(x + \alpha d_x, s + \alpha d_s).$$

Since M is monotone, we have

$$d_x^T d_s = (d_x; d_y)^T (d_s; 0) = (d_x; d_y)^T M(d_x; d_y) \geq 0,$$

where we have $d_x^T d_s = 0$ in the LP case. Nevertheless, we can prove Lemma 4.11 still holds (Exercise 4.7) and from Lemma 4.12

$$\psi_{n+\rho}(x^+, s^+) - \psi_{n+\rho}(x, s)$$

$$\leq -\alpha\sqrt{3/4} + \frac{\alpha^2}{2(1-\alpha)} = -\delta$$

for a constant δ. This leads to a primal-dual potential reduction algorithm with the same complexity bound presented in Theorem 4.13.

4.8 Notes

A similar result to Proposition 4.2 has been proved by Todd [412]. This proposition plays an important role in analyzing several interior-point algorithms.

The Karmarkar projective algorithm in the LP standard form with a lower bound z for z^* was first developed and analyzed by Todd and Burrell [413], Anstreicher [24], Gay [133], and Ye and Kojima [477]. de Ghellinck and Vial [136] developed a projective algorithm that has a unique feature: it does not need to start from a feasible interior point. All of these algorithms have an iteration complexity $O(nL)$. Other extensions and analyses of Karmarkar's algorithm can be found in Akgül [8], Anstreicher [22], Asic, Kovacevic-Vujcic and Radosavljevic-Nikolic [36], Betke and Gritzmann [55], Blair [62], Blum [64], Dennis, Morshedi and Turner [95], Diao [97], Gonzaga [159], Kalantari [211], Kovacevic-Vujcic [238], Lagarias [242], McDiarmid [266], Nemirovskii [319], Nesterov [324], Padberg [335], Sherali [377], Shub [381], Tseng [421], Wei [456], Wu and Wu [460], Zimmermann [489].

The path-following algorithm, described in Section 4.2, is a variant of Renegar [358]. The difference is the analysis used in proving the complexity bound. Renegar measured the duality-gap, while we used the max-potential of the level set. A primal path-following algorithm is independently analyzed by Gonzaga [158]. Both Gonzaga [158] and Vaidya [437] developed a rank-one updating technique in solving the Newton equation of each iteration, and proved that each iteration uses $O(n^{2.5})$ arithmetic operations on average. Kojima, Mizuno and Yoshise [230] and Monteiro and Adler [298] developed a symmetric primal-dual path-following algorithm with the same iteration and arithmetic operation bounds. The algorithm was proposed earlier by Tanabe [400]. Other variants of path-following or homotopy algorithms can be found in Blum [65], Boggs, Domich, Donaldson and Witzgall [67], Nazareth [315, 316].

Recently, Vaidya [438] developed a new center, the volumetric center, for linear inequalities and a path-following algorithm for convex programming. The arithmetic operations complexity bound is identical to that of the ellipsoid method, but its iteration complexity bound is less than that of the ellipsoid method. Also see Anstreicher [26].

The primal potential function with $\rho > 1$ and the affine potential reduction algorithm were developed by Gonzaga [160]. His algorithm has iteration complexity $O(nL)$. The primal-dual potential function and algorithm were analyzed by Anstreicher and Bosch [28], Freund [123], Gonzaga [160], Todd and Ye [415], and Ye [466]. These algorithms possess $O(\sqrt{n}L)$ iteration complexity. Using this function, Ye [468] further developed a projective algorithm with $O(\sqrt{n}L)$ iteration complexity; also see Goldfarb and

4.8. NOTES

Xiao [152].

The affine scaling algorithm was developed by Dikin [98, 99] and was rediscovered by Barnes [43], Cavalier and Soyster [79], Kortanek and Shi [235], Sherali, Skarpness and Kim [378], Vanderbei and Lagarias [448], Vanderbei, Meketon and Freedman [449], Andersen [19], and Monteiro, Adler and Resende [301]. The algorithm also has three forms as the potential algorithm has, and one can view that ρ is chosen as ∞ for the direction of the affine scaling algorithm. The primal or dual algorithm has no polynomial complexity bound yet, but has been proved convergent under very weak conditions as described here, see Tsuchiya [425, 426], Tsuchiya and Muramatsu [429], Monteiro, Tsuchiya and Wang [305], Saigal [369], Sun [396], and Tseng and Luo [424]. Mascarenhas [265] provided a divergence example for the affine scaling algorithm. The primal-dual polynomial affine-scaling-type algorithms can be found in Monteiro, Adler and Resende [301] and Jansen, Roos and Terlaky [199].

The primal-dual potential reduction algorithm described in Section 4.3 is in the primal form. One can develop a potential reduction algorithm in dual form, where z, an upper bound for the optimal objective value z^*, is updated down in each iteration, see Ye [466]. The symmetric primal-dual potential algorithm of Sections 4.4 and 4.7 was developed by Kojima, Mizuno and Yoshise [232]. Other potential reduction algorithms are by Gonzaga and Todd [166], Huang and Kortanek [189], and Tunçel [432]. Todd [408] proposed an extremely simple and elegant $O(\sqrt{n}L)$ algorithm.

The adaptive primal-dual algorithms were developed by Mizuno, Todd and Ye [292], also see Barnes, Chopra and Jensen [44]. A more practical predictor-corrector algorithm was proposed by Mehrotra [276], based on the power series algorithm of Bayer and Lagarias [47] and the primal-dual version of Monteiro, Adler and Resende [301], also see Carpenter, Lustig, Mulvey and Shanno [78] and Zhang and Zhang [486]. His technique has been used in almost all of the LP interior-point implementations. Furthermore, Hung [193] developed a $O(n^{\frac{n+1}{2n}}L)$-iteration variant that uses wider neighborhoods. As n becomes large, this bound approaches the best bound for linear programming algorithms that use the small neighborhood (which are not practical). Other polynomial wide-neighborhood algorithms can be found in Jansen [198] and Sturm and Zhang [391, 390].

There was another polynomial interior-point algorithm, a multiplicative barrier function method, which was developed by Iri and Imai [195]; also see Sturm and Zhang [389].

A modified (shifted) barrier function theory and methods were developed by Polyak [343]; also see Pan [336], and Polak, Higgins and Mayne [341].

Interior-point algorithm computational results can be found in Adler, Karmarkar, Resende and Veiga [4], Altman and Gondzio [13], Bixby, Gregory, Lustig, Marsten and Shanno [59], Choi, Monma and Shanno [85, 86], Christiansen and Kortanek [87], Czyzyk, Mehrotra and Wright [92], Domich, Boggs, Rogers and Witzgall [101], Fourer and Mehrotra [119], Gondzio [155, 156], Lustig, Marsten and Shanno [255, 259, 257, 258], McShane, Monma and Shanno [269], Mehrotra [276], Monma [294], Ponnambalam, Vannelli and Woo [345], Vanderbei [445], and Xu, Hung and Ye [461].

In addition to those mentioned earlier, there are several comprehensive books which cover interior-point linear programming algorithms. They are Bazaraa, Jarvis and Sherali [48], Bertsekas [53], Fang and Puthenpura [111], Saigal [370], and Murty [310].

Many researchers have applied interior-point algorithms to solving convex QP and monotone LCP problems. The algorithms can be divided into three groups: the primal scaling algorithm, see Anstreicher, den Hertog, Roos and Terlaky [29], Ben–Daya and Shetty [50], Goldfarb and Liu [149], Ponceleon [344], and Ye [480, 467]; the dual scaling algorithm, see Jarre [202], Mehrotra and Sun [279], Nesterov and Nemirovskii [326], and Renegar and Shub [360]; and the primal-dual scaling algorithm, see Kojima, Mizuno and Yoshise [231, 232, 230], Mizuno [288], Monteiro and Adler [299], Monteiro, Adler and Resende [301], and Vanderbei [444].

Relations among these algorithms can be seen in den Hertog and Roos [184]. Given an interior feasible point (x, s), the following is a summary of directions generated by the three potential algorithms. They all satisfy

$$Ad_x = 0, \quad d_s = -A^T d_y \quad \text{for LP},$$

$$Ad_x = 0, \quad d_s = Qd_x - A^T d_y \quad \text{for QP},$$

and

$$d_s = M d_x \quad \text{for LCP}.$$

Furthermore, they satisfy, respectively,

Primal:

$$d_s + \frac{x^T s}{(n+\rho)} X^{-2} d_x = -s + \frac{x^T s}{(n+\rho)} X^{-1} e,$$

Dual:

$$d_x + \frac{x^T s}{(n+\rho)} S^{-2} d_s = -x + \frac{x^T s}{(n+\rho)} S^{-1} e,$$

and

Primal-dual:

$$Xd_s + Sd_x = -Xs + \frac{x^T s}{(n+\rho)}e,$$

where $\rho \geq \sqrt{n}$. These algorithms will reduce the primal-dual potential function by a constant, leading to $O(\rho \log(1/\epsilon))$ iteration complexity.

4.9 Exercises

4.1 In Algorithm 4.1, prove that $\|p(z^{k+1})\| \geq \frac{c^T x^k - z^{k+1}}{n+1}$. Then prove Lemma 4.1.

4.2 Compare Karmarkar's algorithm presented in this chapter to that described in Chapter 3. What will happen if $b = 0$ and $z = 0$ in (LP) and (LD) solved here?

4.3 Let $x(z) \in \mathcal{F}_p$ be on the central path associated with $\Omega(z)$ in Proposition 4.2. Prove $z^* > z \geq z^0$ implies $\mu(z) \leq \mu(z^0)$.

4.4 If $\hat{x} \in \overset{\circ}{\mathcal{F}}_p$ satisfies condition (4.9), prove

$$\sum_{j=1}^{n} \left(\frac{\hat{x}_j}{x_j^0}\right) \leq O(n).$$

Use this inequality to prove Corollary 4.3.

4.5 Choose several appropriate constants η for Algorithm 4.2 and calculate how many dual Newton steps are necessary to compute y^{k+1} from y^k. Does the choice of constants affect the complexity of the algorithm?

4.6 Develop a potential reduction algorithm in dual form using the dual potential function $\mathcal{B}(y, s, z)$ of (2.16), where z is a upper bound for the optimal objective value z^*.

4.7 Using the technique in the proof of Lemma 3.12 to prove Lemma 4.11. Moreover, show Lemma 4.11 holds even if $d_x^T d_s \geq 0$ instead of $d_x^T d_s = 0$ as in the LP case.

4.8 Let $v \in \mathcal{R}^n$ be a positive vector and $\rho \geq \sqrt{n}$. Prove

$$\sqrt{\min(v)}\|V^{-1/2}(e - \frac{(n+\rho)}{e^T v}v)\| \geq \sqrt{3/4}.$$

4.9 Use $p+q=r$ and $p^T q = 0$ to prove Lemma 4.14.

4.10 Describe the primal affine scaling algorithm mentioned at the end of Section 4.6. Starting from $x = e$, use it to complete the first three iterations for solving
$$\begin{aligned} \text{minimize} \quad & x_1 + 3x_2 \\ \text{s.t.} \quad & x_1 + x_2 + x_3 = 3, \\ & x_1, x_2, x_3 \geq 0. \end{aligned}$$

4.11 Show that if $\max(p^k) \leq 0$ and $\|p^k\| \neq 0$ in the affine scaling iteration, then the LP problem is unbounded.

4.12 In both updates (4.29) and (4.29), show that $x^{k+1} \in \mathring{\mathcal{F}}_p$.

4.13 Let (A, c) be given. Prove there exists a constant $M(A, c) > 0$ with the property that for any diagonal matrix $D > 0$, the (unique) optimal solution x^* of
$$\begin{aligned} \text{maximize} \quad & c^T x - \tfrac{1}{2}\|Dx\|^2 \\ \text{s.t.} \quad & Ax = 0 \end{aligned}$$
satisfies
$$\|x^*\| \leq M(A,c) c^T x^*.$$

4.14 Show that the sequence of the dual estimates $\{(y^k, s^k)\}$ generated by the affine scaling algorithm is bounded.

Chapter 5

Worst-Case Analysis

There are several remaining key issues concerning interior-point algorithms for LP. The first is the arithmetic operation complexity. In the previous chapters, we have analyzed the total number of iterations needed to solve a LP problem approximately. Since it solves a KKT system of linear equations with dimension $m + n$ and $n \geq m$, each iteration of all interior-point algorithms uses $O(n^3)$ arithmetic operations. Thus, their best operation complexity bound is $O(n^{3.5} \log(R/\epsilon))$, when the initial gap $(x^0)^T s^0 \leq R$. (We assume $(x^0)^T s^0 \leq R$ throughout this section.) The question is whether or not the arithmetic operations can be reduced in each iteration.

The second issue involves termination. Unlike the simplex method for linear programming which terminates with an exact solution, interior-point algorithms are continuous optimization algorithms that generate an infinite solution sequence converging to the optimal solution set. If the data of an LP instance are integral or rational, an argument is made that, after the worst-case time bound, an exact solution can be "rounded" from the latest approximate solution. Thus, several questions arise. First, under the real number computation model (i.e., the LP data consists of real numbers), how do we argue termination to an exact solution? Second, regardless of the data's status, can we utilize a practical test, one which can be computed cost-effectively during the iterative process, to identify an exact solution so that the algorithm can be terminated before the worse-case time bound? Here, the exact solution means a mathematical solution form using exact arithmetic, such as the solution of a system of linear equations or the solution of a least-squares problem, which can be computed in a number of arithmetic operations bounded by a polynomial in n.

The third issue involves initialization. Almost all interior-point algorithms solve the LP problem under the regularity assumption that $\overset{\circ}{\mathcal{F}} \neq \emptyset$. A related issue is that interior-point algorithms have to start at a strictly feasible point. Since no prior knowledge is usually available, one way is to explicitly bound the feasible region of the LP problem by a big M number. If the LP problem has integral data, this number can be set to 2^L in the worst case, where L is the length of the LP data in binary form. This setting is impossible in solving large problems. Moreover, if the LP problem has real data, no computable bound is known to set up the big M.

5.1 Arithmetic Operation

The primary computation cost of each iteration of interior-point algorithms is to inverse a normal matrix AX^2A^T in the primal form, $AS^{-2}A^T$ in the dual form, or $AXS^{-1}A^T$ in the primal-dual form. However, an approximation of these matrices can be inverted using far fewer computations. In this section, we show how to use a rank-one technique to update the inverse of the normal matrix during the iterative progress. This technique reduces the overall number of operations by a factor $\sqrt{n}$.

Consider the normal matrix $A(X^0)^2 A^T$ where x^0 is the initial point of a primal algorithm. To multiply and invert require $O(m^2 n)$ operations and after one iteration of the algorithm, a new point, x^1, is generated. Instead of inverting $A(X^1)^2 A^T$ for the second iteration, we will use $AD^2 A^T$, where D is a positive diagonal matrix defined by

$$d_j = \begin{cases} x_j^0 & \text{if } 1/1.1 \leq x_j^0/x_j^1 \leq 1.1 \\ x_j^1 & \text{otherwise.} \end{cases}$$

In other words, if x_j^1 did not change significantly, its old value is used; otherwise, its new value will be used. The inverse of the normal matrix $AD^2 A^T$ can be calculated from the previous inverse using the Sherman-Morrison-Woodbury rank-one updating formula (Exercise 1.1) whenever d_j is assigned a current value x_j^1. Each update requires $O(m^2)$ arithmetic operations. The procedure can be extended by letting $D^0 = X^0$ and for iterations $k \geq 1$,

$$d_j^k = \begin{cases} d_j^{k-1} & \text{if } 1/1.1 \leq d_j^{k-1}/x_j^k \leq 1.1 \\ x_j^k & \text{otherwise.} \end{cases}$$

Now we estimate the total number of updates needed before the algorithm terminates. Let

$$I^t = \left\{ i : \frac{d_j^t}{x_j^t} \notin [1/1.1, 1.1] \right\}.$$

5.1. ARITHMETIC OPERATION

Then, the computation work in the tth iteration is $O(m^2|I^t|)$ operations, where $|I|$ denotes the number of elements in set I. Thus, the total operations up to the kth iteration is $O(m^2n + m^2\sum_{t=1}^{k}|I^t|)$, where m^2n in the estimate is the amount of work at the initial iteration $t=0$. We have the following lemma to bound this estimate.

Lemma 5.1

$$\|(D^t)^{-1}(x^t - d^t)\|_\infty \leq 0.1 \quad \text{for any} \quad t = 0, 1, ...$$

and

$$\sum_{t=1}^{k}|I^t| \leq 11\sqrt{n}\sum_{t=1}^{k}\|(D^{t-1})^{-1}(x^t - x^{t-1})\|.$$

Proof. The proof of the first inequality is straightforward. Let

$$\sigma^t = \|(D^t)^{-1}(x^t - d^t)\|_1 \quad \text{for} \quad t = 0, 1, ...$$

Then, for $t = 1, 2, ...$

$$\begin{aligned}
\sigma^t &= \sum_{i \in I^t}\frac{|x_i^t - d_i^t|}{d_i^t} + \sum_{i \notin I^t}\frac{|x_i^t - d_i^t|}{d_i^t} \\
&= \sum_{i \notin I^t}\frac{|x_i^t - d_i^{t-1}|}{d_i^{t-1}} \\
&\leq \sum_{i \in I^t}(\frac{|x_i^t - d_i^{t-1}|}{d_i^{t-1}} - \frac{1}{11}) + \sum_{i \notin I^t}\frac{|x_i^t - d_i^{t-1}|}{d_i^{t-1}} \\
&= \|(D^{t-1})^{-1}(x^t - d^{t-1})\|_1 - |I^t|/11 \\
&= \|(D^{t-1})^{-1}(x^t - x^{t-1} + x^{t-1} - d^{t-1})\|_1 - |I^t|/11 \\
&\leq \|(D^{t-1})^{-1}(x^t - x^{t-1})\| + \|(D^{t-1})^{-1}(x^{t-1} - d^{t-1})\|_1 - |I^t|/11 \\
&= \|(D^{t-1})^{-1}(x^t - x^{t-1})\|_1 + \sigma^{t-1} - |I^t|/11.
\end{aligned}$$

Thus, we have

$$\begin{aligned}
\sum_{t=1}^{k}|I^t|/11 &\leq \sum_{t=1}^{k}\left(\|(D^{t-1})^{-1}(x^t - x^{t-1})\|_1 + \sigma^{t-1} - \sigma^t\right) \\
&= \sigma^0 - \sigma^k + \sum_{t=1}^{k}\|(D^{t-1})^{-1}(x^t - x^{t-1})\|_1.
\end{aligned}$$

Since $\sigma^0 = 0$, $\sigma^k \geq 0$, and $\|\cdot\|_1 \leq \|\cdot\|_2$, we have the desired result. $\square$

Furthermore, when
$$\|(D^{t-1})^{-1}(x^t - x^{t-1})\| \leq \alpha,$$
for $t = 1, 2, \ldots$ (as it will be in Algorithm 5.1 below),
$$\sum_{t=1}^{k} |I^t| \leq 11k\alpha\sqrt{n}.$$

If k is about $O(\sqrt{n}\log(R/\epsilon))$, then the total number of rank-one updates is bounded by $O(n\log(R/\epsilon))$ and the total number of operations is bounded by $O(m^2 n \log(R/\epsilon))$.

It is important to note, however, that the sequence of points generated using the approximate normal matrices is different than the sequence that would be generated by using the exact matrices. Therefore, it is necessary to study the convergence of the new sequence. As an example, consider primal potential reduction algorithm in Section 4.3. Replacing X^k in the normal matrix by a positive diagonal matrix D such that
$$\frac{1}{1.1} \leq \frac{d_j}{x_j^k} \leq 1.1 \quad \text{for} \quad j = 1, \ldots, n,$$
we now have
$$x^{k+1} - x^k = -\alpha \frac{D\hat{p}(z^k)}{\|\hat{p}(z^k)\|},$$
where
$$\hat{p}(z^k) = (I - DA^T(AD^2A^T)^{-1}AD)D\nabla \mathcal{P}_{n+\rho}(x^k, z^k).$$
The point x^{k+1} exactly solves the problem
$$\begin{array}{ll} \text{minimize} & \nabla \mathcal{P}_{n+\rho}(x^k, z^k)^T(x - x^k) \\ \text{s.t.} & A(x - x^k) = 0, \ \|D(x - x^k)\| \leq \alpha. \end{array}$$

Following the analysis in Section 4.3, $\hat{p}(z^k)$ can be rewritten as
$$\hat{p}(z^k) = \frac{n+\rho}{c^T x^k - z^k} Ds(z^k) - D(X^k)^{-1}e = D(X^k)^{-1}p(z^k), \tag{5.1}$$
where $p(z^k)$ is given by (4.11),
$$\begin{array}{rl} y(z^k) = & (AD^2A^T)^{-1}AD(Dc - \frac{(s^k)^T x^k}{n+\rho}D(X^k)^{-1}e) \\ s(z^k) = & c - A^T y(z^k). \end{array} \tag{5.2}$$
Thus, we have
$$\|\hat{p}(z^k)\| = \|D(X^k)^{-1}p(z^k)\| \geq \|p(z^k)\|/\|D^{-1}X^k\| \geq \|p(z^k)\|/1.1.$$

5.1. ARITHMETIC OPERATION

Also useful is the inequality

$$\begin{aligned}
\|(X^k)^{-1}(x^{k+1}-x^k)\| &= \|(X^k)^{-1}DD^{-1}(x^{k+1}-x^k)\| \\
&\leq \|(X^k)^{-1}D\|\|D^{-1}(x^{k+1}-x^k)\| \\
&\leq 1.1\|D^{-1}(x^{k+1}-x^k)\| = 1.1\alpha.
\end{aligned}$$

Combining these facts,

$$\nabla \mathcal{P}_{n+\rho}(x^k, z^k)^T (x^{k+1} - x^k) = -\alpha \|\hat{p}(z^k)\|,$$

and the reduction of the potential function is

$$\mathcal{P}_{n+\rho}(x^{k+1}, z^k) - \mathcal{P}_{n+\rho}(x^k, z^k) \leq -\alpha \|\hat{p}(z^k)\| + \frac{(1.1\alpha)^2}{2(1 - 1.1\alpha)}.$$

Since Lemma 4.8 still holds for $p(z^k)$, we only need to modify the potential reduction inequality in the proof of Theorem 4.9 by

$$\mathcal{P}_{n+\rho}(x^{k+1}, z^k) - \mathcal{P}_{n+\rho}(x^k, z^k) \leq \frac{\alpha}{1.1} \min\left(\eta \sqrt{\frac{n}{n+\eta^2}}, 1-\eta\right) + \frac{(1.1\alpha)^2}{2(1 - 1.1\alpha)}.$$

Therefore, upon choosing $\eta = 0.43$ and $\alpha = 0.25$, Theorem 4.9 is still valid for $\delta > 0.04$. As a result, the following modified primal algorithm can be developed.

Algorithm 5.1 *Given $x^0 \in \overset{\circ}{\mathcal{F}}_p$ and $(y^0, s^0) \in \overset{\circ}{\mathcal{F}}_d$. Let $z^0 = b^T y^0$ and $D^0 = X^0$. Set $\alpha = 0.25$ and $k := 0$;*
  While $(s^k)^T x^k \geq \epsilon$ do

1. *For $j = 1, \ldots, n$, if $d_j^k / x_j^k \notin [1/1.1, 1.1]$ then $d_j^k = x_j^k$. Let $D := D^k$. Then, compute $(y(z^k), s(z^k))$ of (5.2) and $\hat{p}(z^k)$ of (5.1).*

2. *If there exists z such that $s(z) > 0$, compute*

$$\bar{z} = \arg\min_z \psi_{n+\rho}(x^k, s(z)),$$

   and if $\psi_{n+\rho}(x^k, s(\bar{z})) < \psi_{n+\rho}(x^k, s^k)$ then $y^{k+1} = y(\bar{z})$, $s^{k+1} = s(\bar{z})$ and $z^{k+1} = b^T y^{k+1}$; otherwise, $y^{k+1} = y^k$, $s^{k+1} = s^k$ and $z^{k+1} = z^k$.

3. *Let $x^{k+1} = x^k - \alpha D \hat{p}(z^{k+1}) / \|\hat{p}(z^{k+1})\|$.*

4. *Let $D^{k+1} = D^k$ and $k := k + 1$, and return to Step 1.*

To summarize, we have

Theorem 5.2 *Let $\rho = \sqrt{n}$ and $\psi_{n+\rho}(x^0, s^0) \leq O(\sqrt{n}\log R)$. Then, Algorithm 5.1 terminates in $O(\sqrt{n}\log(R/\epsilon))$ iterations and uses $O(n^3 \log(R/\epsilon))$ total arithmetic operations.*

The normal matrix in dual or primal-dual algorithms can be updated in a similar manner and most algorithms have their operation complexity improved by the same factor $O(\sqrt{n})$. Up to this point, the most efficient LP algorithms use $O(n^3 \log((x^0)^T s^0/\epsilon))$ operations in the worst case. The KKT matrix in solving convex QP and LCP can be approximated as well and the most efficient algorithms also use $O(n^3 \log((x^0)^T s^0/\epsilon))$ operations in the worst case.

In practice, the rank-one technique is virtually unused. There are two reasons. First, A is "sparse" (containing very few nonzero components) in practice so that sparse numerical linear algebra procedures will be used to factorize the normal matrix, see Chapter 10. The total number of operations in a complete factorization is significantly below $O(n^3)$, so the saving of using rank-one update is insignificant or may be even negative due to its overhead cost. Secondly, the practical step-size α is much larger than the theoretical limit, so that often x^{k+1} is dramatically changed from x^k.

5.2 Termination

We now turn our attention to the termination of interior-point algorithms, the object of a great deal of research efforts. These efforts resulted in four basic approaches.

- The first is a "purification" procedure to find a feasible vertex whose objective value is at least as good as the current interior point. This approach can be done in strongly polynomial time when using the simplex method, and it works for LP with real number data. One difficulty which arises with this method is that many non-optimal vertices may be close to the optimal face, and the simplex method still requires many pivot steps for some "bad" problems. Moreover, the (simplex) purification procedure is *sequential* in nature.

- The second is a theoretical result to identify an optimal basis. A test procedure was also developed to show that, if the LP problem is nondegenerate, the unique optimal basis can be identified before the worst-case time bound. The procedure seemed to work fine for some LP problems but it has difficulty with degenerate LP problems. Unfortunately, most real LP problems are degenerate. The difficulty arises simply because any degenerate LP problem has multiple optimal bases.

5.2. TERMINATION

- The third approach is to slightly randomly perturb the data such that the new LP problem is nondegenerate and its optimal basis remains one of the optimal bases of the original LP problem. Questions remain on how and when to randomize the data during the iterative process, decisions which significantly affect the success of the effort.

- The fourth approach is to guess the optimal face and to find a feasible solution on the face. It consists of two phases: the first phase uses interior point algorithms to identify the complementarity partition (P^*, Z^*), and the second phase solves two linear feasibility problems to find the optimal primal and dual solutions. One can use the solution resulting from the first phase as a starting point for the second phase.

In this section we develop a termination procedure to obtain an exact solution on the interior of the optimal face. We shall see that (i) the termination criterion is guaranteed to work in finite time, (ii) the projection procedure (solving a least-squares problem) is strongly polynomial and can be efficiently performed in *parallel*, and (iii) the approach identifies the optimal face, which is useful in sensitivity analysis.

It has been noticed in practice that many interior-point algorithms generate a sequence of solutions converging to a strictly complementary solution for linear programming. It was subsequently proved that numerous interior-point algorithms for linear programming indeed generate solution sequences that converge to strictly complementary solutions, or interior solutions on the optimal face. Recall that the primal optimal face is

$$\Omega_p = \{x_{P^*} : A_{P^*} x_{P^*} = b, \ x_{P^*} \geq 0\},$$

and the one for the dual is

$$\Omega_d = \{(y, s_{Z^*}) : A_{P^*}^T y = c_{P^*}, \ s_{Z^*} = c_{Z^*} - A_{Z^*}^T y \geq 0\},$$

where (P^*, Z^*) is the strict complementarity partition of the LP problem. Note that these faces have strictly feasible solutions. Define

$$\begin{aligned}
\xi_p(A, b, c) &:= \min_{j \in P^*} \{\max_{x_{P^*} \in \Omega_p} x_j\} > 0, \\
\xi_d(A, b, c) &:= \min_{j \in Z^*} \{\max_{(y, s_{Z^*}) \in \Omega_d} s_j\} > 0, \\
\xi(A, b, c) &:= \min\{\xi_p(A, b, c), \xi_d(A, b, c)\} > 0.
\end{aligned} \qquad (5.3)$$

5.2.1 Strict complementarity partition

To measure the magnitude of positivity of a point $x \in \mathcal{R}_+^n$, we let $\sigma(x)$ be the support, i.e., index set of positive components in x, that is,

$$\sigma(x) = \{i : x_i > 0\}.$$

CHAPTER 5. WORST-CASE ANALYSIS

We first prove the following theorem.

Theorem 5.3 *Given an interior solution x^k and s^k in the solution sequence generated by any of interior-point algorithms possessing property (5.7) below, define*

$$\sigma^k = \{j : x_j^k \geq s_j^k\}.$$

Then, for $K = O(\sqrt{n}(\log(R/\xi^2(A,b,c)) + \log n))$ we have

$$\sigma^k = P^* \quad \text{for all} \quad k \geq K.$$

Proof. For simplicity, we use $\xi = \xi(A,b,c)$. For a given $j \in P^*$, let (x^*, s^*) be a complementarity solution such that x_j^* is maximized on the primal optimal face Ω_p, i.e, $x_j^* \geq \xi_p(A,b,c) \geq \xi$. Since

$$(x^k - x^*)^T(s^k - s^*) = 0,$$

$$\sum_{i \in P^*} x_i^* s_i^k + \sum_{i \in Z^*} s_i^* x_i^k = (x^k)^T s^k. \tag{5.4}$$

Thus, if $(x^k)^T s^k < \epsilon$, then

$$s_j^k \leq \frac{(x^k)^T s^k}{x_j^*} < \epsilon/\xi. \tag{5.5}$$

On the other hand, inequality (5.4) can be written as

$$\sum_{i \in P^*} \frac{x_i^*}{x_i^k}(x_i^k s_i^k) + \sum_{i \in Z^*} \frac{s_i^*}{s_i^k}(x_i^k s_i^k) = (x^k)^T s^k. \tag{5.6}$$

Almost all interior-point algorithms generate a solution sequence (or subsequence) (x^k, s^k) such that

$$\frac{\min(X^k s^k)}{(x^k)^T s^k} > n^{-\eta}, \tag{5.7}$$

where η is a positive constant. Thus, from (5.6) and (5.7) we have

$$\frac{x_j^*}{x_j^k}(x_j^k s_j^k) < (x^k)^T s^k,$$

or

$$\frac{x_j^k}{x_j^*} > \frac{x_j^k s_j^k}{(x^k)^T s^k} \geq n^{-\eta},$$

5.2. TERMINATION

or
$$x_j^k > n^{-\eta} x_j^* \geq n^{-\eta} \xi. \tag{5.8}$$
Thus, if $\epsilon \leq n^{-\eta}\xi^2$, recalling (5.5) we must have
$$s_j^k < \epsilon/\xi \leq n^{-\eta}\xi < x_j^k.$$

Similarly, we can prove this result for each $j \in P^*$. Moreover, for each $j \in Z^*$,
$$x_j^k < \epsilon/\xi \leq n^{-\eta}\xi < s_j^k.$$
Due to the polynomial bound of these interior-point algorithms, in $O(\sqrt{n}(\log(R/\xi^2) + \log n))$ iterations we shall have
$$(x^k)^T s^k \leq \epsilon = n^{-\eta}\xi^2.$$

This concludes the proof of the theorem.

$\square$

5.2.2 Project an interior point onto the optimal face

In practice, we don't wait to see if $\sigma^k = P^*$. In the following we develop a projection procedure to test if $\sigma^k = P^*$ and if an exactly optimal solution can be reached. For simplicity, let $P = \sigma^k$ and the rest be Z. Then we solve
$$(PP) \quad \text{minimize} \quad \|(X_P^k)^{-1}(x_P - x_P^k)\|$$
$$\text{s.t.} \quad A_P x_P = b,$$
and
$$(DP) \quad \text{minimize} \quad \|(S_Z^k)^{-1} A_Z^T(y - y^k)\|$$
$$\text{s.t.} \quad A_P^T y = c_P.$$

Without loss of generality we assume that A_P has full row rank. These two problems can be solved as two least-squares problems. The amount of work is equivalent to the computation in one iteration of interior-point algorithms. Furthermore, if the resulting solutions x_P^* and y^* satisfy
$$x_P^* > 0 \quad \text{and} \quad s_Z^* = c_Z - A_Z^T y^* > 0,$$
then obviously $x^* = (x_P^*, 0)$ and y^* are (exact) optimal solutions for the original LP problems, and $\sigma^k = P^*$ (Figure 5.1).

Let $d_x = (X_P^k)^{-1}(x_P - x_P^k)$ and $d_y = y - y^k$ and $d_s = (S_Z^k)^{-1} A_Z^T(y - y^k)$. Then, the two problems can be rewritten as
$$(PP) \quad \text{minimize} \quad \|d_x\|$$
$$\text{s.t.} \quad A_P X_P^k d_x = b - A_P x_P^k = A_Z x_Z^k,$$

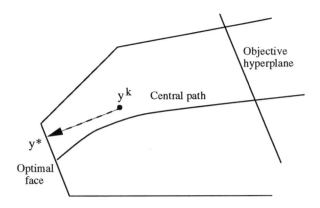

Figure 5.1. Illustration of the projection of y^k onto the (dual) optimal face.

and
$$(DP) \quad \text{minimize} \quad \|(S_Z^k)^{-1} A_Z^T d_y\|$$
$$\text{s.t.} \quad A_P^T d_y = c_P - A_P^T y^k = s_P^k.$$

Thus, if both $\|d_x\|_\infty$ and $\|d_s\|_\infty$ are less than 1, we must have $x_P > 0$ and $s_Z = c_Z - A_Z^T y > 0$. Since $\sigma^k \to P^*$ from the proof of Theorem 5.3 and the right-hand sides of the constraints of both problems converge to zero as $(x^k)^T s^k \to 0$, $\|d_x\|$ and $\|d_s\|$ must converge to zero. After $\sigma^k = P = P^*$, both (PP) and (PD) are feasible. Note that the solutions of (PP) and (PD) are

$$d_x^* = X_P^k A_P^T (A_P (X_P^k)^2 A_P^T)^{-1} A_Z x_Z^k \quad \text{and} \quad d_s^* = (S_Z^k)^{-1} A_Z^T (A_P A_P^T)^{-1} A_P s_P^k.$$

Thus,

$$\|d_x^*\|_\infty$$
$$= \|X_P^k A_P^T (A_P (X_P^k)^2 A_P^T)^{-1} A_P X_P^k (X_P^k)^{-1} A_P^T (A_P A_P^T)^{-1} A_Z x_Z^k\|_\infty$$
$$\leq \|X_P^k A_P^T (A_P (X_P^k)^2 A_P^T)^{-1} A_P X_P^k\| \|(X_P^k)^{-1} A_P^T (A_P A_P^T)^{-1} A_Z x_Z^k\|_\infty$$
$$\leq \|(X_P^k)^{-1} A_P^T (A_P A_P^T)^{-1} A_Z x_Z^k\|_\infty$$
$$\leq \|(X_P^k)^{-1}\| \|A_P^T (A_P A_P^T)^{-1} A_Z\| \|x_Z^k\|_\infty.$$

Let
$$\zeta(A, b, c) = \max(1, \|A_{P^*}^T (A_{P^*} A_{P^*}^T)^{-1} A_{Z^*}\|).$$

Then, if $\sigma^k = P = P^*$ and $\min(x_P^k) > \zeta(A, b, c) \max(x_Z^k)$,

$$\|d_x^*\|_\infty < 1 \quad \text{which implies that} \quad x_P^* > 0.$$

5.3. INITIALIZATION

Similarly, when $\min(s_Z^k) > \zeta(A,b,c)\max(s_P^k)$ then

$$\|d_s^*\|_\infty \leq 1 \quad \text{which implies that} \quad s_P^* > 0.$$

Recall (5.5) that each component of x_Z^k and s_P^k is less than $\epsilon/\xi(A,b,c)$ and (5.8) that each component of x_P^k and s_Z^k is greater than $\xi(A,b,c)n^{-\eta}$. This essentially shows that in

$$O\left(\sqrt{n}\left(\log\left(\frac{R\zeta(A,b,c)}{\xi^2(A,b,c)}\right) + \log n\right)\right)$$

iterations the above projection procedure will succeed in generating the complementarity partition and an exact optimal solution pair. To summarize, we have a condition-based complexity bound.

Theorem 5.4 *All $O(\sqrt{n}\log(R/\epsilon))$-iteration polynomial-time interior-point algorithms discussed earlier, coupled with the termination procedure, will generate an optimal solution in $O(\sqrt{n}(\log(R\zeta(A,b,c)/\xi^2(A,b,c)) + \log n))$ iterations and $O(n^3(\log(R\zeta(A,b,c)/\xi^2(A,b,c)) + \log n))$ arithmetic operations. If the LP problem has integral or rational data, then*

$$R \leq 2^L, \quad \zeta(A,b,c) \leq 2^L, \quad \text{and} \quad \xi(A,b,c) \geq 2^{-L},$$

where L is the size of the LP data. Thus, $\sigma^k = P^*$ and an exact solution will be generated in $O(\sqrt{n}L)$ iterations and $O(n^3L)$ operations.

When an interior solution x_P^*, $P = P^*$, on the primal optimal face is obtained, it can be cornered to an optimal basic solution in no more than $n - m$ pivot operations. See details in Section 10.5.

5.3 Initialization

Most interior-point algorithms have to start at a strictly feasible point. The complexity of obtaining such an initial point is the same as that of solving the LP problem itself. More importantly, a complete LP algorithm should accomplish two tasks: 1) affirmatively detect the infeasibility or unboundedness status of the LP problem, then 2) generate an optimal solution if the problem is neither infeasible nor unbounded.

Several approaches have been proposed to resolve these issues:

- Combining the primal and dual into a single linear feasibility problem (1.1), then applying LP algorithms to find a feasible point of the problem. Theoretically, this approach can retain the currently best complexity result. Practically, the disadvantage of this approach is the doubled dimension of the system of equations which must be solved at each iteration.

- The big M method, i.e., add one or more artificial column(s) and/or row(s) and a huge penalty parameter M to force solutions to become feasible during the algorithm. Theoretically, this approach holds the best complexity. The major disadvantage of this approach is the numerical problems caused by the addition of coefficients of magnitude. It also makes the algorithms slow to converge. This disadvantage also occurs in the primal-dual "exterior" or "infeasible" algorithm. A polynomial complexity can be established for this approach if the LP problem possesses an optimal solution and if the initial point is set to Me. Thus, the big M difficulty even remains in these polynomial infeasible interior-point algorithms.

- Phase I-then-Phase II method, i.e., first try to find a feasible point (and possibly one for the dual problem), and then start to look for an optimal solution if the problem is feasible and bounded. Theoretically, this approach can maintain the polynomial complexity result. The major disadvantage of this approach is that the two (or three) related LP problems are solved sequentially.

- Combined Phase I-Phase II method, i.e., approach feasibility and optimality simultaneously. To our knowledge, the currently "best" complexity of this approach is $O(n \log(R/\epsilon))$. Other disadvantages of the method include the assumption of non-empty interior and/or the use of the big M lower bound. Also, the method works exclusively in either the primal or the dual form.

In this section, we present a homogeneous and self-dual (HSD) LP algorithm to overcome the difficulties mentioned above. The algorithm possesses the following features:

- It solves the linear programming problem without any regularity assumption concerning the existence of optimal, feasible, or interior feasible solutions, while it retains the currently best complexity result

- It can start at any positive primal-dual pair, feasible or infeasible, near the central ray of the positive orthant (cone), and it does not use any big M penalty parameter or lower bound.

5.3. INITIALIZATION

- Each iteration solves a system of linear equations whose dimension is almost the same as that solved in the standard (primal-dual) interior-point algorithms.

- If the LP problem has a solution, the algorithm generates a sequence that approaches feasibility and optimality simultaneously; if the problem is infeasible or unbounded, the algorithm will produce an infeasibility certificate for at least one of the primal and dual problems.

5.3.1 A HSD linear program

Our algorithm is based on the construction of a homogeneous and self-dual linear program related to (LP) and (LD). We now briefly explain the two major concepts, homogeneity and self-duality, used in our construction.

In the context of interior-point algorithms, the idea of attacking a standard-form LP by solving a related homogeneous artificial linear program can be traced to many earlier works. (By a homogeneous linear program, we do not mean that all constraints must be homogeneous, or equivalently all right-hand sides zero. We allow a single inhomogeneous constraint, often called a *normalizing constraint*.) Karmarkar's original canonical form is a homogeneous linear program. One advantage of working in the homogeneous form is that we don't need to be concerned about the magnitude of solutions, since a solution is represented by a ray whose quality is scale-invariant. A disadvantage is that these related homogeneous problems, especially if they do not use any big M parameters, usually involve combining the primal and dual constraints and thus usually lead to algorithms requiring the solution of linear systems roughly twice as large as other methods.

Self-dual linear programs, meaning that the dual of the problem is equivalent to the primal, were introduced many years ago. We state the form of such problems, with inequality constraints, and their properties in the following result, whose proof is omitted.

Proposition 5.5 *Let $\tilde{A} \in R^{p \times p}$ be skew-symmetric, and let $\tilde{b} = -\tilde{c} \in R^p$. Then the problem*

$$\begin{aligned}(SDP) \quad &\text{minimize} \quad \tilde{c}^T \tilde{u} \\ &\text{s.t.} \quad \tilde{A}\tilde{u} \geq \tilde{b},\ \tilde{u} \geq 0,\end{aligned}$$

is equivalent to its dual. Suppose that (SDP) has a feasible solution $\tilde{u}$. Then $\tilde{u}$ is also feasible in the dual problem, and the two objective values sum to zero. Moreover, in this case (SDP) has an optimal solution, and its optimal value is zero.

The advantage of self-duality is that we can apply a primal-dual interior-point algorithm to solve the self-dual problem *without* doubling the dimension of the linear system solved at each iteration.

We now present a homogeneous and self-dual (artificial) linear program (HSDP) relating (LP) and (LD). Given any $x^0 > 0$, $s^0 > 0$, and y^0, we let $n^0 = (x^0)^T s^0 + 1$ and formulate

$$
\begin{array}{rlrrrl}
(HSDP) & \min & & & n^0 \theta & \\
& \text{s.t.} & Ax & -b\tau & +\bar{b}\theta & = 0, \\
& & -A^T y & +c\tau & -\bar{c}\theta & \geq 0, \\
& & b^T y & -c^T x & +\bar{z}\theta & \geq 0, \\
& & -\bar{b}^T y & +\bar{c}^T x & -\bar{z}\tau & = -n^0, \\
& & y \text{ free}, & x \geq 0, & \tau \geq 0, & \theta \text{ free},
\end{array}
$$

where

$$\bar{b} = b - Ax^0, \quad \bar{c} = c - A^T y^0 - s^0, \quad \bar{z} = c^T x^0 + 1 - b^T y^0. \tag{5.9}$$

Here $\bar{b}$, $\bar{c}$, and $\bar{z}$ represent the "infeasibility" of the initial primal point, dual point, and primal-dual "gap," respectively.

Note that the top three constraints in (HSDP), with $\tau = 1$ and $\theta = 0$, represent primal and dual feasibility (with $x \geq 0$) and reversed weak duality, so that they define primal and dual optimal solutions. Making τ a homogenizing variable adds the required dual variable to the third constraint. Then, to achieve feasibility for $x = x^0, (y, s) = (y^0, s^0)$, we add the artificial variable θ with appropriate coefficients, and then the fourth constraint is added to achieve self-duality.

Denote by s the slack vector for the second constraint and by κ the slack scalar for the third constraint. Denote by $\mathcal{F}_h$ the set of all points $(y, x, \tau, \theta, s, \kappa)$ that are feasible for (HSDP). Denote by $\mathcal{F}_h^0$ the set of strictly feasible points with $(x, \tau, s, \kappa) > 0$ in $\mathcal{F}_h$. Note that by combining the constraints, we can write the last (equality) constraint as

$$(s^0)^T x + (x^0)^T s + \tau + \kappa - n^0 \theta = n^0, \tag{5.10}$$

which serves as a normalizing constraint for (HSDP). Also note that the constraints of (HSDP) form a skew-symmetric system, which is basically why it is a self-dual linear program.

With regard to the selection of (x^0, y^0, s^0), note that if x^0 (respectively, (y^0, s^0)) is feasible in (LP) ((LD)), then $\bar{b}$ ($\bar{c}$) is zero, and then every feasible solution to (HSDP) with $\tau > 0$ has x/τ feasible in (LP) (($(y, s)/\tau$ feasible in (LD)). Conversely, if $\bar{z} < 0$, then every feasible solution to (HSDP) with $\theta > 0$ and $\tau > 0$ has $c^T x - b^T y \leq \bar{z}\theta < 0$, so either x/τ or $(y, s)/\tau$ must be infeasible.

5.3. INITIALIZATION

Now let us denote by (HSDD) the dual of (HSDP). Denote by y' the dual multiplier vector for the first constraint, by x' the dual multiplier vector for the second constraint, by τ' the dual multiplier scalar for the third constraint, and by θ' the dual multiplier scalar for the fourth constraint. Then, we have the following result.

Theorem 5.6 *Consider problems (HSDP) and (HSDD).*

i) *(HSDD) has the same form as (HSDP), i.e., (HSDD) is simply (HSDP) with (y, x, τ, θ) being replaced by (y', x', τ', θ').*

ii) *(HSDP) has a strictly feasible point*

$$y = y^0, \quad x = x^0 > 0, \quad \tau = 1, \quad \theta = 1, \quad s = s^0 > 0, \quad \kappa = 1.$$

iii) *(HSDP) has an optimal solution and its optimal solution set is bounded.*

iv) *The optimal value of (HSDP) is zero, and*

$$(y, x, \tau, \theta, s, \kappa) \in \mathcal{F}_h \quad \text{implies that} \quad n^0 \theta = x^T s + \tau \kappa.$$

v) *There is an optimal solution $(y^*, x^*, \tau^*, \theta^* = 0, s^*, \kappa^*) \in \mathcal{F}_h$ such that*

$$\begin{pmatrix} x^* + s^* \\ \tau^* + \kappa^* \end{pmatrix} > 0,$$

which we call a strictly self-complementary solution. (Similarly, we sometimes call an optimal solution to (HSDP) a self-complementary solution; the strict inequalities above need not hold.)

Proof. In what follows, denote the slack vector and scalar in (HSDD) by s' and κ', respectively. The proof of (i) is based on the skew-symmetry of the linear constraint system of (HSDP). We omit the details. Result (ii) can be easily verified. Then (iii) is due to the self-dual property: (HSDD) is also feasible and it has non-empty interior. The proof of (iv) can be constructed as follows. Let $(y, x, \tau, \theta, s, \kappa)$ and $(y', x', \tau', \theta', s', \kappa')$ be feasible points for (HSDP) and (HSDD), respectively. Then the primal-dual gap is

$$n^0 (\theta + \theta') = x^T s' + s^T x' + \tau \kappa' + \kappa \tau'.$$

Let $(y', x', \tau', \theta', s', \kappa') = (y, x, \tau, \theta, s, \kappa)$, which is possible since any feasible point $(y', x', \tau', \theta', s', \kappa')$ of (HSDD) is a feasible point of (HSDP) and vice versa. Thus, we have (iv). Note that (HSDP) and (HSDD) possess a strictly complementary solution pair: the primal solution is the solution for

(HSDP) in which the number of positive components is maximized, and the dual solution is the solution for (HSDD) in which the number of positive components is maximized. Since the supporting set of positive components of a strictly complementary solution is invariant and since (HSDP) and (HSDD) are identical, the strictly complementary solution $(y^*, x^*, \tau^*, \theta^* = 0, s^*, \kappa^*)$ for (HSDP) is also a strictly complementary solution for (HSDD) and vice versa. Thus, we establish (v).

$\square$

Henceforth, we simply choose

$$y^0 = 0, \quad x^0 = e, \quad \text{and} \quad s^0 = e. \tag{5.11}$$

Then, $n^0 = n + 1$ and (HSDP) becomes

$$(HSDP) \quad \min \quad (n+1)\theta$$
$$\text{s.t.} \quad Ax \quad -b\tau \quad +\bar{b}\theta = 0,$$
$$-A^T y \quad +c\tau \quad -\bar{c}\theta \geq 0,$$
$$b^T y \quad -c^T x \quad +\bar{z}\theta \geq 0,$$
$$-\bar{b}^T y \quad +\bar{c}^T x \quad -\bar{z}\tau \quad = -(n+1),$$
$$y \text{ free}, \quad x \geq 0, \quad \tau \geq 0, \quad \theta \text{ free},$$

where

$$\bar{b} = b - Ae, \quad \bar{c} = c - e, \quad \text{and} \quad \bar{z} = c^T e + 1. \tag{5.12}$$

Again, combining the constraints we can write the last (equality) constraint as

$$e^T x + e^T s + \tau + \kappa - (n+1)\theta = n + 1. \tag{5.13}$$

Since $\theta^* = 0$ at every optimal solution for (HSDP), we can see the normalizing effect of equation (5.13) for (HSDP).

We now relate optimal solutions to (HSDP) to those for (LP) and (LD).

Theorem 5.7 *Let $(y^*, x^*, \tau^*, \theta^* = 0, s^*, \kappa^*)$ be a strictly self complementary solution for (HSDP).*

i) *(LP) has a solution (feasible and bounded) if and only if $\tau^* > 0$. In this case, x^*/τ^* is an optimal solution for (LP) and $(y^*/\tau^*, s^*/\tau^*)$ is an optimal solution for (LD).*

ii) *(LP) has no solution if and only if $\kappa^* > 0$. In this case, x^*/κ^* or s^*/κ^* or both are certificates for proving infeasibility: if $c^T x^* < 0$ then (LD) is infeasible; if $-b^T y^* < 0$ then (LP) is infeasible; and if both $c^T x^* < 0$ and $-b^T y^* < 0$ then both (LP) and (LD) are infeasible.*

5.3. INITIALIZATION

Proof. If (LP) and (LD) are both feasible, then they have a complementary solution pair $\bar{x}$ and $(\bar{y}, \bar{s})$ for (LP) and (LD), such that
$$(\bar{x})^T \bar{s} = 0.$$
Let
$$\alpha = \frac{n+1}{e^T \bar{x} + e^T \bar{s} + 1} > 0.$$
Then one can verify (see (5.13)) that
$$\tilde{y}^* = \alpha \bar{y}, \quad \tilde{x}^* = \alpha \bar{x}, \quad \tilde{\tau}^* = \alpha, \quad \tilde{\theta}^* = 0, \quad \tilde{s}^* = \alpha \bar{s}, \quad \tilde{\kappa}^* = 0$$
is a self-complementary solution for (HSDP). Since the supporting set of a strictly complementary solution for (HSDP) is unique, $\tau^* > 0$ at any strictly complementary solution for (HSDP).

Conversely, if $\tau^* > 0$, then $\kappa^* = 0$, which implies that
$$Ax^* = b\tau^*, \quad A^T y^* + s^* = c\tau^*, \quad \text{and} \quad (x^*)^T s^* = 0.$$
Thus, x^*/τ^* is an optimal solution for (LP) and $(y^*/\tau^*, s^*/\tau^*)$ is an optimal solution for (LD). This concludes the proof of the first statement in the theorem.

Now we prove the second statement. If one of (LP) and (LD) is infeasible, say (LD) is infeasible, then we have a certificate $\bar{x} \geq 0$ such that $A\bar{x} = 0$ and $c^T \bar{x} = -1$. Let $(\bar{y} = 0, \bar{s} = 0)$ and
$$\alpha = \frac{n+1}{e^T \bar{x} + e^T \bar{s} + 1} > 0.$$
Then one can verify (see (5.13)) that
$$\tilde{y}^* = \alpha \bar{y}, \quad \tilde{x}^* = \alpha \bar{x}, \quad \tilde{\tau}^* = 0, \quad \tilde{\theta}^* = 0, \quad \tilde{s}^* = \alpha \bar{s}, \quad \tilde{\kappa}^* = \alpha$$
is a self-complementary solution for (HSDP). Since the supporting set of a strictly complementary solution for (HSDP) is unique, $\kappa^* > 0$ at any strictly complementary solution for (HSDP).

Conversely, if $\tau^* = 0$, then $\kappa^* > 0$, which implies that $c^T x^* - b^T y^* < 0$, i.e., at least one of $c^T x^*$ and $-b^T y^*$ is strictly less than zero. Let us say $c^T x^* < 0$. In addition, we have
$$Ax^* = 0, \quad A^T y^* + s^* = 0, \quad (x^*)^T s^* = 0 \quad \text{and} \quad x^* + s^* > 0.$$
From Farkas' lemma, x^*/κ^* is a certificate for proving dual infeasibility. The other cases hold similarly.

□

From the proof of the theorem, we deduce the following:

Corollary 5.8 *Let $(\bar{y}, \bar{x}, \bar{\tau}, \bar{\theta} = 0, \bar{s}, \bar{\kappa})$ be any optimal solution for (HSDP). Then if $\bar{\kappa} > 0$, either (LP) or (LD) is infeasible.*

5.3.2 Solving (HSD)

The following theorem resembles the central path analyzed for (LP) and (LD).

Theorem 5.9 *Consider problem (HSDP).*

i) *For any $\mu > 0$, there is a unique $(y, x, \tau, \theta, s, \kappa)$ in $\mathcal{F}_h^0$, such that*

$$\begin{pmatrix} Xs \\ \tau\kappa \end{pmatrix} = \mu e.$$

ii) *Let $(d_y, d_x, d_\tau, d_\theta, d_s, d_\kappa)$ be in the null space of the constraint matrix of (HSDP) after adding surplus variables s and κ, i.e.,*

$$\begin{array}{rlrlrlr}
 & Ad_x & -bd_\tau & +\bar{b}d_\theta & & = & 0, \\
-A^T d_y & & +cd_\tau & -\bar{c}d_\theta & -d_s & = & 0, \\
b^T d_y & -c^T d_x & & +\bar{z}d_\theta & & -d_\kappa & = 0, \\
-\bar{b}^T d_y & +\bar{c}^T d_x & -\bar{z}d_\tau & & & = & 0.
\end{array} \quad (5.14)$$

Then

$$(d_x)^T d_s + d_\tau d_\kappa = 0.$$

Proof. For any $\mu > 0$, there is a unique feasible point $(y, x, \tau, \theta, s, \kappa)$ for (HSDP) and a unique feasible point $(y', x', \tau', \theta', s', \kappa')$ for (HSDD) such that

$$Xs' = \mu e, \quad Sx' = \mu e, \quad \tau\kappa' = \mu, \quad \kappa\tau' = \mu.$$

However, if we switch the positions of $(y, x, \tau, \theta, s, \kappa)$ and $(y', x', \tau', \theta', s', \kappa')$ we satisfy the same equations. Thus, we must have

$$(y', x', \tau', \theta', s', \kappa') = (y, x, \tau, \theta, s, \kappa),$$

since (HSDP) and (HSDD) have the identical form. This concludes the proof of (i) in the theorem.

The proof of (ii) is simply due to the skew-symmetry of the constraint matrix. Multiply the first set of equations by d_y^T, the second set by d_x^T, the third equation by d_τ and the last by d_θ and add. This leads to the desired result.

□

We see that Theorem 5.9 defines an endogenous path within (HSDP):

$$\mathcal{C} = \left\{ (y, x, \tau, \theta, s, \kappa) \in \mathcal{F}_h^0 : \begin{pmatrix} Xs \\ \tau\kappa \end{pmatrix} = \frac{x^T s + \tau\kappa}{n+1} e \right\},$$

5.3. INITIALIZATION

which we may call the *(self-)central path* for (HSDP). Obviously, if $X^0 s^0 = e$, then the initial interior feasible point proposed in Theorem 5.6 is on the path with $\mu = 1$. Our choice (5.11) for x^0 and s^0 satisfies this requirement. We can define a neighborhood of the path as

$$\mathcal{N}(\eta) = \left\{ (y, x, \tau, \theta, s, \kappa) \in \mathcal{F}_h^0 : \left\| \begin{pmatrix} Xs \\ \tau\kappa \end{pmatrix} - \mu e \right\| \leq \eta \mu, \right.$$

$$\left. \text{where} \quad \mu = \frac{x^T s + \tau \kappa}{n+1} \right\}$$

for some $\eta \in (0, 1)$. Note that from statement (iv) of Theorem 5.6 we have $\theta = \mu$ for any feasible point in $\mathcal{F}_h$.

Since the (HSDP) model constructed and analyzed does not rely on any particular algorithm for solving it, we may use any interior-point algorithm, as long as it generates a strictly complementary solution. Given an interior feasible point $(y, x, \tau, \theta, s, \kappa) \in \mathcal{F}_h^0$, consider solving the following system of linear equations for $(d_y, d_x, d_\tau, d_\theta, d_s, d_\kappa)$:

$$\begin{aligned}(d_y, d_x, d_\tau, d_\theta, d_s, d_\kappa) & \quad \text{satisfies (5.14)}, \\ \begin{pmatrix} X d_s + S d_x \\ \tau^k d_\kappa + \kappa^k d_\tau \end{pmatrix} &= \gamma \mu e - \begin{pmatrix} Xs \\ \tau\kappa \end{pmatrix}. \end{aligned} \quad (5.15)$$

Let $d := (d_y, d_x, d_\tau, d_\theta, d_s, d_\kappa)$. To show the dependence of d on the current pair (x, τ, s, κ) and the parameter γ, we write $d = d(x, \tau, s, \kappa, \gamma)$.

In what follows we apply the predictor-corrector algorithm in Chapter 4 to solving (HSDP):

Predictor Step. At iteration k, we have $(y^k, x^k, \tau^k, \theta^k, s^k, \kappa^k) \in \mathcal{N}(\eta)$ with $\eta = 1/4$. Set $(x, \tau, s, \kappa) = (x^k, \tau^k, s^k, \kappa^k)$ and compute $d = d(x, \tau, s, \kappa, 0)$ from (5.14,5.15). Let

$$\begin{aligned} y(\alpha) &:= y^k + \alpha d_y, & x(\alpha) &:= x^k + \alpha d_x, \\ \tau(\alpha) &:= \tau^k + \alpha d_\tau, & \theta(\alpha) &:= \theta^k + \alpha d_\theta, \\ s(\alpha) &:= s^k + \alpha d_s, & \kappa(\alpha) &:= \kappa^k + \alpha d_\kappa. \end{aligned}$$

We determine the step size using

$$\bar{\alpha} := \max \{ \alpha : (y(\alpha), x(\alpha), \tau(\alpha), \theta(\alpha), s(\alpha), \kappa(\alpha)) \in \mathcal{N}(2\eta) \}. \quad (5.16)$$

Corrector Step. Set

$$(y', x', \tau', \theta', s', \kappa') = (y(\bar{\alpha}), x(\bar{\theta}), \tau(\bar{\alpha}), \theta(\bar{\alpha}), s(\bar{\theta}), \kappa(\bar{\alpha}))$$

and compute $d' = d(x', \tau', s', \kappa', 1)$ from (5.14,5.15). Then let $y^{k+1} = y' + d'_y$, $x^{k+1} = x' + d'_x$, $\tau^{k+1} = \tau' + d'_\tau$, $\theta^{k+1} = \theta' + d'_\theta$, $s^{k+1} = s' + d'_s$, and $\kappa^{k+1} = \kappa' + d'_\kappa$. We have

$$(y^{k+1}, x^{k+1}, \tau^{k+1}, \theta^{k+1}, s^{k+1}, \kappa^{k+1}) \in \mathcal{N}(\eta).$$

Termination. We use the termination technique described earlier to terminate the algorithm. Define σ^k be the index set $\{j : x_j^k \geq s_j^k, \ j = 1, 2, ..., n\}$, and let $P = \sigma^k$ and the rest be Z. Then, we again use a least-squares projection to create an optimal solution that is strictly self-complementary for (HSDP).

Case 1: If $\tau^k \geq \kappa^k$, we solve for y, x_P, and τ from

$$\begin{aligned}
\min \quad & \|(S_Z^k)^{-1} A_Z^T (y^k - y)\|^2 + \|(X_P^k)^{-1}(x_P^k - x_P)\|^2 \\
\text{s.t.} \quad & \qquad\qquad\qquad\qquad\qquad A_P x_P = b\tau^k, \\
& -A_P^T y \qquad\qquad\qquad\qquad = -c_P \tau^k,
\end{aligned}$$

otherwise,

Case 2: $\tau^k < \kappa^k$, and we solve for y and x_P from

$$\begin{aligned}
\min \quad & \|(S_Z^k)^{-1} A_Z^T (y^k - y)\|^2 + \|(X_P^k)^{-1}(x_P^k - x_P)\|^2 \\
\text{s.t.} \quad & \qquad\qquad\qquad\qquad\qquad A_P x_P = 0, \\
& -A_P^T y \qquad\qquad\qquad\qquad = 0, \\
& b^T y \qquad\qquad -c_P^T x_P = \kappa^k.
\end{aligned}$$

This projection guarantees that the resulting x_P^* and s_Z^* ($s_Z^* = c_Z \tau^k - A_Z^T y^*$ in Case 1 or $s_Z^* = -A_Z^T y^*$ in Case 2) are positive, as long as $(x^k)^T s^k + \tau^k \kappa^k$ is reduced to be sufficiently small by the algorithm according to our discussion in the preceding section on termination.

Theorem 5.10 *The $O(\sqrt{n} \log(R/\epsilon))$ interior-point algorithm, coupled with the termination technique described above, generates a strictly self-complementary solution for (HSDP) in $O(\sqrt{n}(\log(c(A,b,c)) + \log n))$ iterations and $O(n^3(\log(c(A,b,c)) + \log n))$ operations, where $c(A,b,c)$ is a positive number depending on the data (A,b,c). If (LP) and (LD) have integer data with bit length L, then by the construction, the data of (HSDP) remains integral and its length is $O(L)$. Moreover, $c(A,b,c) \leq 2^L$. Thus, the algorithm terminates in $O(\sqrt{n}L)$ iterations and $O(n^3 L)$ operations.*

Now using Theorem 5.7 we obtain

5.3. INITIALIZATION

Corollary 5.11 *Within $O(\sqrt{n}(\log(c(A,b,c)) + \log n))$ iterations and $O(n^3(\log(c(A,b,c)) + \log n))$ operations, where $c(A,b,c)$ is a positive number depending on the data (A,b,c), the $O(\sqrt{n}\log(R/\epsilon))$ interior-point algorithm, coupled with the termination technique described above, generates either optimal solutions to (LP) and (LD) or a certificate that (LP) or (LD) is infeasible. If (LP) and (LD) have integer data with bit length L, then $c(A,b,c) \leq 2^L$.*

Again, $c(A,b,c)$ plays the condition number for the data set (A,b,c).

Note that the algorithm may not detect the infeasibility status of both (LP) and (LD).

Example 5.1 *Consider the example where*
$$A = \begin{pmatrix} -1 & 0 & 0 \end{pmatrix}, \quad b = 1, \quad \text{and} \quad c = \begin{pmatrix} 0 & 1 & -1 \end{pmatrix}.$$
Then,
$$y^* = 2, \quad x^* = (0,2,1)^T, \quad \tau^* = 0, \quad \theta^* = 0, \quad s^* = (2,0,0)^T, \quad \kappa^* = 1$$
could be a strictly self-complementary solution generated for (HSDP) with
$$c^T x^* = 1 > 0, \quad b^T y^* = 2 > 0.$$
Thus (y^*, s^*) demonstrates the infeasibility of (LP), but x^* doesn't show the infeasibility of (LD). Of course, if the algorithm generates instead $x^* = (0,1,2)^T$, then we get demonstrated infeasibility of both.

5.3.3 Further analysis

In practice, we may wish to stop the algorithm at an approximate solution. Thus, we wish to analyze the asymptotic behavior of τ^k vs. θ^k.

Theorem 5.12 . *If (LP) possesses an optimal solution, then*
$$\tau^k \geq \frac{1-2\eta}{(e^T\bar{x} + e^T\bar{s} + 1)} \quad \text{for all} \quad k,$$
where $\bar{x}$ and $(\bar{y},\bar{s})$ are any optimal solution pair for (LP) and (LD); otherwise,
$$\kappa^k \geq \frac{1-2\eta}{(e^T\bar{x} + e^T\bar{s} + 1)} \quad \text{for all} \quad k,$$
where $\bar{x}$ and $(\bar{y},\bar{s})$ are any certificate for proving the infeasibility of (LP) or (LD), and moreover,
$$\frac{1-2\eta}{2(n+1)} \leq \frac{1-2\eta}{\kappa^k} \leq \frac{\tau^k}{\theta^k} \leq \frac{1+2\eta}{\kappa^k} \leq \frac{(e^T\bar{x} + e^T\bar{s} + 1)(1+2\eta)}{(1-2\eta)} \quad \text{for all} \quad k.$$

Proof. Note that the sequence generated by the predictor-corrector algorithm is in $\mathcal{N}(2\eta)$. Note that

$$y^* = \alpha\bar{y}, \quad x^* = \alpha\bar{x}, \quad \tau^* = \alpha, \quad \theta^* = 0, \quad s^* = \alpha\bar{s}, \quad \kappa^* = 0,$$

where

$$\alpha = \frac{n+1}{e^T\bar{x} + e^T\bar{s} + 1} > 0,$$

is a self-complementary solution for (HSDP). Now we use

$$(x^k - x^*)^T(s^k - s^*) + (\tau^k - \tau^*)(\kappa^k - \kappa^*) = 0,$$

which follows by subtracting the constraints of (HSDP) for $(y^*, \ldots, \kappa^*)$ from those for $(y^k, \ldots, \kappa^k)$ and then multiplying by $((y^k - y^*)^T, \ldots, \kappa^k - \kappa^k)$. This can be rewritten as

$$(x^k)^T s^* + (s^k)^T x^* + \kappa^k \tau^* = (n+1)\mu^k = (n+1)\theta^k.$$

Thus,

$$\tau^k \geq \frac{\tau^k \kappa^k}{(n+1)\mu^k}\tau^* \geq \frac{1-2\eta}{n+1}\tau^* = \frac{1-2\eta}{(e^T\bar{x} + e^T\bar{s} + 1)}.$$

The second statement follows from a similar argument. We know that there is an optimal solution for (HSDP) with

$$\kappa^* \geq \frac{n+1}{(e^T\bar{x} + e^T\bar{s} + 1)} > 0.$$

Thus

$$\kappa^k \geq \frac{1-2\eta}{(e^T\bar{x} + e^T\bar{s} + 1)}$$

for all k. In addition, from relation (5.13) we have $\kappa^k \leq (n+1)+(n+1)\theta^k \leq 2(n+1)$ for all k.

$\square$

Theorem 5.12 indicates that either τ^k stays bounded away from zero for all k, which implies that (LP) has an optimal solution, or τ^k and θ^k converge to zero at the same rate, which implies that (LP) does not have an optimal solution. In practice, for example, we can adopt the following two convergence criteria:

1. $(x^k/\tau^k)^T(s^k/\tau^k) \leq \epsilon_1$, and $(\theta^k/\tau^k)\|(\bar{b}, \bar{c})\| \leq \epsilon_2$,

2. $\tau^k \leq \epsilon_3$.

5.4. INFEASIBLE-STARTING ALGORITHM

Here ϵ_1, ϵ_2, and ϵ_3 are small positive constants. Since both $(x^k)^T s^k + \tau^k \kappa^k$ and θ^k decrease by at least $(1 - 1/\sqrt{n+1})$ in every two iterations, one of the above convergence criteria holds in $O(\sqrt{n}t)$ iterations for $t = \max\{\log((x^0)^T s^0/(\epsilon_1 \epsilon_3^2)), \log(\|\bar{b}, \bar{c}\|/(\epsilon_2 \epsilon_3))\}$. If the algorithm terminates by the first criterion then we get approximate optimal solutions of (LP) and (LD); otherwise we detect that (LP) and (LD) have no optimal solutions such that $\|(\bar{x}, \bar{s})\|_1 \leq (1 - 2\eta)/\epsilon_3 - 1$ from Theorem 5.12.

5.4 Infeasible-Starting Algorithm

Other popular and effective algorithms to overcome the initialization difficulty are the so-called *infeasible-starting algorithms*. These algorithms start from some $x^0 > 0$, $s^0 > 0$ and y^0, but $Ax^0 - b$ may not be 0 and s^0 may not equal $c - A^T y^0$, that is, (x^0, s^0, y^0) may not be feasible for the primal and dual and it is an interior but infeasible-starting point. The algorithms are based on the following theorem which resembles the central path and its proof is omitted here.

Theorem 5.13 *Consider problem (LP) and (LD). For any given $x^0 > 0$ and $s^0 > 0$ and y^0 and any μ, $\nu > 0$, there is a bounded and unique $(y, x > 0, s > 0)$ in such that*

$$\begin{pmatrix} Xs \\ Ax - b \\ A^T y + s - c \end{pmatrix} = \begin{pmatrix} \mu e \\ \nu(Ax^0 - b) \\ \nu(A^T y^0 + s^0 - c) \end{pmatrix}.$$

We see that Theorem 5.13 defines an interior but infeasible "central" path:

$$\mathcal{C} = \left\{ (y, x > 0, s > 0) : \begin{pmatrix} Xs \\ Ax \\ A^T y + s \end{pmatrix} = \begin{pmatrix} \frac{x^T s}{n} e \\ b + \frac{x^T s}{(x^0)^T s^0}(Ax^0 - b) \\ c - \frac{x^T s}{(x^0)^T s^0}(c - A^T y^0 - s^0) \end{pmatrix} \right\},$$

and its neighborhood

$$\mathcal{N}(\eta) = \left\{ (y, x > 0, s > 0) : \min(Xs) \geq \eta \mu, \frac{\|Ax - b\|}{\|Ax^0 - b\|} \leq \frac{\mu}{\mu^0}, \right.$$

$$\left. \frac{\|A^T y + s - c\|}{\|A^T y^0 + s^0 - c\|} \leq \frac{\mu}{\mu^0}, \text{ where } \mu = \frac{x^T s}{n}, \mu^0 = \frac{(x^0)^T s^0}{n} \right\}$$

for some $\eta \in (0, 1)$. The first inequality is identical to the $\mathcal{N}_\infty^-$ inequality discussed in Section 4.5.2, and the next two inequalities make sure that the (relative) feasibility error decreases to zero faster than the (relative) complementarity gap.

Let

$$\rho_0 \geq \min\{\|(u,w)\|_\infty : Au = b, \ A^Tv + w = c \ \text{ for some } \ v\}, \quad (5.17)$$

and let $\rho \geq \rho_0$ be a number for which we want to find an optimal solution pair (y^*, x^*, y^*), if it exists, such that

$$\rho \geq \|(x^*, s^*)\|_\infty.$$

Algorithm 5.2 *Let $\eta \in (0,1)$ and $\gamma \in (0,1)$. Given $(y^0, x^0, s^0) = \rho(0, e, e)$. Set $k := 0$ and $\theta^0 = 1$.*
While $(x^k)^T s^k > \epsilon$ and $\|(x^k, s^k)\|_1 < \frac{2}{\theta^k \rho}(x^k)^T s^k$ **do:**

1. *Compute directions from*

$$\begin{aligned} X^k d_s + S^k d_x &= \gamma\mu^k e - X^k s^k, \\ A d_x &= b - Ax^k, \\ -A^T d_y - d_s &= s^k - c + A^T y^k. \end{aligned}$$

2. *Compute the largest $\bar{\alpha}$ so that*

$$(y^k, x^k, s^k) + \alpha(d_y, d_x, d_s) \in \mathcal{N}(\eta)$$

and

$$(x^k + \alpha d_x)^T (s^k + \alpha d_s) \leq (1 - \alpha(1-\gamma))(x^k)^T s^k$$

for every $\alpha \in [0, \bar{\alpha}]$. Set

$$(y^{k+1}, x^{k+1}, s^{k+1}) = (y^k, x^k, s^k) + \bar{\alpha}(d_y, d_x, d_s)$$

and

$$\theta^{k+1} = (1 - \bar{\alpha})\theta^k.$$

3. *Let $k := k + 1$ and return to Step 1.*

One can verify that

$$\begin{aligned} Ax^k - b &= \theta^k (Ax^0 - b) \\ A^T y^k + s^k - c &= \theta^k (A^T y^0 + s^0 - c), \end{aligned} \quad (5.18)$$

that is, the $Ax^k - b$ and $A^T y^k + s^k - c$ approach to zero on the same rays, respectively.

We now present the following theorem:

5.4. INFEASIBLE-STARTING ALGORITHM

Theorem 5.14 *Let*
$$\rho \geq \|(x^*, s^*)\|_\infty,$$
where (y^*, x^*, y^*), if it exists, is an optimal solution pair for (LP) and (LD). Then, Algorithm 5.2 terminates in $O(n^2 t)$ iterations, where
$$t = \log\left(\frac{\max\{(x^0)^T s^0, \|Ax^0 - b\|, \|c - A^T y^0 - s^0\|\}}{\epsilon}\right).$$
At the termination, if $(x^k)^T s^k \leq \epsilon$, then (y^k, x^k, s^k) is an ϵ-approximate solution for (LP) and (LD); otherwise there is no optimal solution (y^*, x^*, s^*) such that $\|(x^*, s^*)\|_\infty \leq \rho$.

If (LP) and (LD) have integer data and its binary data length is L, we may set $\rho = 2^L$. Then, no optimal solution (y^*, x^*, s^*), such that $\|(x^*, s^*)\|_\infty \leq \rho$ actually implies that at least one of (LP) and (LD) is infeasible.

To prove the above theorem, we need a couple of lemmas, whose proofs are left as exercises.

Lemma 5.15 *Suppose that at the kth iteration, for all $j = 1, \ldots, n$,*
$$|(d_x)_j (d_s)_j - \eta d_x^T d_s / n| \leq \eta \quad \text{and} \quad |d_x^T d_s| \leq \eta.$$
Then,
$$\bar{\alpha} \geq \min\left\{1, \frac{\gamma(1-\eta)\mu^k}{\eta}, \frac{\gamma \mu^k n}{\eta}\right\}.$$

Note that if (y^k, x^k, s^k) is a feasible pair, then $d_x^T d_s = 0$ and the conditions of the above lemma hold for $\eta = O(n\mu^k)$. Nevertheless, we can prove that the conditions hold in general for $\eta = O(n^2 \mu^k)$, if (LP) and (LD) have solutions. Thus, from the lemma $\bar{\alpha} \geq \Omega(1/n^2)$, which implies that μ^k converges to zero at the rate of $(1 - \Omega(1/n^2))$.

Lemma 5.16 *At the kth iteration, we have*
$$D^{-1} d_x = -\theta^k P D^{-1}(x^0 - u^0) + \theta^k (I - P) D(s^0 - w^0)$$
$$- (I - P)(X^k S^k)^{-.5}(X^k s^k - \gamma \mu^k e),$$
$$d_y = -\theta^k (y^0 - v^0) - (AD^2 A^T)^{-1} AD(\theta^k D^{-1}(x^0 - u^0) + \theta^k D(s^0 - w^0)$$
$$- (X^k S^k)^{-.5}(X^k s^k - \gamma \mu^k e)),$$
$$D d_s = \theta^k P D^{-1}(x^0 - u^0) - \theta^k (I - P) D(s^0 - w^0) - P(X^k S^k)^{-.5}(X^k s^k - \gamma \mu^k e),$$
where
$$D = (X^k)^{.5}(S^k)^{-.5}, \quad P = DA^T (AD^2 A^T)^{-1} AD,$$
and (u^0, v^0, w^0) is the minimizer of (5.17).

From the definition of ρ and ρ^0, we have
$$\|(u^0, w^0)\|_\infty \le \rho^0 \le \rho.$$
Thus,
$$-\rho e \le x^0 - u^0 \le 2\rho e \quad \text{and} \quad -\rho e \le s^0 - w^0 \le 2\rho e.$$
Therefore,
$$\begin{aligned} & \|D^{-1}d_x\| \\ \le\ & \theta^k \|D^{-1}e\| + \theta^k 2\rho\|De\| + \|(X^k S^k)^{-.5}(X^k s^k - \gamma\mu^k e)\| \\ \le\ & \theta^k 2\rho \|(X^k S^k)^{-.5}\|(\|s^k\| + \|x^k\|) + \|(X^k S^k)^{-.5}(X^k s^k - \gamma\mu^k e)\| \\ \le\ & 4\theta^k \rho\|(X^k S^k)^{-.5}\|\|(x^k, s^k)\| + \|(X^k S^k)^{-.5}(X^k s^k - \gamma\mu^k e)\|. \end{aligned}$$

Since $x_j^k s_j^k \ge \eta\mu^k$ for each j and
$$\|(x^k, s^k)\| \le \|(x^k, s^k)\|_1 \le \frac{2}{\theta^k \rho}(x^k)^T s^k,$$
we have
$$4\theta^k \rho\|(X^k S^k)^{-.5}\|\|(x^k, s^k)\| \le 4\theta^k \rho \frac{1}{\sqrt{\eta\mu^k}} \frac{2}{\theta^k \rho}(x^k)^T s^k \le \frac{8n\sqrt{\mu^k}}{\sqrt{\eta}}$$
and
$$\|(X^k S^k)^{-.5}(X^k s^k - \gamma\mu^k e)\| \le \sqrt{n\mu^k(1 - 2\gamma + \gamma^2/\eta)}\ .$$
This implies that $\|D^{-1}d_x\| \le O(n\sqrt{\mu^k})$. Similarly, we have $\|Dd_s\| \le O(n\sqrt{\mu^k})$. Combining them, we have
$$|(d_x)^T d_s| = O(n^2 \mu^k) \quad \text{and} \quad (d_x)_j (d_s)_j = O(n^2 \mu^k) \quad \text{for each} \quad j,$$
implying that the conditions of Lemma 5.15 hold.

The question now is what happens if
$$\|(x^k, s^k)\|_1 > \frac{2}{\theta^k \rho}(x^k)^T s^k?$$

Suppose that there is an optimal solution pair such that $\|(x^*, s^*)\|_\infty \le \rho$. Then, from (5.18) we see
$$\begin{aligned} A(\theta^k x^0 + (1-\theta^k)x^* - x^k) &= 0 \\ A^T(\theta^k y^0 + (1-\theta^k)y^* - y^k) + (\theta^k s^0 + (1-\theta^k)s^* - s^k) &= 0 \end{aligned}.$$

So we have
$$(\theta^k x^0 + (1-\theta^k)x^* - x^k)^T(\theta^k s^0 + (1-\theta^k)s^* - s^k) = 0,$$
which means
$$\begin{aligned}&(\theta^k x^0 + (1-\theta^k)x^*)^T s^k + (\theta^k s^0 + (1-\theta^k)s^*)^T x^k \\ &= (\theta^k x^0 + (1-\theta^k)x^*)^T(\theta^k s^0 + (1-\theta^k)s^*) + (x^k)^T s^k.\end{aligned}$$

Using this inequality together with
$$x^0 = s^0 = \rho e, \quad x^* \leq \rho e, \quad s^* \leq \rho e, \quad \text{and} \quad (x^*)^T s^* = 0,$$
we have
$$\begin{aligned}&\theta^k \rho \|(x^k, s^k)\|_1 \\ &= \theta^k((x^k)^T s^0 + (x^0)^T s^k) \\ &\leq (\theta^k x^0 + (1-\theta^k)x^*)^T s^k + (\theta^k s^0 + (1-\theta^k)s^*)^T x^k \\ &= (\theta^k x^0 + (1-\theta^k)x^*)^T(\theta^k s^0 + (1-\theta^k)s^*) + (x^k)^T s^k \\ &\leq \theta^k n \rho^2 + (x^k)^T s^k.\end{aligned}$$

Furthermore, $(y^k, x^k, s^k) \in \mathcal{N}(\eta)$ means that
$$(x^k)^T s^k \geq \theta^k (x^0)^T s^0 = \theta^k n \rho^2.$$

Hence we must have
$$\theta^k \rho \|(x^k, s^k)\|_1 \leq 2(x^k)^T s^k,$$
which leads to a contradiction.

5.5 Notes

Using the rank-one updating technique to improve the arithmetic operation complexity of interior-point algorithms by a factor of $\sqrt{n}$ was first due to Karmarkar [217]. Gonzaga [158] and Vaidya [437] used this technique to obtain the current-best LP arithmetic complexity bound $O(n^3 L)$; see also Mizuno [287] for a general treatment. Specific computation issues of the rank-one technique were presented in Shanno [375].

Parallel worst-case complexity results on interior-point algorithms can be seen in Goldberg, Plotkin, Shmoys and Tardos [147, 148] and Nesterov and Nemirovskii [326].

The convergence behavior of various interior-point trajectories was studied by Adler and Monteiro [6], Güler [173], Lagarias [241], McLinden [267], Megiddo and Shub [274], Monteiro [295] [296], and Monteiro and Tsuchiya [304]. The analysis of identifying the optimal partition of variables at a strictly complementary solution was due to Güler and Ye [177]. Adler and Monteiro [7], Jansen, Roos and Terlaky [198], Greenberg [167, 168], and Monteiro and Mehrotra [302] provided a post-optimality analysis based on the optimal partition of variables.

The termination procedure described here was developed by Mehrotra and Ye [281]. They also reported effective computational results for solving Netlib problems. A more recent termination or cross-over procedure for obtaining a basic optimal solution is developed by Andersen and Ye [17], Bixby and Saltzman [60], Kortanek and Zhu [236], and Megiddo [272]. Andersen and Ye proved a polynomial bound and showed its practical effectiveness. For a comprehensive survey on identifying an optimal basis and the optimal partition, see El–Bakry, Tapia and Zhang [108].

The homogeneous and self-dual algorithm is due to Mizuno, Todd and Ye [479], which is based on the homogeneous model of Goldman and Tucker [153, 431]. The algorithm is simplified and implemented by Xu, Hung and Ye [461]; also see Tutuncu [435]. A combined phase I and phase II algorithm was proposed by Anstreicher [25], also see Freund [124].

Infeasible-starting algorithms, which are very popular and effective, were developed and analyzed by Lustig [254], Kojima, Megiddo and Mizuno [226], Mizuno [289], Mizuno, Kojima and Todd [291], Potra [347], Tanabe [401], Wright [458], and Zhang [482]. Under certain conditions for choosing the initial point, these algorithms have polynomial iteration bounds (e.g. Zhang [482] and Mizuno [289]). In fact, using a smaller neighborhood and the predictor-corrector type algorithm in the preceding chapter one can reduce the iteration bound in Theorem 5.14 to $O(nt)$ with the same selection of the initial point. A surface theory of all infeasible-starting interior-point algorithms can be seen in Mizuno, Todd and Ye [293].

There have also been efforts to look for *lower* bounds on the number of iterations required; see Anstreicher [27], Bertsimas and Luo [54], Ji and Ye [207], Powell [352], Sonnevend, Stoer and Zhao [384, 385], and Zhao and Stoer [488]. One important recent result is due to Todd [411], who obtains a bound of at least $n^{1/3}$ iterations to achieve a constant factor decrease in the duality gap. The algorithm he studies is the primal-dual affine-scaling algorithm, which is close to methods used in practical implementations. He allows almost any reasonable step size rule, such as going 99.5 percent of the way to the boundary of the feasible region, again as used in practical codes; such step size rules definitely do not lead to iterates lying close to the central path. The weakness of the primal-dual affine-scaling algorithm

5.5. NOTES

is that no polynomiality or even global convergence has been established for it, except for the case of very small step sizes, and practical experiments indicate that the algorithm alone may not perform well.

Todd also showed that his lower bound extends to other polynomial primal-dual interior-point methods that use directions, including some centering component if the iterates are restricted to a certain neighborhood of the central path. Todd and Ye [416] further extended his result to longstep primal-dual variants that restrict the iterates to a wider neighborhood. This neighborhood seems the least restrictive while also guaranteeing polynomiality for primal-dual path-following methods, and the variants are even closer to what is implemented in practice.

Recently, Atkinson and Vaidya [439] used a combined logarithmic and volumetric potential function to derive an algorithm for LP in $O(n^{1/4}m^{1/4}L)$ iterations. Their algorithm is simplified and improved by Anstreicher [26] and Ramaswamy and Mitchell [356].

Condition-based complexity analyses could be found in Renegar [359], who developed a general condition number and ill-posedness theory for the generalized linear programming. See also Filipowski [115], Freund and Vera [125], and Todd and Ye [417] for a related discussion. Their discussion addresses the issue that interior-point algorithms do not provide clear-cut information for a particular linear programming problem being infeasible or unbounded. They provide general tools, approximate Farkas' lemma (Exercise 5.12) and the "gauge duality" theory (e.g., Freund [122]), for concluding that a problem or its dual is likely to be infeasible, and apply them to develop stopping rules for a homogeneous self-dual algorithm and for a generic infeasible-starting method. These rules allow precise conclusions to be drawn about the linear programming problem and its dual: either near-optimal solutions are produced, or we obtain "certificates" that all optimal solutions, or all feasible solutions to the primal or dual, must have large norm. Their rules thus allow more definitive interpretation of the output of such an algorithm than previous termination criteria. They have given bounds on the number of iterations required before these rules apply.

More recently, Vavasis and Ye [452] proposed a primal-dual "layered-step" interior point (LIP) algorithm for linear programming. This algorithm follows the central path, either with short steps or with a new type of step called a *layered least-squares* (LLS) step. The algorithm returns an exact optimum after a finite number of steps; in particular, after $O(n^{3.5}c(A))$ iterations, where $c(A)$ is a function of the coefficient matrix, which is independent of b and c. One consequence of the new method is a new characterization of the central path: we show that it composed of at most n^2 alternating straight and curved segments. If the LIP algorithm is applied to integer data, we get as another corollary a new proof of a well-known theorem of Tardos that linear programming can be solved in strongly polynomial time provided that A contains small-integer entries. Megiddo, Mizuno and Tsuchiya [273] further proposed an enhanced version of the LIP algorithm.

5.6 Exercises

5.1 *Verify inequality*

$$\mathcal{P}_{n+\rho}(x^{k+1}, z^k) - \mathcal{P}_{n+\rho}(x^k, z^k) \leq \frac{\alpha}{1.1} \min\left(\alpha\sqrt{\frac{n}{n+\alpha^2}}, 1-\alpha\right) + \frac{(1.1\alpha)^2}{2(1-1.1\alpha)}$$

in Section 5.1

5.2 *In the termination section, prove if both $\|d_x\|_\infty$ and $\|d_s\|_\infty$ are less than 1, then $x_P > 0$ and $s_Z = c_Z - A_Z^T y > 0$, which imply that $\sigma^k = P^*$.*

5.3 *Analyze the complexity bound of the termination procedure if A_P is not of full row rank.*

5.4 *Prove that if the LP problem has integral data, then*

$$\zeta(A, b, c) \leq 2^L \quad \text{and} \quad \xi(A, b, c) \geq 2^{-L},$$

where L is the size of the binary LP data.

5.5 *Prove that the total number of required pivots in the process described at the end of Section 5.2.2 is at most $|\sigma(x_{P*}^*)| - m \leq n - m$.*

5.6 *Prove Proposition 5.5.*

5.7 *Prove Theorem 5.10 for the predictor-corrector algorithm described in Section 5.3.2.*

5.8 *Similar to $\xi(A, b, c)$ and $\zeta(A, b, c)$, derive an expression for the condition number $c(A, b, c)$ in Theorem 5.10. Prove that if the LP problem has integral data, then*

$$c(A, b, c) \leq 2^L,$$

where L is the size of the binary LP data.

5.9 *Prove that in Algorithm 5.2*

$$Ax^k - b = \theta^k(Ax^0 - b)$$

and

$$c - A^T y^k - s^k = \theta^k(c - A^T y^0 - s^0).$$

5.6. EXERCISES

5.10 *Verify Lemma 5.16*

5.11 *(Gauge Duality Theorem) Show that*

$$\alpha := \min\{\|z\| : Mz \geq d\}$$

and

$$\eta := \min\{\|t\|^* : M^T r = t,\ d^T r = 1,\ r \geq 0\}$$

satisfy $\alpha\eta = 1$. Here, $\|.\|$ is a p norm, $p = 1, 2, ...,$ and $\|.\|^*$ is the corresponding dual norm.

5.12 *Using the "gauge duality" result (Exercise 5.11) to show the following approximate Farkas' lemma. Let $A \in \mathcal{R}^{m \times n}$, $b \in \mathcal{R}^m$, and $c \in \mathcal{R}^n$. Let*

$$\begin{aligned}
\alpha_x &:= \min\{\|x\| : Ax = b,\ x \geq 0\}, \\
\alpha_y &:= \min\{\|y\| : A^T y \leq c\},\ and \\
\alpha_s &:= \min\{\|s\|^* : A^T y + s = c,\ s \geq 0\}.
\end{aligned}$$

Let

$$\begin{aligned}
\beta_u &:= \min\{\|u\|^* : A^T y \leq u,\ b^T y = 1\}, \\
\beta_v &:= \min\{\|v\|^* : Ax = v,\ c^T x = -1,\ x \geq 0\},\ and \\
\beta_w &:= \min\{\|w\| : Ax = 0,\ c^T x = -1,\ x \geq -w\}.
\end{aligned}$$

Then

$$\alpha_x \beta_u = \alpha_y \beta_v = \alpha_s \beta_w = 1.$$

Here, $\|.\|$ is a p norm and $\|.\|^*$ is the corresponding dual norm.

5.13 *If every feasible solution of an LP problem is large, i.e., $\|x\|$ is large, then the problem is near infeasible. Prove this statement using Exercise 5.12.*

Chapter 6

Average-Case Analysis

The idea of average-case analysis is to obtain rigorous probabilistic bounds on the number of iterations required by an iterative algorithm to reach some termination criterion. Although many interior point algorithms devised in the last several years are polynomial time methods, in practice they generally perform much better than worst-case bounds would indicate. A "gap" between theory and practice exists that average-case analysis might (at least partially) close.

There are two main viewpoints in the probabilistic analysis of algorithms. First one can develop randomized algorithms, and show that, on a worst-case instance of a problem, the average running time of the algorithm has a certain bound, or the running time satisfies a certain bound with high probability, or the running time always satisfies a certain bound and the algorithm gives a correct answer with high probability, meaning converging to 1 as the dimension of the problem goes to ∞.

Second one can consider the expected running time of a deterministic algorithm when applied to problem instances generated according to some probability distribution (or class of such distributions). For linear programming, researchers have provided some theoretical justification for the observed practical efficiency of the simplex method, despite its exponential worst-case bound. Of course, this viewpoint might be less compelling, since one can always argue that the distribution chosen for problem instances is inappropriate.

Another minor viewpoint is the so called *one-step analysis*: at an iteration we make an nonrigorous but plausible assumption concerning the current data generated by the algorithm, and then address the expected behavior or behavior which occurs with high probability at that iteration. The anticipated number of iterations is then defined to be the number of iterations required if this behavior actually occurs at every iteration (or at least once every, say, ten iterations). This analysis is distinct from the two just described. As the reader will see, the assumptions we make at each iteration can be inconsistent with one another. Nevertheless, such an approach might add insight in the case where a more rigorous analysis seems intractable.

In this chapter, we first develop a one-step analysis for several adaptive interior-point algorithms described in Section 4.5, which all have complexities of $O(n^{1/2}\log(1/\epsilon))$ or $O(n\log(1/\epsilon))$ iterations to attain precision ϵ. (Here we assume, without loss of generality, that $(x^0)^T s^0 = R = 1$.) Based on the one-step analysis, we anticipate that these algorithms would only require $O(n^{1/4}\log(1/\epsilon))$ or $O((\log n)\log(1/\epsilon))$ iterations, where n is the number of variables in (LP).

We then develop a rigorous analysis, based on the second main viewpoint of probabilistic analysis, of interior-point algorithms coupled with the termination procedure described in Chapter 5. We will first show that a random linear feasibility problem can be solved in $O(\sqrt{n}\log n)$ iterations with high probability. Using the homogeneous and self-dual algorithm described in Chapter 5, we then show that the expected number of iterations required to solve a random LP problem is bounded above by $O(\sqrt{n}\log n)$.

Let us formally define high probability: an event in n-dimensional space is true with probability approaching one as $n \to \infty$. Such an event is called a *high probability event*. Note that a result based on high probability may be stronger than the one based on the standard expected or average analysis. We first derive some *Observations*:

1) Let events E_1 and E_2 be true with high probability. Then the event $E_1 \cap E_2$ is also true with high probability.

2) Let the event E_1 be true with high probability, and let E_1 imply E_2. Then the event E_2 is also true with high probability.

Observation (1) cannot be induced to m events where m is proportional to n. However, we have the following lemma:

Lemma 6.1 *Let $\bar{E}_j$ be the complement of E_j, $j = 1, 2, \ldots$. Then, if the probability*

$$\lim_{n \to \infty} \left(\sum_{j=1}^{n} \Pr(\bar{E}_j) \right) = 0,$$

then $E_1 \cap E_2 \cap \ldots \cap E_m$ is true with high probability for any $1 \le m \le n$.

Proof. The proof simply follows

$$\Pr(E_1 \cap E_2 \cap ... \cap E_m) \geq \Pr(E_1 \cap E_2 \cap ... \cap E_n) \geq 1 - \sum_{j=1}^{n} \Pr(\bar{E}_j)$$

for any $1 \leq m \leq n$. Since $\lim_{n \to \infty}(1 - \sum_{j=1}^{n} \Pr(\bar{E}_j)) = 1$, we have the lemma proved.

□

6.1 One-Step Analysis

Consider two adaptive interior-point algorithms in Section 4.5: the predictor-corrector and wide-neighborhood algorithms, with worst-case complexities $O(n^{1/2} \log(1/\epsilon))$ and $O(n \log(1/\epsilon))$ iterations to attain precision ϵ—to generate $(x^k, y^k, s^k) \in \overset{\circ}{\mathcal{F}}$ from (x^0, y^0, s^0) such that $\mu^k/\mu^0 \leq \epsilon$, respectively. The progress will be far faster in the predictor-corrector algorithm if $\|Pq\|$ is typically smaller than the worst-case bound given by Lemma 4.14. From (4.23) and (4.24), the corrector step will be far better centered than is guaranteed by Lemma 4.16, and the predictor step will be much longer than $O(n^{-1/2})$ by Lemma 4.17.

On the other hand, Lemma 4.14 shows that, for the wide-neighborhood algorithm, $\|Pq\|_\infty$ and $\|Pq\|_\infty^-$ can only be bounded by a multiple of $\|r\|^2$, not $\|r\|_\infty^2$, unless an extra factor of n is introduced. But $\|r\|$ may be large compared to $\|r\|_\infty$, which is related to η with $(x, s) \in \mathcal{N}_\infty(\eta)$. Again, if $\|Pq\|_\infty$ and $\|Pq\|_\infty^-$ are typically much smaller than the bound given by Lemma 4.14, then the duality gap reduction will be far greater.

For now, we note the following results which follow immediately from Lemmas 4.17 and 4.19, and inequalities (4.26) and (4.27).

Corollary 6.2 *Consider the predictor-corrector and wide-neighborhood algorithms in Section 4.5.*

i) *If at a particular iteration we have $\|Pq\| \leq n^{1/2}\mu$ in the predictor step of the predictor-corrector algorithm, then the duality gap at that iteration will decrease at least by a factor of $(1 - 2/(1 + \sqrt{1 + 8\sqrt{n}}))$.*

ii) *Let η and γ be as in Theorem 4.20. If at a particular iteration of the wide-neighborhood algorithm we have $\|Pq\|_\infty \leq \mu^k \log n$ for $\mathcal{N} = \mathcal{N}_\infty(\eta)$ and $\|Pq\|_\infty^- \leq \mu^k \log n$ for $\mathcal{N} = \mathcal{N}_\infty^-(\eta)$, then the duality gap at that iteration will decrease at least by a factor $(1 - \eta\gamma(1 - \gamma)/\log n)$, with either $\mathcal{N}_\infty(\eta)$ or $\mathcal{N}_\infty^-(\eta)$.*

6.1.1 High-probability behavior

In this section we provide heuristic arguments as to why we might expect $\|Pq\|$, $\|Pq\|_\infty$, and $\|Pq\|_\infty^-$ to be of the sizes stated in the above corollary. Recall that p and q are the projections of $r \in \mathcal{R}^n$ onto the subspaces U and $U^\perp$ respectively. In this section we suppose r is fixed, but assume that

Assumption 6.1 U *is a random subspace of* $\mathcal{R}^n$ *of dimension* $d := n - m$, *drawn from the unique distribution on such subspaces that is invariant under orthogonal transformations.*

Given that U is the null space of $AX^{1/2}S^{-1/2} =: \tilde{A}$, this assumption would hold, for example, if each component of the matrix $\tilde{A}$ were independently drawn from a standard normal distribution. Note that such assumptions, made at different iterations and hence values of X and S, are not consistent with one another. Further, for several interior-point algorithms the asymptotic behavior of (x^k, s^k) is known, and this behavior is also inconsistent with our assumption, see the next chapter. On the other hand, it is also known that the asymptotic behavior is in favor of the convergence of the duality gap to 0 (see Chapter 7). Thus, such inconsistency may not deny our result which, under the assumption, establishes a faster convergence rate for the duality gap to 0. We will comment further on our approach at the end of the chapter.

For now let us examine the consequences on Pq of our assumption. Note that to compensate for the deficiencies of our assumption, the results we obtain hold with probability approaching 1 as $n \to \infty$. We establish the following theorem:

Theorem 6.3 *Let* $\xi = \|r\|_\infty / \|r\|$. *Then, with the assumption above,*

i)
$$\Pr\left(\|Pq\| \le \frac{\|r\|^2}{4}\left(2\xi^2 + \frac{6.5}{n}\right)^{1/2}\right) \to 1 \quad \text{as} \quad n \to \infty;$$

ii)
$$\Pr\left(\|Pq\|_\infty^- \le \left(\frac{\log(n)}{n}\right) \|r\|^2\right) \to 1 \quad \text{as} \quad n \to \infty.$$

Before we show how these results are proved, we indicate how they relate to the bounds on $\|Pq\|$ that form the hypotheses of Corollary 6.2. In Corollary 6.2(i), we are analyzing the predictor step, so $r = -(XS)^{1/2}e$ and $(x, s) \in \mathcal{N}_2(1/4)$. Hence $\|r\|^2 = x^T s = n\mu$ and $\|r\|_\infty^2 = \|Xs\|_\infty \le$

6.1. ONE-STEP ANALYSIS

$\mu + \|Xs - \mu e\| \leq 5\mu/4$. Thus $\xi^2 \leq 5/(4n)$ and by Theorem 6.3(i), with probability approaching 1

$$\|Pq\| \leq \frac{\|r\|^2}{4}\left(2\frac{5}{4n} + \frac{6.5}{n}\right)^{1/2} = \frac{3\|r\|^2}{4n^{1/2}} < n^{1/2}\mu,$$

which is the hypothesis of Corollary 6.2(i).

For Corollary 6.2(ii), we consider first the case where $\mathcal{N} = \mathcal{N}_\infty^-(\eta)$. Then by Theorem 6.3(ii) and Lemma 4.15(iii), with probability approaching 1

$$\|Pq\|_\infty^- \leq \frac{\log(n)}{n}\|r\|^2 \leq \log(n)\mu^k,$$

which gives the hypothesis of Corollary 6.2(ii) in this case. Now suppose $\mathcal{N} = \mathcal{N}_\infty(\eta)$. Then with high probability

$$\|Pq\|_\infty^- \leq \log(n)\mu^k$$

as above. Also, by Lemma 4.14(iii) and Lemma 4.15(iii),

$$\|Pq\|_\infty^+ \leq \frac{\|r\|_\infty^2}{4} \leq \frac{(1+\eta)\mu^k}{4} \leq \frac{\mu^k}{2}.$$

Hence $\|Pq\|_\infty \leq \log(n)\mu^k$ with probability approaching 1, which gives the hypothesis for Corollary 6.2(ii) with $\mathcal{N} = \mathcal{N}_\infty(\eta)$.

6.1.2 Proof of the theorem

Now we sketch the proof of Theorem 6.3. The proof of (i) is long and technical and hence we omit it here. However, we will prove a slightly weaker version of (i) at the end of this section.

Because p and q are homogeneous of degree 1 in $\|r\|$, we assume henceforth without loss of generality that r is scaled so that

$$g = r/2 \text{ satisfies } \|g\| = 1.$$

Let $F = (g, H)$ be an orthogonal $n \times n$ matrix. If we express the vector p in terms of the basis consisting of the columns of F, we get

Lemma 6.4 *We can write*

$$p = (1 + \zeta)g + \eta H v, \tag{6.1}$$

where $\frac{1+\zeta}{2}$ has a beta distribution with parameters $\frac{d}{2}$ and $\frac{m}{2}$; $\eta = \sqrt{1 - \zeta^2}$; and v is uniformly distributed on the unit sphere in $\mathcal{R}^{n-1}$.

184   CHAPTER 6. AVERAGE-CASE ANALYSIS

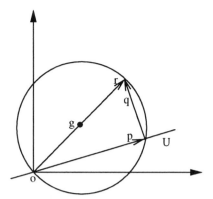

Figure 6.1. Illustration of the projection of r onto a random subspace.

Proof. Since p and q are orthogonal with $p + q = r$, p lies on the sphere of center $r/2 = g$ and radius $\|g\| = 1$ (Figure 6.1). Thus p can be written in the form (6.1), with $\eta = \sqrt{1 - \zeta^2}$ and $\|v\| = 1$. We need to establish that ζ and v have the given distributions.

Note that $\|p\|^2 = (1 + \zeta)^2 + \eta^2 = 2(1 + \zeta)$. However, we can obtain the distribution of $\|p\|^2$ directly. The invariance under orthogonal transformations implies that we can alternatively take U as a fixed d-subspace, say, $\{x \in \mathcal{R}^n : x_{d+1} = \cdots = x_n = 0\}$, and r uniformly distributed on a sphere of radius 2. Then r can be generated as

$$\left(\frac{2\lambda_1}{\|\lambda\|}, \frac{2\lambda_2}{\|\lambda\|}, \ldots, \frac{2\lambda_n}{\|\lambda\|} \right)^T,$$

where $\lambda \sim N(0, I)$ in $\mathcal{R}^n$ (i.e., the components of λ are independent normal random variables with mean 0 and variance 1, denoted by $N(0, 1)$). But then

$$p = \left(\frac{2\lambda_1}{\|\lambda\|}, \frac{2\lambda_2}{\|\lambda\|}, \ldots, \frac{2\lambda_d}{\|\lambda\|}, 0, \ldots, 0 \right)^T,$$

and $\|p\|^2 = 4(\lambda_1^2 + \cdots + \lambda_d^2)/(\lambda_1^2 + \cdots + \lambda_n^2)$. This has the distribution of four times a beta random variable with parameter $\frac{d}{2}$ and $\frac{m}{2}$, which confirms the distribution of ζ.

Now let W be an orthogonal matrix with $Wg = g$. W can be thought of as rotating the sphere with center g around its diameter from 0 to $2g = r$. We can view the random d-subspace U as the null space of an $m \times n$ random matrix $\bar{A}$ with independent standard normal entries. The fact that p is the projection of r onto U is then equivalent to $\bar{A}p = 0$, $r - p = \bar{A}^T v$ for some v.

6.1. ONE-STEP ANALYSIS

But then $(\bar{A}W^T)Wp = 0$ and $r - Wp = Wr - Wp = (\bar{A}W^T)^T v$, so that Wp is the projection of r onto $U' = \{x : (\bar{A}W^T)x = 0\}$. If $\bar{A}$ has independent standard normal entries, so does $\bar{A}W^T$, so U' is also a random d-subspace. Thus Wp has the same distribution as p. But writing W as $HW'H^T + gg^T$, where W' is an arbitrary orthogonal matrix of order $n-1$, we see that v has the same distribution as $W'v$. Since $\|v\| = 1$, v is uniformly distributed on the unit sphere $\mathcal{R}^{n-1}$.

□

Since $p + q = r = 2g$, relation (6.1) implies

$$\begin{aligned} q &= (1-\zeta)g - \eta Hv, \text{ so that} \\ Pq &= \eta^2 g^2 - 2\zeta\eta GHv - \eta^2(Hv)^2 \quad (6.2) \\ &= -(Hv)^2 + (\eta g - \zeta Hv)^2 \\ &\geq -\|Hv\|_\infty^2 e, \quad (6.3) \end{aligned}$$

where $G := \text{diag}(g)$, and g^2, $(Hv)^2$, and $(\eta g - \zeta Hv)^2$ denote the vectors whose components are the squares of those of g, Hv, and $\eta g - \zeta Hv$, respectively.

The proof of Theorem 6.3(i) proceeds by using (6.2) to evaluate $\|Pq\|^2$, and then analyzing all the terms in the resulting expression. The proof of Theorem 6.3(ii) follows from (6.3) (which gives $\|Pq\|_\infty^- \leq \|Hv\|_\infty^2$) and the following result:

Lemma 6.5 *Let $F = [g, H]$ be an orthogonal matrix. If v is uniformly distributed on the unit sphere in $\mathcal{R}^{n-1}$,*

$$\Pr\left(\|Hv\|_\infty \leq \sqrt{3\frac{\log n}{n}}\right) \to 1 \quad as \quad n \to \infty.$$

Proof. Since v is uniformly distributed on the unit sphere in $\mathcal{R}^{n-1}$, it can be generated as follows: $v = \lambda/\|\lambda\|$, where $\lambda \sim N(0, I)$ (the standard normal distribution in $\mathcal{R}^{n-1}$). Hence we wish to obtain an upper bound on $\|H\lambda\|_\infty$ and a lower bound on $\|\lambda\|$, both of which hold with high probability. Now $\|\lambda\|^2$ is a χ^2 random variable with $n-1$ degrees freedom, so

$$\begin{aligned} E(\|\lambda\|^2) &= n-1, \\ \text{Var}(\|\lambda\|^2) &= 2(n-1). \end{aligned}$$

From Chebychev's inequality, we have

$$\Pr(\|\lambda\| \geq (1-\epsilon)\sqrt{n-1}) \to 1 \quad as \quad n \to \infty \quad (6.4)$$

for any $\epsilon > 0$.

Let λ_0 be a standard normal variable, and let $\lambda' = (\lambda_0, \lambda)$, also $N(0, I)$ but in $\mathcal{R}^n$. Then $\|\lambda'\|_\infty = \max\{\nu_j : j = 0, 1, 2, \cdots, n - 1\}$ where $\nu_j = |\lambda_j|$ has the positive normal distribution. Then $1 - N_+(x) = 2(1 - N(x))$ where N_+ is the distribution function of ν, and N is the normal distribution function. It now follows from results in extreme value theory [1] that

$$\Pr\left(\|\lambda'\|_\infty \leq \sqrt{2\log(2n)}\right) \to 1 \text{ as } n \to \infty.$$

Since $F\lambda'$ is also $N(0, I)$,

$$\Pr\left(\|F\lambda'\|_\infty \leq \sqrt{2\log(2n)}\right) \to 1 \text{ as } n \to \infty.$$

Now we have

$$\|H\lambda\|_\infty \leq \|F\lambda'\|_\infty + \|\lambda_0 g\|_\infty.$$

Since $\|g\| = 1$,

$$\Pr\left(\|\lambda_0 g\|_\infty \leq \epsilon\sqrt{\log n}\right) \to 1 \text{ as } n \to \infty$$

for any $\epsilon > 0$. From the above relations and (6.4), we get the result of the lemma.

□

We conclude this section by showing how (6.2) and Lemma 6.5 imply a slightly weaker form of Theorem 6.3(i). Indeed, (6.2) yields

$$\begin{aligned}\|Pq\| &\leq \eta^2\|g^2\| + 2|\zeta\eta|\|g\|_\infty\|Hv\| + \eta^2\|(Hv)^2\| \\ &\leq \|g\|_\infty\|g\| + 2\|g\|_\infty + \|Hv\|_\infty\|Hv\| \\ &= 3\xi + \|Hv\|_\infty.\end{aligned}$$

By Lemma 6.5, this is at most $3\xi + \sqrt{3\log(n)/n}$ with probability approaching 1 as $n \to \infty$. This bound would lead one to hope that $\|Pq\|$ would be at most $(n\log(n))^{1/2}\mu$ at a typical predictor step. The predictor-corrector algorithm, with the worst-case bound $O(\sqrt{n}\log(1/\epsilon))$, would require at most $O((n\log(n))^{1/4}\log(1/\epsilon))$ iterations of this type, while the wide-neighborhood algorithm, with the worst-case bound $O(n\log(1/\epsilon))$, would require at most $O((\log n)\log(1/\epsilon))$ iterations of this type.

---

[1] S. I. Resnick, *Extreme Values, Regular Variation, and Point Processes*, Springer-Verlag (1987), pp. 42 and 71.

6.2 Random-Problem Analysis I

We now develop a rigorous analysis, based on the second main viewpoint introduced at the beginning of the chapter, of interior-point algorithms coupled with the termination procedure described in Chapter 5. We use a simple problem, the homogeneous linear feasibility problem, to illustrate the analysis and show that a random such problem can be solved in $O(\sqrt{n}\log n)$ iterations with high probability.

Consider finding a feasible point for the homogeneous linear system

$$\mathcal{X} = \{x : Ax = 0, \quad x \geq 0, \quad x \neq 0\}. \tag{6.5}$$

We assume that $A \in \mathcal{R}^{m \times n}$ has full row-rank. Let us reformulate the problem as a Phase I linear program:

$$\begin{array}{ll} \text{minimize} & z \\ \text{s.t.} & Ax + (-Ae)z = 0, \quad e^T x = 1, \quad (x, z) \geq 0, \end{array} \tag{6.6}$$

and its dual

$$\begin{array}{ll} \text{maximize} & \lambda \\ \text{s.t.} & s = -A^T y - e\lambda \geq 0, \quad s_z = 1 + e^T A^T y \geq 0. \end{array} \tag{6.7}$$

Obviously, LP problem (6.6) has nonempty interior, its optimal solution set is bounded, and its optimal objective value is 0 if and only if $\mathcal{X}$ is nonempty. In fact, we can select an initial feasible point as $x^0 = e/n$, $z^0 = 1/n$ for the primal, and $y^0 = 0$, $\lambda^0 = -1$, $s^0 = e$ and $s_z^0 = 1$ for the dual. Thus, (x^0, z^0) and (y^0, λ^0) are "centered," and the initial primal-dual gap is $(1 + 1/n)$.

We now specialize the termination procedure proposed in Chapter 5 for solving problem (6.6). Suppose $A = (A_P, A_Z)$, where

$$P = \{j : x_j^k \geq s_j^k\} \quad \text{and} \quad Z = \{j : x_j^k < s_j^k\}.$$

We solve the least-squares problem

$$\begin{array}{rl} (PP) \quad \text{minimize} & \|(X_P^k)^{-1}(x_P - x_P^k)\| \\ \text{s.t.} & A_P(x_P - x_P^k) = A_Z x_Z^k + (-Ae)z^k \end{array}$$

and

$$\begin{array}{rl} (DD) \quad \text{minimize} & \|(S_Z^k)^{-1} A_Z^T (y - y^k)\| \\ \text{s.t.} & A_P^T (y - y^k) = s_P^k. \end{array}$$

Here, we have ignored variable z and the last (normalization) equality constraint $e^T x = 1$ in the problem, when we apply the termination projection.

In particular, if $A = A_P$, then the minimizer $x^* = x_P^*$ of (PP) satisfies

$$(X^k)^{-1}(x^* - x^k) = X^k A^T (A(X^k)^2 A^T)(-Ae)z^k.$$

Thus,

$$\begin{aligned}
&\|(X^k)^{-1}(x^* - x^k)\| \\
&= \|X^k A^T (A(X^k)^2 A^T)(-Ae)z^k\| \\
&= \|X^k A^T (A(X^k)^2 A^T) A X^k (X^k)^{-1} A^T (AA^T)^{-1}(-Ae)z^k\| \\
&\leq \|X^k A^T (A(X^k)^2 A^T) A X^k\| \|(X^k)^{-1} A^T (AA^T)^{-1}(-Ae)z^k\| \\
&\leq \|(X^k)^{-1} A^T (AA^T)^{-1} Ae(-z^k)\| \\
&\leq \|(X^k)^{-1}\| \|A^T (AA^T)^{-1} Ae\| |z^k| \\
&\leq \sqrt{n} z^k \|(X^k)^{-1}\|.
\end{aligned}$$

(Note that $\zeta(A, b, c)$ defined in Section 5.2.2 is less than or equal to $\sqrt{n}$ here.) This implies that if $\min(x^k) > \sqrt{n} z^k$, then the projection x^* satisfies

$$\|(X^k)^{-1} x^* - e\| \leq \sqrt{n} z^k \|(X^k)^{-1}\| < 1, \qquad (6.8)$$

and x^* must be a point in $\overset{\circ}{\mathcal{X}}$.

Let the optimal partition of problem (6.6) be (P^*, Z^*). If system (6.5) has an interior feasible point, then $A = A_{P^*}$ and $z^* = 0$. Using Theorem 5.4 with $\eta = 1$ in (5.7), we have, when the duality gap $z^k - \lambda^k \leq \xi^2/n^2$,

$$s_j^k < \xi/n^2 < \xi/n < x_j^k, \ j \in P^* \quad \text{and} \quad s_j^k > \xi/n > \xi/n^2 > x_j^k, \ j \in Z^*,$$

or

$$n s_j^k < x_j^k, \ j \in P^* \quad \text{and} \quad s_j^k > n x_j^k, \ j \in Z^*,$$

where, recall from (5.3), that for the standard LP problem

$$\xi := \xi(A, b, c) = \min\{\xi_p, \xi_d\}.$$

Thus, in $O(\sqrt{n}(|\log \xi| + \log n))$ iterations we have $A_P = A_P^* = A$ and (6.8), and therefore we generate an interior-point in $\overset{\circ}{\mathcal{X}}$.

Consider the case that system (6.5) is empty, then the optimal value $z^* = \lambda^*$ of problem (6.6) is positive and we can choose $y = y^k$ in (DD) and have

$$s = -A^T y^k = s^k + e\lambda^k.$$

Thus, if $\lambda^k \geq 0$, then we must have $s > 0$, which proves that $\mathcal{X}$ is empty from the Farkas lemma. Note that in $O(\sqrt{n}(|\log z^*| + \log n))$ iterations we have the duality gap $z^k - \lambda^k \leq z^*$ or $\lambda^k \geq z^k - z^* \geq 0$.

6.2. RANDOM-PROBLEM ANALYSIS I

Let us estimate ξ for problem (6.6) if system (6.5) has an interior feasible point $\bar{x}$ such that

$$p(1/n) \leq \bar{x}_j \leq p(n) \quad \text{for} \quad j = 1, 2, ..., n+1, \qquad (6.9)$$

where $p(\alpha)$ is a polynomial α^d for a constant $d \geq 1$. Then, for problem (6.6) we must have

$$\xi_p \geq p(1/n)/(np(n)) \quad \text{and} \quad \xi_d = 1,$$

since $(\bar{x}/e^T\bar{x}, 0)$ is a primal optimal solution and $\bar{y} = 0$ is a dual optimal solution with $\bar{s} = (0, 1)^T$. Thus, $\xi \geq p(1/n)/(np(n))$.

On the other hand, if system (6.5) is empty then $\{s : s = -A^T y \geq 0\}$ has an interior feasible point $(\bar{y}, \bar{s})$. Let $(\bar{y}, \bar{s})$ satisfy

$$p(1/n) \leq \bar{s}_j \leq p(n) \quad \text{for} \quad j = 1, 2, ..., n+1. \qquad (6.10)$$

Then, the dual LP problem (6.7) has a feasible point $y = \bar{y}/e^T\bar{s}$, $s = \bar{s}/e^T\bar{s}$, $\lambda = \min(s)$, and $s_z = 0$. Thus, $z^* \geq \lambda \geq p(1/n)/(np(n))$.

To summarize, we have

Theorem 6.6 *Let $p(\alpha) = \alpha^d$ for a constant $d \geq 1$ and let the homogeneous system (6.5) either have a feasible point $\bar{x}$ satisfying (6.9) or be empty with an $\bar{s} = -A^T\bar{y}$ satisfying (6.10). Then, finding a feasible point for system (6.5) or proving it empty can be completed in $O(\sqrt{n}\log n)$ iterations by an $O(\sqrt{n}\log(1/\epsilon))$ interior-point algorithm, where each iteration solves a system of linear equations.*

We emphasize that ξ or z^* is a non-combinatorial measure of the feasible region $\mathcal{X}$ or its dual. For an LP problem (as opposed to a feasibility problem), ξ or z^* is determined by the geometry of the optimal face.

6.2.1 High-probability behavior

From Lemma 6.1 we can derive several propositions.

Proposition 6.7 *Let $\hat{x}_j$, $j = 1, 2, ..., n$, have the identical standard Gauss distribution $N(0, 1)$ and condition on the event that $\hat{x}_j \geq 0$ for $j = 1, ..., n$. Then, with high probability,*

$$p(1/n) \leq \hat{x}_j \leq p(n) \quad \text{for} \quad j = 1, 2, ..., n,$$

where $p(n) = n^d$ for some constant $d \geq 1$.

Proposition 6.8 Let $\hat{x}_j$, $j = 1, 2, ..., n$, have the identical Cauchy distribution, i.e., the quotient of two independent $N(0,1)$ random variables, and condition on the event that $\hat{x}_j \geq 0$ for $j = 1, ..., n$. Then, with high probability,
$$p(1/n) \leq \hat{x}_j \leq p(n) \quad for \quad j = 1, 2, ..., n,$$
where $p(n) = n^d$ for some constant $d \geq 1$.

Proposition 6.9 Let λ_0, λ_1,..., λ_m be independent $N(0,1)$ random variables and condition on the event that $\lambda_i \leq |\lambda_0|/\sqrt{d}$ for $i = 1, ..., m$ ($d = n - m \geq 1$). Then, the non-negative random variables, $\hat{x}_i := 1 - \sqrt{d}\lambda_i/|\lambda_0|$ for $i = 1, ..., m$, satisfy
$$p(1/n) \leq \hat{x}_i \leq p(n) \quad for \quad i = 1, ..., m,$$
with high probability, where $p(n) = n^d$ for some constant d.

The first two propositions are relatively easy to prove. To prove the third, we first prove a similar proposition:

Proposition 6.10 Let λ_0, λ_1,..., λ_m be independent $N(0,1)$ random variables and condition on the event that $\lambda_i \leq |\lambda_0|/\sqrt{d}$ for $i = 1, ..., m$ ($d = n - m \geq 1$). Then, the non-negative random variables, $x_i := |\lambda_0|/\sqrt{d} - \lambda_i$ for $i = 1, ..., m$, satisfy
$$p(1/n) \leq x_i \leq p(n) \quad for \quad i = 1, ..., m,$$
with high probability, where $p(n) = n^d$ for some constant $d \geq 1$.

Proof. In proving Proposition 6.10, we fix $p(n) = n^4$. Let $f(\lambda)$ be the probability density function of $N(0,1)$,
$$P(m) := \Pr(x_1 \geq 0, x_2 \geq 0, ..., x_m \geq 0)$$
and
$$P(m-1) := \Pr(x_2 \geq 0, ..., x_m \geq 0).$$
Also note that $|N(0,1)|$ has the probability density function $2f(\lambda)$ in $[0, \infty)$. Then, we have

$P(m)$
$= \Pr(x_1 \geq 0, x_2 \geq 0, ..., x_m \geq 0)$
$= \int_0^\infty 2f(\lambda_0) \int_{-\infty}^{\lambda_0/\sqrt{d}} f(\lambda_1) \int_{-\infty}^{\lambda_0/\sqrt{d}} f(\lambda_2) \cdots \int_{-\infty}^{\lambda_0/\sqrt{d}} f(\lambda_m) \prod_{i=0}^{m} d\lambda_i$
$\geq \int_0^\infty 2f(\lambda_0) \int_{-\infty}^{0} f(\lambda_1) \int_{-\infty}^{0} f(\lambda_2) \cdots \int_{-\infty}^{0} f(\lambda_m) \prod_{i=0}^{m} d\lambda_i$
$= (1/2)^m.$

6.2. RANDOM-PROBLEM ANALYSIS I

We also have

$$
\begin{aligned}
P(m) &= \Pr(x_1 \geq 0, x_2 \geq 0, \ldots, x_m \geq 0) \\
&= \int_0^\infty 2f(\lambda_0) \int_{-\infty}^{\lambda_0/\sqrt{d}} f(\lambda_1) \int_{-\infty}^{\lambda_0/\sqrt{d}} f(\lambda_2) \cdots \int_{-\infty}^{\lambda_0/\sqrt{d}} f(\lambda_m) \prod_{i=0}^m d\lambda_i \\
&\geq \int_0^\infty 2f(\lambda_0) \int_{-\infty}^0 f(\lambda_1) \int_{-\infty}^{\lambda_0/\sqrt{d}} f(\lambda_2) \cdots \int_{-\infty}^{\lambda_0/\sqrt{d}} f(\lambda_m) \prod_{i=0}^m d\lambda_i \\
&= (1/2) \int_0^\infty 2f(\lambda_0) \int_{-\infty}^{\lambda_0/\sqrt{d}} f(\lambda_2) \cdots \int_{-\infty}^{\lambda_0/\sqrt{d}} f(\lambda_m) \prod_{i=0, i \neq 1}^m d\lambda_i \\
&= P(m-1)/2.
\end{aligned}
$$

Consider the probability

$$P_1^- := \Pr(x_1 \leq p(1/n) | x_1 \geq 0, \ldots, x_m \geq 0).$$

We have

$$
\begin{aligned}
P_1^- &= \frac{1}{P(m)} \int_0^\infty 2f(\lambda_0) \int_{\lambda_0/\sqrt{d}-n^{-4}}^{\lambda_0/\sqrt{d}} f(\lambda_1) \int_{-\infty}^{\lambda_0/\sqrt{d}} f(\lambda_2) \cdots \prod_{i=0}^m d\lambda_i \\
&\leq \frac{1}{P(m)} \int_0^\infty 2f(\lambda_0) \int_{\lambda_0/\sqrt{d}-n^{-4}}^{\lambda_0/\sqrt{d}} (\frac{1}{\sqrt{2\pi}}) \int_{-\infty}^{\lambda_0/\sqrt{d}} f(\lambda_2) \cdots \prod_{i=0}^m d\lambda_i \\
&= \frac{n^{-4}}{\sqrt{2\pi}} \frac{P(m-1)}{P(m)} \\
&\leq \frac{2n^{-4}}{\sqrt{2\pi}} = O(n^{-2}).
\end{aligned}
$$

Now consider the probability

$$P_1^+ := \Pr(x_1 \geq p(n) | x_1 \geq 0, \ldots, x_m \geq 0).$$

We have

$$
P_1^+ = \frac{1}{P(m)} \int_0^\infty 2f(\lambda_0) \int_{-\infty}^{\lambda_0/\sqrt{d}-n^4} f(\lambda_1) \int_{-\infty}^{\lambda_0/\sqrt{d}} f(\lambda_2) \cdots \prod_{i=0}^m d\lambda_i
$$

$$
\begin{aligned}
&= \frac{1}{P(m)} \int_0^{n^2} 2f(\lambda_0) \int_{-\infty}^{\lambda_0/\sqrt{d}-n^4} f(\lambda_1) \int_{-\infty}^{\lambda_0/\sqrt{d}} f(\lambda_2) \cdots \prod_{i=0}^{m} d\lambda_i \\
&\quad + \frac{1}{P(m)} \int_{n^2}^{\infty} 2f(\lambda_0) \int_{-\infty}^{\lambda_0/\sqrt{d}-n^4} f(\lambda_1) \int_{-\infty}^{\lambda_0/\sqrt{d}} f(\lambda_2) \cdots \prod_{i=0}^{m} d\lambda_i \\
&:= P_1'^+ + P_1''^+.
\end{aligned}
$$

For $P_1''^+$, we have

$$
\begin{aligned}
P_1''^+ &= \frac{1}{P(m)} \int_{n^2}^{\infty} 2f(\lambda_0) \int_{-\infty}^{\lambda_0/\sqrt{d}-n^4} f(\lambda_1) \int_{-\infty}^{\lambda_0/\sqrt{d}} f(\lambda_2) \cdots \prod_{i=0}^{m} d\lambda_i \\
&\leq \frac{1}{P(m)} \int_{n^2}^{\infty} 2f(\lambda_0) d\lambda_0 \\
&= \frac{1}{P(m)} \int_{n^2}^{\infty} \frac{2}{\sqrt{2\pi}} \exp(-x^2/2) dx \\
&= \frac{1}{P(m)} \int_{n^2}^{\infty} \frac{2n^2}{\sqrt{2\pi}} [\exp(-x^2/2n^4)]^{n^4} d(x/n^2) \\
&\leq \frac{1}{P(m)} \int_{n^2}^{\infty} \frac{2n^2}{\sqrt{2\pi}} \exp(-(n^4-1)/2) \exp(-x^2/2n^4) d(x/n^2) \\
&= \frac{n^2 \exp(-(n^4-1)/2)}{P(m)} \int_1^{\infty} \frac{2}{\sqrt{2\pi}} \exp(-x^2/2) d(x) \\
&\leq 2^m n^2 \exp(-(n^4-1)/2) \\
&= O(n^{-2})
\end{aligned}
$$

for n large enough. For $P_1'^+$, we have

$$
\begin{aligned}
P_1'^+ &= \frac{1}{P(m)} \int_0^{n^2} 2f(\lambda_0) \int_{-\infty}^{\lambda_0/\sqrt{d}-n^4} f(\lambda_1) \int_{-\infty}^{\lambda_0/\sqrt{d}} f(\lambda_2) \cdots \prod_{i=0}^{m} d\lambda_i \\
&\leq \frac{1}{P(m)} \int_0^{n^2} 2f(\lambda_0) \int_{-\infty}^{n^2/\sqrt{d}-n^4} f(\lambda_1) \int_{-\infty}^{\lambda_0/\sqrt{d}} f(\lambda_2) \cdots \prod_{i=0}^{m} d\lambda_i \\
&\leq \frac{P(m-1)}{P(m)} \int_{-\infty}^{n^2/\sqrt{d}-n^4} f(\lambda_1) d\lambda_1 \\
&\leq 2 \int_{-\infty}^{n^2/\sqrt{d}-n^4} f(\lambda_1) d\lambda_1
\end{aligned}
$$

6.2. RANDOM-PROBLEM ANALYSIS I

$$\begin{aligned} &= 2\int_{n^4-n^2/\sqrt{d}}^{\infty} f(\lambda_1)d\lambda_1 \\ &= O(n^{-2}). \end{aligned}$$

Thus, the probability, P_1, that either $0 < x_1 \leq p(1/n)$ or $x_1 \geq p(n)$ satisfies

$$P_1 := P_1^- + P_1^+ = P_1^- + P_1'^+ + P_1''^+ \leq O(n^{-2}).$$

The same result holds for P_i, $i = 2, ..., m$, the probability that either $0 < x_i \leq p(1/n)$ or $x_i \geq p(n)$. Thus, we shall have

$$\sum_{i=1}^{m} P_i \leq O(n^{-1}),$$

which approaches zero as $n \to \infty$. Using Lemma 6.1 we prove Proposition 6.10.

□

In a similar manner, we can prove that $\Pr(n^{-2} \leq |\lambda_0| \leq n^2 | x_1 \geq 0, ..., x_m \geq 0)$ approaches 1 as $n \to \infty$. This leads to the final proof of Proposition 6.9 for $d = 7$.

In the next section, we prove that the conditions in Theorem 6.6 are satisfied with high probability for some random LP problems.

6.2.2 Random linear problems

Let $A \in \mathcal{R}^{m \times n}$ of the homogeneous linear system (6.5) be standard Gaussian: Each component of A is independently distributed as a standard Gauss random variable.

Since each column, a_j, of A is also standard Gaussian, $a_j/\|a_j\|$ is uniformly distributed on the unit sphere in $\mathcal{R}^m$. This is a special case of a rotationally-symmetric distribution. Denote by $d = n - m$. It is shown that the bases of the null space of A are standard Gaussian random vectors.

Corollary 6.11 *The probability that system (6.5) has a feasible point is*

$$p_{nd} = \frac{\binom{n-1}{0} + ... + \binom{n-1}{d-1}}{2^{n-1}}.$$

Corollary 6.12 *The probability that system (6.5) has an interior feasible point is*

$$p_{nd} = \frac{\binom{n-1}{0} + ... + \binom{n-1}{d-1}}{2^{n-1}}.$$

Proof. From the strictly complementary property of linear systems, this probability is the probability that the dual

$$s = A^T y \geq 0, \qquad s \neq 0 \qquad (6.11)$$

is infeasible. However, the latter probability is exact

$$1 - p_{nm} = p_{nd}.$$

□

Theorem 6.13 *With probability one, exactly one of systems (6.5) and (6.11) is empty and the other has an interior feasible point.*

Proof. The probability that (6.11) has an interior feasible solution is p_{nm}. Note that these two events are exclusive, and $p_{nd} + p_{nm} = 1$.

□

We now prove another lemma.

Lemma 6.14 *System (6.5) is feasible if and only if there is a partition $A = (A_B, A_N)$, where A_B is $m \times m$, such that*

$$A_B x_B + (A_N e) x_{m+1} = 0, \quad x \neq 0 \quad \text{and} \quad x \geq 0 \qquad (6.12)$$

is feasible.

Proof. It is obvious that system (6.12) being feasible implies that (6.5) is feasible. Conversely, if (6.5) is feasible, then it implies that 0 belongs to the polytope P defined as the convex hull of the columns of A. Let $(a_1, a_2, ..., a_d)$ be a minimal affinely independent set of columns of A such that 0 belongs to the convex hull of $(a_1, a_2, ..., a_d)$. By Carathéodory's theorem $d \leq m+1$, if $d < m+1$, take as columns of A_B as $(a_1, a_2, ..., a_d)$ plus $m-d$ any other columns of A. If $d = m+1$, then there is an $(m+1)$-vector $\bar{u}$ such that

$$(a_1, a_2, ..., a_{m+1})\bar{u} = 0 \quad \text{and} \quad \bar{u} > 0.$$

Let b be the sum of the rest of the columns in A, and let $(\bar{v}, 1)$ be a vector in the null space of the matrix $(a_1, a_2, ..., a_{m+1}, b)$ (since b can be expressed as a linear combination of $(a_1, a_2, ..., a_{m+1})$). Then, for scalars α and η,

$$x(\alpha, \beta) = \alpha \begin{pmatrix} \bar{u} \\ 0 \end{pmatrix} + \beta \begin{pmatrix} \bar{v} \\ 1 \end{pmatrix}$$

6.2. RANDOM-PROBLEM ANALYSIS I

is also in the null space of $(a_1, a_2, ..., a_{m+1}, b)$. Let

$$k = \arg\min\left\{\frac{\bar{v}_j}{\bar{u}_j} : j = 1, 2, ..., (m+1)\right\}.$$

If $\bar{v}_k = 1$, then select $\alpha^* = 0$ and $\beta^* = 1$; if $\bar{v}_k < 1$, then select $\alpha^* = 1$ and $\beta^* = \bar{u}_k/(1 - \bar{v}_k)$; else select $\alpha^* = -1$ and $\beta^* = \bar{u}_k/(\bar{v}_k - 1)$. In all of the three cases, $x^* = x(\alpha^*, \beta^*) > 0$ and $x_k^* = x_{m+2}^*$. In other words, the lemma is true by selecting $A_B = (a_1, ..., a_{k-1}, a_{k+1}, ..., a_{m+1})$.

□

Let us call the partition satisfying the condition in Lemma 6.14 a *basic feasible partition*. We now analyze a feasible solution of (6.5) or (6.11). We develop the following result:

Theorem 6.15 *With high probability, a feasible point of system (6.5) or (6.11) satisfies condition (6.9) or (6.10), respectively.*

Proof. Let (A_B, A_N) be any partition of A and $a = A_N e/\sqrt{n-m}$. Consider the system

$$(A_B, a)x = 0, \quad x \neq 0 \quad \text{and} \quad x \geq 0. \tag{6.13}$$

Since (A_B, a) is standard Gaussian, the vector $\hat{x}$ in the null space of (A_B, a) is the line generated by a standard Gauss random vector $(\lambda_1, \lambda_2, ..., \lambda_{m+1})$, that is,

$$\hat{x}_i = \delta\lambda_i \quad \text{for} \quad i = 1, 2, ..., m+1,$$

where δ is a scalar. Without loss of generality, we can let $\delta > 0$. Hence, (A_B, A_N) is a *basic feasible* partition or $\hat{x}$ is feasible for system (6.13) if and only if $\hat{x}_i = \lambda_i \geq 0$ for $i = 1, 2, ..., m+1$. Thus, each component of a feasible solution of (6.13) has the identical distribution $|N(0,1)|$. Thus, due to Proposition 6.7, with high probability

$$p(1/n) \leq \hat{x}_i \leq p(n) \quad \text{for} \quad i = 1, 2, ..., m+1.$$

Note that $\hat{x}$ induces a feasible solution for system (6.5) by assigning

$$x_B = (\hat{x}_1, ..., \hat{x}_m)^T \quad \text{and} \quad x_N = (\hat{x}_{m+1}/\sqrt{n-m})e.$$

This completes the proof for (6.5).

The same result applies to the dual slack vector s of system (6.11) when it is feasible, where m is replaced by $n - m$.

□

Based on Theorems 6.13 and 6.15, we have the final result

Theorem 6.16 *With high probability the homogeneous random linear feasibility problem (6.5) can be solved in $O(\sqrt{n}\log n)$ iterations.*

Proof. From Theorem 6.13, with probability one either $A = A_P^*$ or $z^* > 0$ for problem (6.6) associated with (6.5). Then from Theorem 6.15, with high probability, there exists positive primal variables or positive dual slacks (not both) satisfying condition (6.9) or (6.10). Thus, the theorem follows from Theorem 6.6.

□

Note that the non-homogeneous linear system

$$Ax = b, \quad x \geq 0, \qquad (6.14)$$

where $(A, -b)$ is standard Gaussian, can be solved by solving the system (6.5) with $A := (A, -b)$, which remains standard Gaussian. Note that system (6.14) is feasible if and only if $b \in P^*$ where (P^*, Z^*) is the optimal partition of problem (6.6). Thus,

Corollary 6.17 *With high probability the random linear feasibility problem (6.14) can be solved in $O(\sqrt{n}\log n)$ iterations.*

6.3 Random-Problem Analysis II

This section analyzes the average complexity of interior-point algorithms for solving the probabilistic LP model of Todd. This model allows for degenerate optimal solutions, and does not provide a feasible starting point. We refer to this model as "Todd's degenerate model." The lack of an initial solution in the degenerate model is problematic for many interior point algorithms, which require an interior solution to start. We obtain a bound of $O(\sqrt{n}\log n)$ iterations for the expected number of iterations before termination with an exact optimal solution, using the homogeneous and self–dual algorithm of Chapter 5 as applied to this model.

Denote by $\mathcal{F}_h$ the set of all points that are feasible for (HSDP). Denote by $\mathcal{F}_h^0$ the set of strictly feasible (interior) points in $\mathcal{F}_h$, with $(x, \tau, s, \kappa) > 0$. It is easily seen that (HSDP) has a strictly feasible point: $y = 0$, $x = e > 0$, $\tau = 1$, $\theta = 1$, $s = e > 0$, $\kappa = 1$.

From Theorem 5.7, it is clear that the key to solving a LP problem, or alternatively detecting its infeasibility or unboundedness, is to find a

6.3. RANDOM-PROBLEM ANALYSIS II

strictly self-complementary solution to (HSDP). Many interior point algorithms might be used to solve (HSDP), as long as they generate a sequence or subsequence of feasible pairs which converge to a strictly complementary solution of the problem being solved, such as the predictor-corrector or wide-neighborhood algorithm described in Chapter 4. By using the analysis employed in Section 5.3, with $\eta = 1/4$, we generate a sequence $(y^k, x^k, \tau^k, \theta^k, s^k, \kappa^k) \in \mathcal{N}(\eta)$, and

$$\frac{\theta^{k+1}}{\theta^k} = \frac{\mu^{k+1}}{\mu^k} \le \frac{2}{1 + \sqrt{1 + 4\sqrt{2}(n+1)}}. \qquad (6.15)$$

6.3.1 Termination scheme

In this section we consider the problem of generating an exact optimal solution to (HSDP). For simplicity, we denote

$$u = \begin{pmatrix} x \\ \tau \end{pmatrix} \in \mathcal{R}^{n+1}, \quad v = \begin{pmatrix} s \\ \kappa \end{pmatrix} \in \mathcal{R}^{n+1}.$$

To begin, let $(u^*, y^*, v^*, \theta^* = 0)$ be any strictly self-complementary solution, i.e., $u^* + v^* > 0$. Note that

$$e^T u^* + e^T v^* = e^T x^* + \tau^* + e^T s^* + \kappa^* = n + 1.$$

Define

$$\sigma_h^* = \{i : 0 \le i \le n+1, \ u_i^* > 0\}, \quad \text{and} \quad \xi_h^* = \min_i(u_i^* + v_i^*).$$

We refer to σ_h^* as the self-complementary partition of (HSDP), and clearly $0 < \xi_h^* \le 1$. Our goal here is to use the iterates (u^k, v^k) of the algorithm to eventually identify the complementary partition, and to generate an exact optimal solution of (HSDP). Using the techniques developed in Chapter 5, we can prove the following result:

Lemma 6.18 *Let $\zeta = (1-\eta)\xi_h^*/(n+1)$. Then in order to obtain $v_j^k < \zeta \le u_j^k$, $j \in \sigma_h^*$, and $u_j^k < \zeta \le v_j^k$, $j \notin \sigma_h^*$, it suffices to have*

$$\theta^k < \frac{1-\eta}{(n+1)^2}(\xi_h^*)^2. \qquad (6.16)$$

Given an iterate $(x^k, \tau^k, y^k, s^k, \kappa^k)$, let A_P denote the columns of A having $x_j^k > s_j^k$, and let x_P denote the corresponding components of x.

Similarly let A_Z and s_Z denote the remaining columns of A, and the corresponding components of s. Note that (6.15) implies that $\theta^k \to 0$, so Lemma 6.18 implies that eventually we always have $P = \sigma_h^* \setminus \{n+1\}$. In what follows, we assume that k is in fact large enough so that $P = \sigma_h^* \setminus \{n+1\}$. We employ the following projections to generate an exact optimal solution. We distinguish two cases:

Case 1. ($\tau^k > \kappa^k$). Find the solution $\hat{x}_P^k$ of

$$(PP1) \quad \min \quad \|x_P - x_P^k\|$$
$$\text{s.t.} \quad A_P x_P = b\tau^k.$$

If $\hat{x}_P^k \geq 0$, then compute the solution $\hat{y}^k$ of

$$(DP1) \quad \min \quad \|s_Z - s_Z^k\|$$
$$\text{s.t.} \quad A_P^T y = c_P \tau^k, \quad c_Z \tau^k - A_Z^T y = s_Z,$$

and set

$$\hat{s}_Z^k = c_Z \tau^k - A_Z^T \hat{y}^k = s_Z^k - A_Z^T (\hat{y}^k - y^k).$$

Case 2. ($\tau^k \leq \kappa^k$). Find the solution $\hat{x}_P^k$ of

$$(PP2) \quad \min \quad \|x_P - x_P^k\|$$
$$\text{s.t.} \quad A_P x_P = 0.$$

If $\hat{x}_P^k \geq 0$, then compute the solution $\hat{y}^k$ of

$$(DP2) \quad \min \quad \|s_Z - S_Z^k\|$$
$$\text{s.t.} \quad A_P^T y = 0, \quad -A_Z^T y = s_Z,$$

and set $\hat{s}_Z^k = -A_Z^T \hat{y}^k$, and $\hat{\kappa}^k = b^T \hat{y}^k - c_P^T \hat{x}_P^k$.

According to Lemma 6.18, exactly one of the above cases occurs for all sufficiently large k. Also, Case 1 eventually occurs exactly when (LP) has an optimal solution, in which case (PP1) and (DP1) are both clearly feasible. In what follows, we consider Case 1 only, since our random model always has a solution.

It is easily seen from the definition of (HSDP) that:
(PP1) is equivalent to

$$\min \quad \|x_P - x_P^k\|$$
$$\text{s.t.} \quad A_P(x_P - x_P^k) = A_Z x_Z^k + \bar{b}\theta^k; \qquad (6.17)$$

6.3. RANDOM-PROBLEM ANALYSIS II

(DP1) is equivalent to

$$\begin{array}{ll} \min & \|\bar{c}_Z \theta^k - A_Z^T(y - y^k)\| \\ \text{s.t.} & A_P^T(y - y^k) = \bar{c}_P \theta^k + s_P^k, \end{array} \quad (6.18)$$

because

$$s_Z^k - s_Z^k = A_Z^T(y^k - y^k) + \bar{c}_Z \theta^k. \quad (6.19)$$

From (5.4) and $\theta^k \to 0$, we conclude that $(x_Z^k, s_P^k) \to 0$ as $k \to \infty$, and also $\kappa^k \to 0$ if $\{n+1\} \in \sigma_h^*$. Using these facts and (6.17)–(6.18) we can easily deduce that

$$(\hat{x}_P^k - x_P^k) \to 0 \text{ and } (\hat{s}_Z^k - s_Z^k) \to 0 \quad \text{as} \quad k \to \infty.$$

From this relation and Lemma 6.18 it follows that $(\hat{x}_P^k, \tau^k > 0, \hat{s}_Z^k) > 0$ (if k is large enough) is a strictly complementary solution to (HSDP).

The above discussion shows that our projection scheme works provided k is large enough. Below we give a more precise characterization of this fact. Again, for our probabilistic model, to be described below, only Case 1 occurs provided k is large enough. Therefore in what follows we assume that k is large enough and that we are always in Case 1.

A matrix A_P satisfies the Haar condition if every square submatrix of A_P is invertible. It is well known that the standard Gaussian matrix A_P is a Haar matrix with probability one. Thus, for the purposes of studying probabilistic behavior, we only have to deal with matrices that satisfy the Haar condition. Let $\hat{A}_B$ denote any square submatrix of A_P with its full row or column dimension. Also, if A_P has more rows than columns, let $\hat{A}$, $\hat{A}_Z$, and $\hat{b}$ denote the $\hat{A}_B$-corresponding rows of A, A_Z, and b, respectively; Otherwise, $\hat{A} = A$, $\hat{A}_Z = A_Z$, and $\hat{b} = b$. Then we have

Lemma 6.19 Let $(x_P^* > 0, \tau^* > 0, y^*, s_Z^* > 0)$ be any strictly (self) complementary solution for (HSDP). Then Case 1 occurs and (PP1) generates $\hat{x}_P^k > 0$ and $\hat{s}_Z^k > 0$ whenever

$$\theta^k \leq \frac{(1 - \eta)(\xi_h^*)^2}{(n+1)^2(1 + \sqrt{n}\|\hat{A}_B^{-1}\hat{A}_Z\|)} \; . \quad (6.20)$$

Proof. Assume that (6.20) holds. Since (6.20) implies (6.16), we have $\tau^k > \kappa^k$ and P must be the self-complementary partition $\sigma_h^* \setminus \{n+1\}$, by Lemma 6.18. From (6.17), the constraint in (PP1) is clearly equivalent to

$$A_P(x_P - x_P^k) = A_Z x_Z^k + (b - Ae)\theta^k,$$

and it is consistent. Note that $b = A_P x_P^*$. We have

$$A_P(x_P - x_P^k + e\theta^k - x_P^*\theta^k) = A_Z(x_Z^k - e\theta^k). \quad (6.21)$$

One solution to (6.21) is
$$x_P - x_P^k + e\theta^k - x_P^*\theta^k = \hat{A}_B^{-1}\hat{A}_Z(x_Z^k - e\theta^k)$$
if A_P has more rows than columns, or
$$x_P - x_P^k + e\theta^k - x_P^*\theta^k = \begin{pmatrix} \hat{A}_B^{-1}\hat{A}_Z(x_Z^k - e\theta^k) \\ 0 \end{pmatrix}$$
otherwise. Thus, the solution $\hat{x}_P^k$ of (PP1) must satisfy
$$\left\|\hat{x}_P^k - x_P^k\right\| \le \left\|\hat{A}_B^{-1}\hat{A}_Z(x_Z^k - e\theta^k)\right\| + \theta^k \left\|e - x_P^*\right\|. \quad (6.22)$$
For the first term in (6.22), we have
$$\begin{aligned}
\left\|\hat{A}_B^{-1}\hat{A}_Z(x_Z^k - e\theta^k)\right\| &\le \left\|\hat{A}_B^{-1}\hat{A}_Z\right\| \left\|x_Z^k - e\theta^k\right\| \\
&\le \max\{\max(x_Z^k), \theta^k\} \|e\| \left\|\hat{A}_B^{-1}\hat{A}_Z\right\| \\
&\le \frac{(n+1)\theta^k}{\xi_h^*} \sqrt{|Z|} \left\|\hat{A}_B^{-1}\hat{A}_Z\right\|,
\end{aligned}$$
where the last inequality is from (5.5). For the second term of (6.22), we have
$$\begin{aligned}
\theta^k \|e - x_P^*\| &\le \theta^k \sqrt{|P| - 2e^T x_P^* + \|x_P^*\|^2} \\
&\le \theta^k \sqrt{|P| - 2e^T x_P^* + (e^T x_P^*)^2} \\
&\le \theta^k (n+1),
\end{aligned}$$
since $e^T x_P^* \le n + 1$. Substituting the above two inequalities into (6.22) results in
$$\begin{aligned}
\left\|\hat{x}_P^k - x_P^k\right\| &\le \frac{(n+1)\theta^k}{\xi_h^*} \sqrt{|Z|} \left\|\hat{A}_B^{-1}\hat{A}_Z\right\| + (n+1)\theta^k \\
&\le (1 + \sqrt{n} \left\|\hat{A}_B^{-1}\hat{A}_Z\right\|)(n+1)\frac{\theta^k}{\xi_h^*}, \quad (6.23)
\end{aligned}$$
so (6.20), (6.23), and (5.7) imply that $\hat{x}_P^k > 0$.

Now consider (DP1). From (6.18), the constraint in (DP1) is clearly equivalent to
$$A_P^T(y - y^k) = (c_P - e)\theta^k + s_P^k;$$

6.3. RANDOM-PROBLEM ANALYSIS II

and it is consistent. Note that $c_P = P^T y^*$. We have

$$A_P^T(y - y^k - y^*\theta^k) = s_P^k - e\theta^k. \tag{6.24}$$

Similarly, the solution $\hat{y}_P^k$ of (DP1) must satisfy

$$\left\|\hat{s}_Z^k - s_Z^k\right\| \leq \left\|\hat{A}_Z^T(\hat{A}_B^T)^{-1}(s_P^k - e\theta^k)\right\| + \theta^k \left\|e - s_Z^*\right\|. \tag{6.25}$$

For the first term in (6.25), we have

$$\begin{aligned}
\left\|\hat{A}_Z^T(\hat{A}_B^T)^{-1}(s_P^k - e\theta^k)\right\| &\leq \left\|\hat{A}_B^{-1}\hat{A}_Z\right\| \left\|s_P^k - e\theta^k\right\| \\
&\leq \max\{\max(s_P^k), \theta^k\} \|e\| \left\|\hat{A}_B^{-1}\hat{A}_Z\right\| \\
&\leq \frac{(n+1)\theta^k}{\xi_h^*} \sqrt{|P|} \left\|\hat{A}_B^{-1}\hat{A}_Z\right\|,
\end{aligned}$$

where the last inequality is from (5.5). For the second term of (6.25), we have

$$\begin{aligned}
\theta^k \|e - s_Z^*\| &\leq \theta^k \sqrt{|Z| - 2e^T x_P^* + \|s_Z^*\|^2} \\
&\leq \theta^k \sqrt{|Z| - 2e^T x_P^* + (e^T s_Z^*)^2} \\
&\leq \theta^k (n+1).
\end{aligned}$$

Substituting the above two inequalities into (6.25) results in

$$\begin{aligned}
\left\|\hat{s}_Z^k - s_Z^k\right\| &\leq \frac{(n+1)\theta^k}{\xi_h^*} \sqrt{|P|} \left\|\hat{A}_B^{-1}\hat{A}_Z\right\| + (n+1)\theta^k \\
&\leq (1 + \sqrt{n} \left\|\hat{A}_B^{-1}\hat{A}_Z\right\|)(n+1)\frac{\theta^k}{\xi_h^*},
\end{aligned} \tag{6.26}$$

so (6.20), (6.26), and (5.7) imply that $\hat{s}_Z^k > 0$.

□

6.3.2 Random model and analysis

In this section we describe a random LP model proposed by Todd ([410]), and perform a probabilistic analysis of the behavior of the homogeneous and self-dual algorithm, using the finite termination scheme described above. We will refer to the model under consideration as "Todd's degenerate model."

Todd's Degenerate Model. Let $A = (A_1, A_2)$, where A_i is $m \times n_i$, $n_i \geq 1$, $n_1 + n_2 = n$, and each component of A is i.i.d. from the $N(0,1)$ distribution. Let

$$\hat{x} = \begin{pmatrix} \hat{x}_1 \\ 0 \end{pmatrix}, \quad \hat{s} = \begin{pmatrix} 0 \\ \hat{s}_2 \end{pmatrix},$$

where the components of $\hat{x}_1 \in \mathcal{R}^{n_1}$ and $\hat{s}_2 \in \mathcal{R}^{n_2}$ are i.i.d. from the $|N(0,1)|$ distribution. Finally let $b = A\hat{x}, c = \hat{s} + A^T\hat{\pi}$. We assume that either $\hat{\pi} = 0$, or the components of $\hat{\pi}$ are i.i.d. from any distribution with $O(1)$ mean and variance.

Clearly this model has degenerate solutions if $n_1 \neq m$, and produces instances of (LP) having no easy feasible starting point. This presents an obstacle for most interior point methods, which require interior feasible points for initialization. Since an instance of Todd's degenerate model always has an optimal solution, it follows from Theorem 5.7 that $n+1 \in \sigma_h^*$. Therefore, if the homogeneous and self–dual algorithm described in Chapter 5 is applied to an instance of Todd's degenerate model, we are eventually always in Case 1.

Now, we begin a probabilistic analysis of the self–dual algorithm equipped with the termination scheme described in the preceding section. Since our finite termination criterion in Lemma 6.19 depends on ξ_h^*, from a strictly complementary solution (x^*, τ^*, s^*) to (HSDP), we must first infer a valid value of ξ_h^* from the given strictly complementary solution $(\hat{x}, \hat{s})$ for (LP) and (LD).

Let

$$\hat{\xi} = \min(\hat{x} + \hat{s}) = \min\begin{pmatrix} \hat{x}_1 \\ \hat{s}_2 \end{pmatrix}, \quad \hat{\rho} = 1 + e^T\hat{x} + e^T\hat{s}. \tag{6.27}$$

Note that $x^* = (n+1)\hat{x}/\hat{\rho}$, $\tau^* = (n+1)/\hat{\rho}$, $y^* = (n+1)\hat{\pi}/\hat{\rho}$, $\kappa^* = 0$, and $s^* = (n+1)\hat{s}/\hat{\rho}$ is a strictly self-complementary solution to (HSDP). Thus, we have the following proposition:

Proposition 6.20 *Consider Todd's degenerate model with optimal solution $(\hat{x}, \hat{s})$. Then there is a strictly self-complementary solution $(x^*, \tau^*, y^*, s^*, \kappa^*)$ to (HSDP) such that $\xi_h^* \geq \hat{\xi}/\hat{\rho}$.*

This proposition and Lemma 6.19 lead to Lemma 6.21.

Lemma 6.21 *Consider an instance of Todd's degenerate model, and let $\hat{\xi}$ and $\hat{\rho}$ be as in (6.27). Suppose that k is large enough so that the following inequality is satisfied:*

$$\theta^k \leq \frac{(1-\eta)\hat{\xi}^2}{(n+1)^2\hat{\rho}^2(1+\sqrt{n}\|\hat{A}_B^{-1}\hat{A}_Z\|)}, \tag{6.28}$$

6.3. RANDOM-PROBLEM ANALYSIS II

where
$$A_P = A_1 \quad \text{and} \quad A_Z = A_2.$$

Then (PP1) and (DP1) generate solutions $\hat{x}_P^k > 0$ and $\hat{y}^k, \hat{s}_Z^k > 0$, so that $\hat{x} = (\hat{x}_P, 0)$ and $\hat{y}, \hat{s} = (0, \hat{s}_Z)$ solve (LP) and (LD), where $\hat{x}_P = \hat{x}_P^k/\tau^k$, $\hat{y} = \hat{y}^k/\tau^k$, and $\hat{s}_Z = \hat{s}_Z^k/\tau^k$.

Using the criterion in the previous lemma, we can terminate the algorithm once (6.28) holds. From $\theta^0 = 1$, (6.15), and (6.28), this definitely happens if

$$\theta^k \leq \left(1 - \frac{2}{1 + \sqrt{1 + 4\sqrt{2}(n+1)}}\right)^k \leq \frac{(1-\eta)\hat{\xi}^2}{(n+1)^2\hat{\rho}^2(1 + \sqrt{n}\|\hat{A}_B^{-1}\hat{A}_Z\|)},$$

which requires

$$k = O(\sqrt{n})\left(\log n + \log \hat{\rho} + \log(1 + \sqrt{n}\|\hat{A}_B^{-1}\hat{A}_Z\|) - \log \hat{\xi}\right).$$

We now introduce a lemma which is frequently used later and whose straightforward proof is omitted.

Lemma 6.22 *Let ζ and η be two continuous random variables, with sample space $(0, \infty)$. Define the new variables $\xi = \min(\zeta, \eta)$ and $\rho = \max(\zeta, \eta)$. Then, for any $x \geq 0$,*

$$f_\xi(x) \leq f_\zeta(x) + f_\eta(x) \quad \text{and} \quad f_\rho(x) \leq f_\zeta(x) + f_\eta(x),$$

where $f_\chi(\cdot)$ is the probability density function (p.d.f.) of a random variable χ.

Let λ have distribution $|N(0,1)|$ with the p.d.f.

$$f_\lambda(x) = \sqrt{2/\pi}\exp(-x^2/2).$$

Then,

$$\begin{aligned}
\mathrm{E}(\log \hat{\xi}) &= \int_0^{1/n} \log x f_{\hat{\xi}}(x)dx + \int_{1/n}^\infty \log x f_{\hat{\xi}}(x)dx \\
&\geq -\log n + \int_0^{1/n} \log x f_{\hat{\xi}}(x)dx \\
&\geq -\log n - \int_0^{1/n} |\log x| f_{\hat{\xi}}(x)dx.
\end{aligned}$$

Using Lemma 6.22, we have

$$\begin{aligned}
\int_0^{1/n} |\log x| f_{\hat{\xi}}(x) dx &\leq n \int_0^{1/n} |\log x| f_\lambda(x) dx \\
&\leq n\sqrt{2/\pi} \int_0^{1/n} |\log x| \exp(-x^2/2) dx \\
&\leq n\sqrt{2/\pi} \int_0^{1/n} |\log x| dx \\
&= -n\sqrt{2/\pi} \int_0^{1/n} \log x\, dx \\
&= n\sqrt{2/\pi}(1 + \log n)/n \\
&< 1 + \log n.
\end{aligned}$$

Also, we have

$$E(\log \hat{\rho}) = E(\log(1 + e^T \hat{x}_1 + e^T \hat{s}_2)) \leq \log(1 + E(e^T \hat{x}_1 + e^T s_2)) = O(\log n).$$

Moreover, consider

$$\begin{aligned}
&E\left(\log(1 + \sqrt{n}\|\hat{A}_B^{-1} \hat{A}_Z\|)\right) \\
&\leq E\left(\log\left(1 + \sqrt{n}\left(\sum_{j \in Z} \|\hat{A}_B^{-1} \hat{a}_j\|^2\right)^{1/2}\right)\right) \\
&\leq E\left(\log\left((1 + \sqrt{n})\left(|Z| + \sum_{j \in Z} \|\hat{A}_B^{-1} \hat{a}_j\|^2\right)^{1/2}\right)\right) \\
&= \log(1 + \sqrt{n}) + (1/2) E\left(\log \sum_{j \in Z}(1 + \|\hat{A}_B^{-1} \hat{a}_j\|^2)\right),
\end{aligned}$$

where $\hat{a}_j$ is the jth column of $\hat{A}$. Note that $(\hat{A}_B, -\hat{a}_j)$ is a Gaussian matrix, $\hat{A}_B^{-1} \hat{a}_j$ has the distribution of the Cauchy random variables λ_i/λ_0 where λ_i, $i = 0, 1, ..., |\hat{A}_B|$, are independent $N(0,1)$ random variables. Note that $|\hat{A}_B|$, the dimension of $\hat{A}_B$, is less than or equal to m. Without losing generality, we assume $|\hat{A}_B| = m$. Hence

$$1 + \|\hat{A}_B^{-1} \hat{a}_j\|^2 \sim \frac{\lambda_0^2 + \lambda_1^2 + ... + \lambda_m^2}{\lambda_0^2} \sim \frac{\eta_j^2}{\nu_j^2},$$

6.3. RANDOM-PROBLEM ANALYSIS II

where η_j^2 has a chi-square distribution with $m+1$ degrees of freedom, $\chi^2(m+1)$, and ν_j is a $|N(0,1)|$ random variable. Thus,

$$\begin{aligned}
\mathrm{E}\left[\log \sum_{j \in Z}(1+\|\hat{A}_B^{-1}\hat{a}_j\|^2)\right] &\leq \mathrm{E}\left[\log \frac{\max_{j \in Z}\{\eta_j^2\}}{\min_{j \in Z}\{\nu_j^2\}}\right] \\
&= \mathrm{E}\left[\log \max_{j \in Z}\{\eta_j^2\}\right] - \mathrm{E}\left[\log(\min_{j \in Z}\{\nu_j\})^2\right] \\
&:= \mathrm{E}\left[\log \hat{\eta}^2\right] - \mathrm{E}\left[\log(\hat{\nu})^2\right].
\end{aligned}$$

Using Lemma 6.22 again, we have

$$\begin{aligned}
\mathrm{E}[\log \hat{\eta}^2] &= \int_0^\infty \log x f_{\hat{\eta}^2}(x) dx \\
&\leq \log\left(\int_0^\infty x f_{\hat{\eta}^2}(x) dx\right) \\
&\leq \log\left(\int_0^\infty x |Z| f_{\eta_j^2}(x) dx\right) \\
&\leq \log\left(|Z| \int_0^\infty x f_{\eta^2}(x) dx\right) \\
&= \log(|Z|(m+1)),
\end{aligned}$$

where η^2 is a $\chi^2(m+1)$ random variable, whose expected value is $m+1$.
Finally,

$$\begin{aligned}
\mathrm{E}[\log(\hat{\nu})^2] &= 2\mathrm{E}[\log \hat{\nu}] \\
&= 2\int_0^\infty \log x f_{\hat{\nu}}(x) dx \\
&= 2\int_0^{1/n} \log x f_{\hat{\nu}}(x) dx + 2\int_{1/n}^\infty \log x f_{\hat{\nu}}(x) dx \\
&\geq -2\log n + 2\int_0^{1/n} \log x f_{\hat{\nu}}(x) dx \\
&\geq -2\log n - 2\int_0^{1/n} |\log x| f_{\hat{\nu}}(x) dx
\end{aligned}$$

and

$$\begin{aligned}
\int_0^{1/n} |\log x| f_{\hat{\nu}}(x) dx &\leq \int_0^{1/n} |\log x| |Z| f_\lambda(x) dx \\
&\leq |Z|\sqrt{2/\pi} \int_0^{1/n} |\log x| \exp(-x^2/2) dx
\end{aligned}$$

$$\begin{aligned}&\le\ |Z|\sqrt{2/\pi}\int_0^{1/n}|\log x|dx\\&\le\ |Z|\sqrt{2/\pi}(1+\log n)/n\\&<\ 1+\log n.\end{aligned}$$

Therefore, termination occurs on an iteration k, whose expected value is bounded as

$$\mathrm{E}[k]\ \le\ O(\sqrt{n}\log n).$$

Thus we have proved the main result:

Theorem 6.23 *Assume that the homogeneous and self–dual algorithm, using the termination scheme described in the preceding section, is applied to an instance of Todd's degenerate model. Then the expected number of iterations before termination with an exact optimal solution of (LP) is bounded above by* $O(\sqrt{n}\log n)$.

6.4 Notes

For examples in linear programming randomized algorithms, we cite the recent paper of Seidel [374], who gives a simple randomized algorithm whose expected running time for (LP) is $O(m!n)$, and the references therein.

For the expected running time of the simplex method, a deterministic algorithm, when applied to LP instances generated according to some probability distribution (or class of such distributions), see Adler, Karp and Shamir [2], Adler and Megiddo [3], Borgwardt [70], Megiddo [270], Smale [382], Todd [409], and the references cited there.

"One-step analysis" of a variant of Karmarkar's method can be seen in Nemirovsky [319]. Similar analysis of a primal-dual method can be seen in Gonzaga and Todd [166]. The analysis described in Section 6.1 is due to Mizuno et al. [292]. Let us describe a possible program to make one-step analysis rigorous. Suppose we assume that our original problem (LP) is generated probabilistically as follows: the entries of A are independent standard normal random variables, $b = Ae$ and $c = A^T y + e$ for some y. Then $(x, s) = (e, e)$ is an initial point on the central path $\mathcal{C}$. Moreover, for all of our algorithms, r is a multiple of e and U is a random subspace with the orthogonal transformation-invariant distribution. Hence our analysis holds at the initial iteration. We now apply an algorithm that requires that each iterate lies in $\mathcal{C}$ and hence $r = e$ at each iteration. However, the null space U of $AX^{1/2}S^{-1/2}$ will have a different induced distribution at later iterations. We could hope that before (x,s) gets too close to an optimal pair, this induced distribution is somewhat close to what we have assumed in Section 6.1.2, so that its Radon-Nikodym derivative with respect to our distribution is suitably bounded. In this case, the probability that $\|Pq\|$ exceeds $n^{1/2}\mu$, which is small under the distribution we have assumed, will also be small under the distribution induced by the initial probabilistic generation of (LP). Hence, for most iterations, the improvement in the duality gap would be as in Corollary 6.2. A great number of difficulties need to be resolved before such an approach could succeed. We would probably need bounds on how fast the probabilities in Theorem 6.3 approach 1, and clearly as (x,s) approaches the optimum the induced distribution differs drastically from what we have assumed.

In the meantime, we hope that the one-step nonrigorous analysis has lent some insight into the practical behavior of primal-dual algorithms. Our algorithms using $\mathcal{N} = \mathcal{N}_\infty^-(\eta)$ for η close to 1 are quite close to implemented primal-dual methods, and the result of our nonrigorous analysis, that $O((\log n)\log(1/\epsilon))$ iterations typically suffice, is borne out by several large-scale tests.

The properties of the Gaussian matrix in Section 6.2 and the rotationally-symmetric distribution LP model can be found in Borgwardt [70]. In particular, system (6.5) was discussed in Girko [140], Schmidt and Mattheiss [372], and Todd [410]. Our probabilistic analysis is essentially focused on the initialization and termination of interior-point algorithms. In other words, we have focused on the factor $\xi(A,b,c)$ and $\zeta(A,b,c)$ in the worst complexity result of Section 5.2. Essentially, we have proved that, for the above random problem, $\zeta(A,b,c) = \sqrt{n}$ with probability 1 and $\xi(A,b,c) \geq p(1/n)$ with high probability. Possible new topics for further research in this area include whether our analysis will hold for other probability distributions and the expected behavior.

Most of results in Section 6.3 are due to Anstreicher et al. [31, 30], where they proved that Theorem 6.23 holds for a more general degenerate model ([410]):

Todd's Degenerate Model. Let $A = (A_1, A_2, A_3)$, where A_i is $m \times n_i$, $n_i \geq 1$, $n_1 < m$, $n_1 + n_2 + n_3 = n$, and each component of A is i.i.d. from the $N(0,1)$ distribution. Let

$$\hat{x} = \begin{pmatrix} \hat{x}_1 \\ 0 \\ 0 \end{pmatrix}, \quad \hat{s} = \begin{pmatrix} 0 \\ 0 \\ \hat{s}_3 \end{pmatrix},$$

where the components of $\hat{x}_1$ and $\hat{s}_3$ are i.i.d. from the $|N(0,1)|$ distribution. Finally let $b = A\hat{x}, c = \hat{s} + A^T\hat{\pi}$. We assume that either $\hat{\pi} = 0$, or the components of $\hat{\pi}$ are i.i.d. from any distribution with $O(1)$ mean and variance.

6.5 Exercises

6.1 *Prove Proposition 6.7.*

6.2 *Prove Proposition 6.8.*

6.3 *Prove Corollary 6.11.*

6.4 *Let $\lambda_0, \lambda_1,..., \lambda_m$ be independent $N(0,1)$ random variables and condition on the event that $\lambda_i \leq |\lambda_0|/\sqrt{d}$ for $i = 1, \ldots, m$ ($d = n - m \geq 1$). Prove that $\Pr(n^{-2} \leq |\lambda_0| \leq n^2 | x_1 \geq 0, ..., x_m \geq 0)$ approaches 1 as $n \to \infty$, where $x_i := |\lambda_0|/\sqrt{d} - \lambda_i$ for $i = 1, \ldots, m$.*

6.5 *Prove Proposition 6.20.*

6.6 *Prove Lemma 6.22.*

Chapter 7
Asymptotic Analysis

Interior-point algorithms generate a sequence of ever-improving points $x^0, x^1, ..., x^k, ...$ approaching the solution set. For many optimization problems, the sequence never exactly reaches the solution set. One theory of iterative algorithms is referred to as local or asymptotic convergence analysis and is concerned with the rate at which the optimality error, $\{r^k\}$, of the generated sequence converges to zero. Obviously, if each iteration of competing algorithms requires the same amount of work, the speed of the convergence reflects the effectiveness of the algorithm. This convergence rate, although it holds locally or asymptotically, allows quantitative evaluation and comparison among different algorithms. It has been widely used in classical optimization and numerical analyses as an efficiency criterion. Generally, this criterion does explain the practical behavior of many iterative algorithms.

In this chapter we analyze the asymptotic convergence rate of iteration sequences generated by some interior-point algorithms. The asymptotic complexity presented in this chapter has several surprising but pleasing aspects. First, the theory is simple in nature. Second, it partially explains the excellent behavior of interior-point algorithms in practice. Third, it provides a tool to identify the strict complementarity partition for the termination method discussed in Chapter 5.

7.1 Rate of Convergence

The asymptotic convergence rate is a rich and yet elementary theory to predict the relative efficiency of a wide class of algorithms. It consists of two measures: the order and ratio of convergence.

7.1.1 Order of convergence

In Section 1.4.4 we have introduced, p, the order of convergence. To ensure that those definitions are applicable to any sequence, they are usually stated in terms of limit superior rather than just limit and $0/0$ is regarded as a finite number. In optimization, these technicalities are rarely necessary since $\{r^k \geq 0\}$ represents a measure towards optimality, and $r^k = 0$ implies that optimality is exactly reached.

We might say that the order of convergence is a measure of how good the tail of $\{r^k\}$ is in the worst case. Large values of p imply the faster convergence of the tail. The convergence of order equal two is called (sub) quadratic convergence. Indeed, if the sequence has order $p > 1$ and the limit

$$\lim_{k \to \infty} \frac{r^{k+1}}{(r^k)^p} = \beta < \infty$$

exists, then there exists a finite K, such that

$$\frac{r^{k+1}}{(r^k)^p} \leq 2\beta$$

or

$$(2\beta)^{1/(p-1)} r^{k+1} \leq [(2\beta)^{1/(p-1)} r^k]^p$$

and

$$(2\beta)^{1/(p-1)} r^K < 1$$

for all $k \geq K$. Thus, if we wish to reduce

$$(2\beta)^{1/(p-1)} r^k \leq \epsilon,$$

we need only

$$k - K = \frac{\log \log(1/\epsilon) + \log \log \left((2\beta)^{1/(p-1)} r^K\right)^{-1}}{\log p}$$

iterations, since

$$(2\beta)^{1/(p-1)} r^k \leq \left((2\beta)^{1/(p-1)} r^K\right)^{p^{k-K}}.$$

We also have the following proposition:

Proposition 7.1 *Let the positive sequence $\{r^k\}$ converge to zero. Then, the order of convergence equals*

$$\liminf_{k \to \infty} \frac{\log r^{k+1}}{\log r^k}.$$

7.1. RATE OF CONVERGENCE

Example 7.1 *The sequence with $r^k = (\alpha)^k$ where $0 < \alpha < 1$ converges to zero with order unity.*

Example 7.2 *The sequence with $r^k = (\alpha)^{2^k}$ where $0 < \alpha < 1$ converges to zero with order two, and the sequence is quadratically convergent.*

7.1.2 Linear convergence

Most of iterative algorithms have an order of convergence equal to unity, i.e., $p = 1$. It is therefore appropriate to consider this class in greater detail and develop another measure of speed for this class: the ratio of linear convergence, which was introduced in Section 1.4.4.

Linear convergence is the most important type of convergence behavior. A linearly convergence sequence, with convergence ratio β, can be said to have a tail that converges to zero at least as fast as the geometric sequence $M(\beta)^k$ for some fixed positive number M independent of k. Thus, we also call linear convergence geometric convergence.

As a rule, when comparing the relative effectiveness of two competing algorithms both of which produce linearly convergent sequences, the comparison is based on their corresponding convergence ratio—the smaller the ratio the faster the convergence. The ultimate case where $\beta = 0$ is referred to as superlinear convergence. We note immediately that convergence of any order greater than unity is superlinear. It is possible for superlinear convergence to have unity convergence order.

Example 7.3 *The sequence with $r^k = 1/k$ converges to zero. The convergence is of order one but it is not linear, since $\lim(r^{k+1}/r^k) = 1$, that is, β is not strictly less than one.*

Example 7.4 *The sequence with $r^k = (1/k)^k$ is of order unity, and it is superlinearly convergent.*

7.1.3 Average order

In practical optimization, the convergence order at each iteration may not be the same during the iterative process. We now define the average order related to the speed of convergence of such a sequence.

Definition 7.1 *Let the positive sequence $\{r^k\}$ converge to zero. The average order per iteration of convergence of $\{r^k\}$ between k and $k + K$ is defined as*

$$\bar{p} = \left(\prod_{i=1}^{K} p_i\right)^{1/K},$$

where p_i is the convergence order from $k+i-1$ to $k+i$.

In other words, the average convergence order during this period is the geometric mean of the orders of each iteration. Using the average order, from k to $k+K$ we should have

$$r^{k+K} = (r^k)^{\bar{p}^K} = (r^k)^{\prod_{i=1}^{K} p_i}.$$

The right-hand side is precisely the accumulated convergence orders between k and $K+k$.

Example 7.5 *The sequence with $r^0 = \alpha$, $0 < \alpha < 1$, $r^{k+1} = (r^k)^2$ if k is even and $r^{k+1} = r^k$ if k is odd. Then, the average converge order between k to $k+2$ is $\sqrt{2}$.*

7.1.4 Error function

In optimization, the decision variables form a vector in $\mathcal{R}^n$, and iterative algorithms generate a sequence $\{x^k\}$ in $\mathcal{R}^n$ space. Thus, if $\{x^k\}$ converges to the optimal solution set, the convergence properties of such a sequence are defined with respect to some particular error or residual function, $r(x)$, that converts the vector sequence into a real number sequence. Such an error function satisfies the property that $r(x) > 0$ for all non-optimal solutions and $r(x) = 0$ for every optimal solution. Hence, the convergence rate of $\{x^k\}$ is represented by the convergence rate of $\{r^k := r(x^k)\}$.

It is common to choose the error function by which to measure convergence as the same function that defines the objective function of the original optimization problem. This means that we measure convergence by how fast the objective converges to its optimum. Alternatively, we sometimes use the function $\min_{x \in \mathcal{S}} \|x^k - x\|$ which represents the distance from x^k to the solution set $\mathcal{S}$.

Example 7.6 *The primal, dual and primal-dual Newton procedures are presented in Chapter 3. They generate sequences of points $\{x^k\}$, $\{s^k\}$ and $\{x^k, s^k\}$, whose error functions are the proximity measures to the analytic center $\eta_p(x^k)$, $\eta_d(s^k)$ and $\eta(x^k, s^k)$. When the errors are less than 1, 1 and 2/3, respectively, $\eta_p(x^{k+1}) \leq \eta_p(x^k)^2$, $\eta_d(s^{k+1}) \leq \eta_d(s^k)^2$ and $\eta(x^{k+1}, s^{k+1}) \leq \eta(x^k, s^k)^2$. Thus, these errors converge to 0 quadratically.*

In the analysis of interior-point algorithms the error function is chosen as the primal-dual or complementarity gap $x^T s$, which should be zero at an optimal solution pair. For an optimization problem that possesses a strict complementarity solution, the above two error functions will have the same convergence rate.

7.2 Superlinear Convergence: LP

Consider the predictor-corrector algorithm described in Section 4.5. We will show that this $O(\sqrt{n}\log(R/\epsilon))$-iteration algorithm actually forces quadratic convergence of the duality gap $r^k := (x^k)^T s^k > 0$ to zero. In the context of the present work it is important to emphasize that the notions of convergence, superlinear convergence, or quadratic convergence of the duality gap sequence in no way require the convergence of the iteration sequence $\{(x^k, s^k)\}$.

We follow the same notations in Section 4.5. At the kth predictor step, let $\mu^k = (x^k)^T s^k / n$, $(d_x, d_s) := d(x^k, s^k, 0)$, and

$$\delta^k = \frac{D_x d_s}{\mu^k}. \tag{7.1}$$

If $\theta^k = \bar{\theta}$ is the largest step-size chosen in Algorithm 4.5, then from Lemma 4.17 (note $\delta^k = Pq/\mu^k$),

$$\begin{aligned}
1 - \theta^k &\leq 1 - \frac{2}{1 + \sqrt{1 + 4\|\delta^k\|/\eta}} \\
&= \frac{\sqrt{1 + 4\|\delta^k\|/\eta} - 1}{1 + \sqrt{1 + 4\|\delta^k\|/\eta}} \\
&= \frac{4\|\delta^k\|/\eta}{(1 + \sqrt{1 + 4\|\delta^k\|/\eta})^2} \\
&\leq \|\delta^k\|/\eta
\end{aligned} \tag{7.2}$$

(recall $\eta = 1/4$) and

$$(x^{k+1})^T s^{k+1} \leq (1 - \theta^k)(x^k)^T s^k \leq \frac{\|\delta^k\|}{\eta}(x^k)^T s^k. \tag{7.3}$$

Our goal is to prove that $\|\delta^k\| = O((x^k)^T s^k)$. Then, inequality (7.3) guarantees quadratic convergence of $(x^k)^T s^k$ to zero. (In this section, the big "O" notation represents a positive quantity that may depend on n and/or the original problem data, but which is independent of the iteration k.)

7.2.1 Technical results

We first introduce several technical lemmas. For simplicity, we drop the index k and recall the linear system during the predictor step

$$\begin{aligned}
Xd_s + Sd_x &= -Xs \\
Ad_x &= 0 \\
-A^T d_y - d_s &= 0.
\end{aligned} \tag{7.4}$$

Let $\mu = x^T s/n$ and $z = Xs$. Then from $(x,s) \in \mathcal{N}_2(\eta)$ we must have

$$(1-\eta)\mu \le z_j \le (1+\eta)\mu \quad \text{for} \quad j = 1, 2, ..., n. \tag{7.5}$$

We shall estimate $\|d_x\|$ and $\|d_s\|$. Our present objective is to demonstrate that $\|d_x\| = O(\mu)$ and $\|d_s\| = O(\mu)$. We start by characterizing the direction to more general systems, including LP (7.4) and monotone LCP (4.39).

Lemma 7.2 *If d_x and d_s satisfy the equation*

$$Xd_s + Sd_x = -Xs$$

and the inequality

$$(d_x)^T d_s \ge 0,$$

then,

$$\|D^{-1}d_x\|^2 + \|Dd_s\|^2 \le x^T s,$$

where $D = X^{1/2} S^{-1/2}$.

Proof. Multiplying the diagonal matrix $(XS)^{-1/2}$ on both sides of the equation, we have

$$D^{-1}d_x + Dd_s = -(XS)^{1/2} e.$$

Taking the norms of both sides,

$$\|D^{-1}d_x + Dd_s\|^2 = x^T s$$

or

$$\|D^{-1}d_x\|^2 + \|Dd_s\|^2 + (d_x)^T d_s = x^T s.$$

Since $d_x^T d_s \ge 0$, we have the lemma proved.

$\square$

Let $(P, Z) = (P^*, Z^*)$ be the strict complementarity partition of the LP problem. For all k, Theorem 2.18 and relation (5.8) imply that

$$\begin{aligned} \xi \le x_j^k \le 1/\xi \quad \text{for} \quad j \in P \\ \xi \le s_j^k \le 1/\xi \quad \text{for} \quad j \in Z, \end{aligned} \tag{7.6}$$

where $\xi < 1$ is a fixed positive quantity independent of k.

Lemma 7.3 *If d_x and d_s are obtained from the linear system (7.4) and $\mu = x^T s/n$, then*

$$\|(d_x)_Z\| = O(\mu) \quad \text{and} \quad \|(d_s)_P\| = O(\mu).$$

7.2. SUPERLINEAR CONVERGENCE: LP

Proof. From Lemma 7.2 and relation (7.6), we obtain

$$\begin{aligned}
\|(d_x)_Z\| &= \|D_Z D_Z^{-1}(d_x)_Z\| \\
&\le \|D_Z\| \|D_Z^{-1}(d_x)_Z\| \\
&\le \|D_Z\| O(\sqrt{\mu}) \\
&= \|(X_Z S_Z)^{1/2} S_Z^{-1}\| O(\sqrt{\mu}) \\
&= O(\sqrt{\mu}) O(\sqrt{\mu}) = O(\mu).
\end{aligned}$$

This proves that $\|(d_x)_Z\| = O(\mu)$. The proof that $\|(d_s)_P\| = O(\mu)$ is similar.

□

The proofs of $\|(d_x)_P\| = O(\mu)$ and $\|(d_s)_Z\| = O(\mu)$ are more involved. Toward this end, we first note

$$\begin{aligned}
x + d_x &\in \mathcal{R}(D^2 A^T), \\
s + d_s &\in \mathcal{N}(AD^2).
\end{aligned} \quad (7.7)$$

This is because from the first equation of (7.4) we have

$$\begin{aligned}
S(x + d_x) &= -X d_s \\
X(s + d_s) &= -S d_x.
\end{aligned}$$

Thus,

$$\begin{aligned}
x + d_x &= -(XS^{-1})d_s = D^2 A^T d_y \\
s + d_s &= -(SX^{-1})d_x = -D^{-2} d_x,
\end{aligned}$$

which gives relation (7.7).

Lemma 7.4 *If d_x and d_s are obtained from the linear system (7.4), then $(d_x)_P$ is the solution to the (weighted) least-squares problem*

$$\begin{aligned}
\min_u \quad & (1/2)\|D_P^{-1} u\|^2 \\
\text{s.t.} \quad & A_P u = -A_Z (d_x)_Z
\end{aligned}$$

and $(d_s)_Z = A_Z^T v$ and v is the solution to the (weighted) least-squares problem

$$\begin{aligned}
\min_v \quad & (1/2)\|D_Z v\|^2 \\
\text{s.t.} \quad & A_P^T v = -(d_s)_P.
\end{aligned}$$

Proof. From (7.7), we see that

$$x_P + (d_x)_P \in R(D_P^2 A_P^T). \tag{7.8}$$

Since $s_P^* = 0$ for all optimal s^*, we must have $c_P \in R(A_P^T)$. Thus,

$$s_P = c_P - A_P^T y \in R(A_P^T),$$

which implies that

$$x_P = D_P^2 s_P \in R(D_P^2 A_P^T). \tag{7.9}$$

From (7.8) and (7.9) we have

$$(d_x)_P \in R(D_P^2 A_P^T).$$

Moreover, $(d_x)_P$ satisfies the equation

$$A_P (d_x)_P = -A_Z (d_x)_Z.$$

Thus, $(d_x)_P$ satisfies the KKT conditions for the first least-squares problem. Since $AD^2(s + d_s) = -A d_x = 0$ and $AD^2 s = Ax = b$, it follows that

$$-b = AD^2 d_s = A_P D_P^2 (d_s)_P + A_Z D_Z^2 (d_s)_Z. \tag{7.10}$$

Also, since $x_Z^* = 0$ for all optimal x^*, we have $A_P x_P^* = b$ implying $b \in \mathcal{R}(A_P)$. Therefore, relation (7.10) implies

$$A_Z D_Z^2 (d_s)_Z \in \mathcal{R}(A_P).$$

Moreover, d_y satisfies the equation

$$A_P^T d_y = -(d_s)_P.$$

Thus, d_y satisfies the KKT conditions for the second least-squares problem. □

7.2.2 Quadratic convergence

Theorem 7.5 *If d_x and d_s are obtained from the linear system (7.4) and $\mu = x^T s/n$, then*

$$\|(d_x)_P\| = O(\mu) \quad \text{and} \quad \|(d_s)_Z\| = O(\mu).$$

7.2. SUPERLINEAR CONVERGENCE: LP

Proof. Since the first least-squares problem is always feasible, there must be a *feasible* $\bar{u}$ such that
$$\|\bar{u}\| = O(\|(d_x)_Z\|),$$
which, together with Lemma 7.3, implies
$$\|\bar{u}\| = O(\mu).$$
Furthermore, from Lemma 7.4 and relations (7.5) and (7.6)
$$\begin{aligned}
\|(d_x)_P\| &= \|D_P D_P^{-1}(d_x)_P\| \\
&\leq \|D_P\| \|D_P^{-1}(d_x)_P\| \\
&\leq \|D_P\| \|D_P^{-1} \bar{u}\| \\
&\leq \|D_P\| \|D_P^{-1}\| \|\bar{u}\| \\
&= \|(X_P S_P)^{-1/2} X_P\| \|(X_P S_P)^{1/2} X_P^{-1}\| \|\bar{u}\| \\
&\leq \|(X_P S_P)^{-1/2}\| \|X_P\| \|(X_P S_P)^{1/2}\| \|X_P^{-1}\| \|\bar{u}\| \\
&= O(\|\bar{u}\|) = O(\mu).
\end{aligned}$$
Similarly, we can prove the second statement of the theorem.
□

Theorem 7.5 indicates that at the kth predictor step, d_x^k and d_s^k satisfy
$$\|(d_x^k)_P\| = O(\mu^k) \quad \text{and} \quad \|(d_s^k)_Z\| = O(\mu^k), \qquad (7.11)$$
where $\mu^k = (x^k)^T s^k / n$. We are now in a position to state our main result.

Theorem 7.6 *Let $\{(x^k, s^k)\}$ be the sequence generated by Algorithm 4.5. Then,*

i) *the Algorithm has iteration complexity $O(\sqrt{n} \log(R/\epsilon))$;*

ii) $1 - \theta^k = O((x^k)^T s^k)$;

iii) $(x^k)^T s^k \to 0$ *quadratically.*

Proof. The proof of (i) is in Theorem 4.18, which also establishes
$$\lim_{k \to \infty} \mu^k = 0.$$
From relation (7.1), Lemma 7.3 and Theorem 7.5 we have
$$\|\delta^k\| = \|D_x d_s / \mu^k\| \leq O((x^k)^T s^k),$$
which, together with inequality (7.2), establishes (ii).

From inequality (7.3) we see that (ii) implies (iii). This proves the theorem.
□

7.3 Superlinear Convergence: Monotone LCP

In this section, we consider the monotone LCP extension of the predictor-corrector LP algorithm. We show that this $O(\sqrt{n}\log(R/\epsilon))$-iteration algorithm for the monotone LCP actually possesses quadratic convergence assuming

Assumption 7.1 *The monotone LCP possesses a strict complementarity solution.*

This assumption is restrictive since in general it does not hold for the monotone LCP. We will actually show by example, however, that Assumption 7.1 appears to be necessary in order to achieve superlinear convergence for the algorithm.

Again, the LCP being monotone means that the iterate direction

$$d_s = M d_x \quad \text{implies} \quad d_x^T d_s \geq 0.$$

Note that for LP, we have $d_x^T d_s = 0$. This is the only difference between LP and LCP analyses. Almost all technical results on iterate directions developed for LP ($d_x^T d_s = 0$) hold for the monotone LCP ($d_x^T d_s \geq 0$).

7.3.1 Predictor-corrector algorithm for LCP

In this section, we briefly describe the predictor-corrector LCP algorithm for solving the monotone LCP given by (1.4). Recall that $\overset{\circ}{\mathcal{F}}$ denote the collection of all strictly feasible points (x, s) and the neighborhood of the central path

$$\mathcal{N}_2(\eta) = \{(x, s) \in \overset{\circ}{\mathcal{F}}: \; \|Xs/\mu - e\| \leq \eta\},$$

where $\mu = x^T s/n$ and η is a constant between 0 and 1.

To begin with choose $0 < \eta \leq 1/4$ (a typical choice would be 1/4). All search directions d_x and d_s will be defined as the solutions of the following system of linear equations

$$\begin{array}{rcl} Xd_s + Sd_x & = & \gamma\mu e - Xs \\ Md_x - d_s & = & 0, \end{array} \qquad (7.12)$$

where $0 \leq \gamma \leq 1$. (This is a special case of system (4.39). There are no free variables here.) To show the dependence of $d = (d_x, d_s)$ on the pair (x, s) and parameter γ, we write $d = d(x, s, \gamma)$.

A typical iteration of the algorithm proceeds as follows. Given $(x^k, s^k) \in \mathcal{N}_2(\eta)$, we solve system (7.12) with $(x, s) = (x^k, s^k)$ and $(d_x, d_s) = d(x^k, s^k, 0)$. For some step length $\theta \geq 0$ let

$$x(\theta) = x^k + \theta d_x, \quad s(\theta) = s^k + \theta d_s,$$

7.3. SUPERLINEAR CONVERGENCE: MONOTONE LCP

and $\mu(\theta) = x(\theta)^T s(\theta)/n$. This is the predictor step.

Again, we can choose the *largest* step length $\bar{\theta} \leq 1$ such that for all $0 \leq \theta \leq \bar{\theta}$ $(x(\theta), s(\theta)) \in \mathcal{N}_2(\eta + \tau)$ where $0 < \tau \leq \eta$, and let

$$x' = x(\bar{\theta}) \quad \text{and} \quad s' = s(\bar{\theta}).$$

We can compute $\bar{\theta}$ by finding the roots of a quartic equation.

Next we solve system (7.12) with $(x, s) = (x', s') \in \mathcal{N}(\eta + \tau)$, $\mu' = (x')^T s'/n$, and $\gamma = 1$, i.e., $(d'_x, d'_s) = d(x', s', 1)$. Let $x^{k+1} = x' + d'_x$ and $s^{k+1} = s' + d'_s$. This is the corrector (or centering) step.

Similar to Lemma 4.16, for all k we can show that

$$(x^k, s^k) \in \mathcal{N}_2(\eta) \tag{7.13}$$

as long as $0 < \eta \leq 1/4$ and $0 < \tau \leq \eta$, and

$$\begin{aligned}(x')^T s' &= (1 - \theta^k)(x^k)^T s^k + (\theta^k)^2 (d_x)^T d_s \\ (x^{k+1})^T s^{k+1} &= (x')^T s' + (d'_x)^T d'_s.\end{aligned} \tag{7.14}$$

One can also show that

$$\begin{aligned}(d_x)^T d_s &\leq (x^k)^T s^k / 4 \\ (d'_x)^T d'_s &\leq (x')^T s'/(8n).\end{aligned} \tag{7.15}$$

Let $\delta^k = D_x d_s / \mu^k$ in the predictor step. Then, we can show that

$$\|\delta^k\| \leq \sqrt{2}n/4, \tag{7.16}$$

and the following lemma, which resembles Lemma 4.17.

Lemma 7.7 *If $\theta^k := \bar{\theta}$ is the largest θ such that $(x(\theta), s(\theta)) \in \mathcal{N}_2(\eta + \tau)$ with $0 < \eta \leq 1/4$ and $0 < \tau \leq \eta$, then*

$$\theta^k \geq \frac{2}{1 + \sqrt{1 + 4\|\delta^k\|/\tau}}.$$

Clearly, this lemma together with (7.14), (7.15) and (7.16) implies that the iteration complexity of the algorithm is $O(\sqrt{n}\log(R/\epsilon))$ for a constant $0 < \tau \leq \eta$. Note again that

$$1 - \theta^k \leq \frac{\|\delta^k\|}{\tau}. \tag{7.17}$$

Relations (7.14), (7.15), (7.16), and (7.17), and Lemma 7.7, imply

$$\begin{aligned}\mu^{k+1} &\leq (1 + 1/8n)(\frac{\|\delta^k\|}{\tau}\mu^k + (d_x)^T d_s / n) \\ &\leq (1 + 1/8n)(\frac{\|D_x d_s\|}{\tau} + \frac{\|D_x d_s\|}{\sqrt{n}}).\end{aligned} \tag{7.18}$$

From (7.18), we see that if

$$\|d_x\| = O(\mu^k) \quad \text{and} \quad \|d_s\| = O(\mu^k),$$

then the complementarity gap converges to zero quadratically.

7.3.2 Technical results

For a LCP possessing a strictly complementary solution, a unique partition P and Z, where $P \cap Z = \{1, 2, ..., n\}$ and $P \cup Z = \emptyset$, exists such that $x_Z^* = 0$ and $s_P^* = 0$ in every complementary solution and at least one complementarity solution has $x_P^* > 0$ and $s_Z^* > 0$. We can also prove that relation (7.6) holds for the sequence generated by the predictor-corrector MCP algorithm. Let $\mu = x^T s/n$ and $z = Xs$. We must also have relation (7.5) if $(x, s) \in \mathcal{N}_2(\eta)$.

We now introduce several technical lemmas. For simplicity, we drop the index k and recall the linear system during the predictor step

$$\begin{array}{rcl} Xd_s + Sd_x & = & -Xs \\ Md_x - d_s & = & 0. \end{array} \quad (7.19)$$

Define $D = X^{1/2}S^{-1/2}$. We now estimate $\|d_x\|$ and $\|d_s\|$. Since M is monotone, i.e., $(d_x)^T d_s \geq 0$, both Lemma 7.2 and the following lemma hold.

Lemma 7.8 *If d_x and d_s are obtained from the linear system (7.19), and $\mu = x^T s/n$, then*

$$\|(d_x)_Z\| = O(\mu) \quad \text{and} \quad \|(d_s)_P\| = O(\mu).$$

The proofs of $\|(d_x)_P\| = O(\mu)$ and $\|(d_s)_Z\| = O(\mu)$ are, again, more involved. We first note

$$\begin{array}{rcl} S(x + d_x) & = & -Xd_s, \\ X(s + d_s) & = & -Sd_x, \end{array}$$

and therefore

$$\begin{array}{rcl} x + d_x & = & -(XS^{-1})d_s = -D^2 d_s \\ s + d_s & = & -(X^{-1}S)d_x = -D^{-2} d_x. \end{array} \quad (7.20)$$

Before proceeding, we need some results regarding (non-symmetric) positive semi-definite (PSD) matrices that may be of independent interest. In what

7.3. SUPERLINEAR CONVERGENCE: MONOTONE LCP

follows, we will consider M to be partitioned (following a re-ordering of rows and columns) as

$$M = \begin{pmatrix} M_{PP} & M_{PZ} \\ M_{ZP} & M_{ZZ} \end{pmatrix}. \tag{7.21}$$

Lemma 7.9 *Let M be a PSD matrix, partitioned as in (7.21). Then $M_{PP}x_P = 0$ if and only if $M_{PP}^T x_P = 0$. Furthermore, $M_{PP}x_P = 0$ implies that $(M_{ZP} + M_{PZ}^T)x_P = 0$.*

Proof. Let $x = (x_P^T, 0^T)^T$. If either $M_{PP}x_P = 0$ or $M_{PP}^T x_P = 0$, then $x^T M x = 0$, so x is a global minimizer of the quadratic form $y^T M y$. Consequently $(M + M^T)x = 0$, which is exactly

$$\begin{aligned} (M_{PP} + M_{PP}^T)x_P &= 0 \\ (M_{ZP} + M_{PZ}^T)x_P &= 0. \end{aligned}$$

$\square$

Lemma 7.10 *Let M be a PSD matrix, partitioned as in (7.21). Then*

$$\mathcal{R}\begin{pmatrix} M_{PP} & M_{PZ} \\ 0 & I \end{pmatrix} = \mathcal{R}\begin{pmatrix} M_{PP}^T & M_{ZP}^T \\ 0 & -I \end{pmatrix}.$$

Proof. From the fundamental theorem of linear algebra, it is equivalent to prove that

$$\mathcal{N}\begin{pmatrix} M_{PP}^T & 0 \\ M_{PZ}^T & I \end{pmatrix} = \mathcal{N}\begin{pmatrix} M_{PP} & 0 \\ M_{ZP} & -I \end{pmatrix},$$

where $\mathcal{N}(\cdot)$ denotes the null space of a matrix. To begin, assume that

$$\begin{pmatrix} M_{PP}^T & 0 \\ M_{PZ}^T & I \end{pmatrix} \begin{pmatrix} x_P \\ x_Z \end{pmatrix} = 0. \tag{7.22}$$

From Lemma 7.9, $M_{PP}x_P = 0$. Also $x_Z = -M_{PZ}^T x_P$, so showing that $M_{ZP}x_P - x_Z = 0$ is equivalent to showing that $(M_{ZP} + M_{PZ}^T)x_P = 0$, which also holds by Lemma 7.9. Thus

$$\begin{pmatrix} M_{PP} & 0 \\ M_{ZP} & -I \end{pmatrix} \begin{pmatrix} x_P \\ x_Z \end{pmatrix} = 0. \tag{7.23}$$

The argument that (7.23) implies (7.22) is similar.

$\square$

7.3.3 Quadratic convergence

Now we can establish

Lemma 7.11 *If d_x and d_s are obtained from the linear system (7.19), and $\mu = x^T s/n$, then $u = (d_x)_P$ and $v = (d_s)_Z$ are the solutions to the (weighted) least-squares problem*

$$\begin{aligned} \min_{u,v} \quad & (1/2)\|D_P^{-1}u\|^2 + (1/2)\|D_Z^1 v\|^2 \\ \text{s.t.} \quad & M_{PP}u = -M_{PZ}(d_x)_Z + (d_s)_P \\ & M_{ZP}u - v = -M_{ZZ}(d_x)_Z. \end{aligned} \quad (7.24)$$

Proof. Note that from (7.19), $u = (d_x)_P$, $v = (d_s)_Z$ is certainly feasible in the problem (7.24). Next, from (7.19) and (7.20), we see that

$$\begin{aligned} x_P + (d_x)_P &= -D_P^2 M_B . d_x \\ s_Z + (d_s)_Z &= -D_Z^{-2}(d_x)_Z. \end{aligned} \quad (7.25)$$

Since $s_P^* = 0$ for all optimal s^*, with $x_Z^* = 0$, we must have $q_P = -M_{PP}x_P^* \in \mathcal{R}(M_{PP})$. Therefore,

$$D_P^{-2} x_P = s_P = M_B . x + q_P = M_{PP}(x_P - x_P^*) + M_{PZ}x_Z.$$

Substituting this into the first equation of (7.25) obtains

$$D_P^{-2}(d_x)_P = -M_{PP}(x_P - x_P^* + (d_x)_P) - M_{PZ}(x_Z + (d_x)_Z). \quad (7.26)$$

Also $s_Z = D_Z^{-2} x_Z$, which substituted into the second equation of (7.25) yields

$$D_Z^2 (d_s)_Z = -x_Z - (d_x)_Z. \quad (7.27)$$

Then (7.26) and (7.27) together imply that

$$\begin{pmatrix} D_P^{-2}(d_x)_P \\ D_Z^2 (d_s)_Z \end{pmatrix} \in \mathcal{R} \begin{pmatrix} M_{PP} & M_{PZ} \\ 0 & I \end{pmatrix}.$$

Applying Lemma 7.10, we conclude that

$$\begin{pmatrix} D_P^{-2}(d_x)_P \\ D_Z^2 (d_s)_Z \end{pmatrix} \in \mathcal{R} \begin{pmatrix} M_{PP}^T & M_{ZP}^T \\ 0 & -I \end{pmatrix},$$

which shows exactly that $u = (d_x)_P$, $v = (d_s)_Z$ satisfies the KKT conditions for optimality in the least-squares problem (7.24). □

7.3. SUPERLINEAR CONVERGENCE: MONOTONE LCP

Theorem 7.12 *If d_x and d_s are obtained from the linear system (7.19), and $\mu = x^T s/n$, then $\|d_x\| = O(\mu)$ and $\|d_s\| = O(\mu)$.*

Proof. Due to Lemma 7.8, we only need to prove

$$\|(d_x)_P\| = O(\mu) \quad \text{and} \quad \|(d_s)_Z\| = O(\mu).$$

Since the least-squares problem (7.24) is always feasible, there must be *feasible $\bar{u}$ and $\bar{v}$* such that

$$\|\bar{u}\| = O(\|(d_x)_Z\| + \|(d_s)_P\|) \quad \text{and} \quad \|\bar{v}\| = O(\|(d_x)_Z\| + \|(d_s)_P\|),$$

which, together with Lemma 7.8, implies $\|\bar{u}\| = O(\mu)$ and $\|\bar{v}\| = O(\mu)$. Furthermore, from Lemma 7.11 and relations (7.5) and (7.6),

$$\begin{aligned}
&\|(d_x)_P\|^2 + \|(d_s)_Z\|^2 \\
&= \|D_P D_P^{-1}(d_x)_P\|^2 + \|D_Z^{-1} D_Z(d_s)_Z\|^2 \\
&\leq \|D_P^2\| \, \|D_P^{-2}(d_x)_P\|^2 + \|D_Z^{-2}\| \, \|D_Z(d_s)_Z\|^2 \\
&= \|(X_P S_P)^{-1} X_P^2\| \, \|D_P^{-1}(d_x)_P\|^2 + + \|(X_Z S_Z)^{-1} S_Z^2\| \, \|D_Z(d_s)_Z\|^2 \\
&\leq \left(\|(X_P S_P)^{-1} X_P^2\| + \|(X_Z S_Z)^{-1} S_Z^2\|\right) \left(\|D_P^{-1}(d_x)_P\|^2 + \|D_Z(d_s)_Z\|^2\right) \\
&\leq \left(\|(X_P S_P)^{-1} X_P^2\| + \|(X_Z S_Z)^{-1} S_Z^2\|\right) \left(\|D_P^{-1}\bar{u}\|^2 + \|D_Z \bar{v}\|^2\right) \\
&\leq \left(\|(X_P S_P)^{-1} X_P^2\| + \|(X_Z S_Z)^{-1} S_Z^2\|\right) \left(\|D_P^{-2}\| \, \|\bar{u}\|^2 + \|D_Z^2\| \, \|\bar{v}\|^2\right) \\
&\leq O(1/\mu) \left(\|D_P^{-2}\| \, \|\bar{u}\|^2 + \|D_Z^2\| \, \|\bar{v}\|^2\right) \\
&= O(\mu) \left(\|D_P^{-2}\| + \|D_Z^2\|\right) \\
&= O(\mu) \left(\|(X_P S_P) X_P^{-2}\| + \|(X_Z S_Z) S_Z^{-2}\|\right) \\
&= O(\mu^2).
\end{aligned}$$

□

The above theorem leads to the result described in Theorem 7.6 for the predictor-corrector LCP algorithm. The following proposition concerns Assumption 7.1.

Proposition 7.13 *There is a monotone LCP problem, where a strict complementarity solution does not exist, for which the predictor-corrector algorithm or affine scaling algorithm possesses no superlinear convergence.*

Proof. Consider the simple monotone LCP with $n = 1$, $M = 1$ and $q = 0$. The unique complementarity solution is $s = x = 0$, which is not strictly complementary. Note that the feasible solution $s = x = \epsilon$ is a perfectly centered pair for any $\epsilon > 0$. The direction in the predictor step (or affine scaling algorithm) is

$$d_x = -x/2 \quad \text{and} \quad d_s = -s/2.$$

Thus, even taking the step size $\theta = 1$, the new solution will be $s = x = \epsilon/2$. Thus, the complementarity slackness sequence is reduced at most linearly, with constant $1/4$, which proves the proposition.

□

7.4 Quadratically Convergent Algorithms

The predictor-corrector algorithm described in previous sections needs to solve two systems of linear equations or two least-squares problems—one in the predictor step and one in the corrector step. If one counts each iteration as solving one system of linear equations, as is usually done in the classical analysis of interior-point algorithms, the average order of convergence of this algorithm is only $\sqrt{2}$. In this section we further show how to construct an algorithm for solving LP and monotone LCP whose order of convergence exactly equals 2. We also show that the solution sequence generated by the algorithm is a Cauchy, and therefore convergent, sequence.

7.4.1 Variant 1

An iteration of the variant proceeds as follows. Given $(x^k, s^k) \in \mathcal{N}_2(\eta)$, we perform $T (\geq 1)$ successive predictor steps followed by one corrector step, where in tth predictor step of these T steps we choose $\tau = \tau_t > 0$, where

$$\sum_{t=1}^{T} \tau_t = \eta. \tag{7.28}$$

In other words, at the tth predictor step of these T steps, we solve system (7.12) with $\mu' = (x')^T s'/n$ and $(x, s) = (x', s') \in \mathcal{N}(\eta + \tau_1 + ... + \tau_{t-1})$ (the initial $(x', s') = (x^k, s^k) \in \mathcal{N}(\eta)$) and $\gamma = 0$, i.e., $(d_x, d_s) = d(x', s', 0)$. For some $\theta > 0$ let

$$x(\theta) = x' + \theta d_x, \quad s(\theta) = s' + \theta d_s \quad \text{and} \quad \mu(\theta) = (x(\theta))^T s(\theta)/n.$$

Our specific choice, $\bar{\theta}$, for θ is similar as before: the largest θ such that

$$(x(\theta), s(\theta)) \in \mathcal{N}_2(\eta + \tau_1 + ... + \tau_{t-1} + \tau_t).$$

7.4. QUADRATICALLY CONVERGENT ALGORITHMS

From the first inequality in (7.14), the fact $\bar{\theta} \leq 1$, (7.17), and Theorem 7.12 we have

$$\mu(\bar{\theta}) \leq \frac{\|D_x d_s\|}{\tau_t} + \frac{\|D_x d_s\|}{\sqrt{n}} \leq \frac{R(\mu')^2}{\tau_t} \tag{7.29}$$

for some fixed positive quantity R independent of k and t. Now update $x' := x(\bar{\theta})$ and $s' := s(\bar{\theta})$.

After T predictor steps we have $(x', s') \in \mathcal{N}_2(2\eta)$. Now we perform one corrector step as before to generate

$$(x^{k+1}, s^{k+1}) \in \mathcal{N}_2(\eta).$$

Based on the previous lemmas and results, each predictor step within an iteration achieves quadratic convergence order for any positive constant sequence $\{\tau_t\}$ satisfying (7.28). For example, one natural choice would be $\tau_t = \eta/T$ for $t = 1, 2, ..., T$. Since each iteration solves $T+1$ systems of linear equations, the average order of the convergence of the complementary gap to zero in Variant 1 is $2^{T/(T+1)}$ per linear system solver for any constant $T \geq 1$.

Theorem 7.14 *Variant 1 generates a sequence $\{x^k, s^k\}$ such that the average convergence order is $2^{T/(T+1)}$ per linear system solver for any constant $T \geq 1$.*

7.4.2 Variant 2

Now we develop a new variant where we let $T = \infty$, that is, no corrector step is needed anymore in the rest of the iterations of the algorithm. The algorithm becomes the pure Newton method or the primal-dual affine scaling algorithm.

After $(x^K, s^K) \in \mathcal{N}(\eta)$ for some finite K, we perform only the predictor step, where we choose $\tau = \tau_t > 0$ satisfying (7.28). One natural choice will be

$$\tau_t = \eta(1/2)^t \quad \text{for} \quad t = 1, 2, ...$$

For simplicity, let us reset $K := 1$. Then, in the kth iteration we solve system (7.12) with

$$(x, s) = (x^k, s^k) \in \mathcal{N}(\eta + \sum_{t=1}^{k-1} \tau_t) \quad \left(\text{where } \sum_{t=1}^{0} \tau_t := 0\right)$$

and $\gamma = 0$, i.e., $(d_x, d_s) = d(x^k, s^k, 0)$. For some $\theta > 0$ let

$$x(\theta) = x^k + \theta d_x, \quad s(\theta) = s^k + \theta d_s. \tag{7.30}$$

Our specific choice for θ is $\bar{\theta}$, the largest θ such that $(x(\theta), s(\theta)) \in \mathcal{N}_2(\eta + \sum_{t=1}^{k} \tau_t)$. Now directly update

$$x^{k+1} := x(\bar{\theta}) \quad \text{and} \quad s^{k+1} := s(\bar{\theta}). \tag{7.31}$$

Theorem 7.15 *Let $(x^K)^T s^K$ be small enough. Then, Variant 2 generates a sequence $\{x^k, s^k\}$ with $k \geq K$ such that*

i) *the order of the convergence of the complementary gap to zero equals at least 2,*

ii) *$\{x^k, s^k\}$ is a Cauchy, and therefore convergent, sequence.*

Proof. At the kth iteration ($k \geq K := 1$) we have from (7.29)

$$(x^{k+1})^T s^{k+1} \leq \frac{R((x^k)^T s^k)^2}{\tau_k} = R((x^k)^T s^k)^2 2^k/\eta,$$

or

$$\log_2((x^{k+1})^T s^{k+1}) \leq 2\log_2((x^k)^T s^k) + \log_2(R/\eta) + k. \tag{7.32}$$

For $(x^K)^T s^K$ small enough, the inequality (7.32) implies that $\{\log_2((x^k)^T s^k)\}$ is a geometric sequence (with base close to 2, say, 1.5) tending to $-\infty$. Since k is just an arithmetic sequence and $\log_2(R/\eta)$ is fixed, we should have

$$\lim_{k \to \infty} \frac{k + \log_2(R/\eta)}{\log_2((x^k)^T s^k)} \to 0, \tag{7.33}$$

geometrically. This implies that

$$\liminf_{k \to \infty} \frac{\log((x^{k+1})^T s^{k+1})}{\log((x^k)^T s^k)} \geq 2,$$

which from Proposition 7.1 proves (i).

Now from Theorem 7.12, (7.30) and (7.31)

$$\|x^{k+1} - x^k\| = \bar{\theta}\|d_x^k\| < \|d_x^k\| = O(\mu^k) = O((x^k)^T s^k/n)$$

and

$$\|s^{k+1} - s^k\| = \bar{\theta}\|d_s^k\| < \|d_s^k\| = O(\mu^k) = O((x^k)^T s^k/n).$$

Hence, $\{x^k, s^k\}$ must be a Cauchy sequence, since $\{(x^k)^T s^k\}$ converges to zero superlinearly from (i). This proves (ii). □

To actually achieve the order 2 of convergence of the primal-dual gap, we need to decide when to start the primal-dual affine scaling procedure described in Variant 2. Note from (7.32) that as long as $\{\log_2((x^k)^T s^k)\}$ is a geometric sequence with base close to 1.5 tending to $-\infty$, we shall have the order 2 of convergence of $\{(x^k)^T s^k / (x^0)^T s^0\}$ to zero. Thus, we can start the procedure at any time when $(x^K)^T s^K < 1$. Again for simplicity, let $K := 1$. Then we add a *safety check* to see if for $k = 1, 2, \ldots$

$$\begin{aligned} |\log((x^{k+1})^T s^{k+1})|/|\log((x^k)^T s^k)| &\geq 1.5 \\ (x^{k+1})^T s^{k+1}/(x^k)^T s^k &\leq 1 - \Omega(1/\sqrt{n}). \end{aligned} \quad (7.34)$$

If both inequalities in (7.34) are satisfied, we *continue* the predictor step. Otherwise we conclude that $(x^K)^T s^K$ was not "small enough," and we do one *corrector* step and then *restart* the predictor procedure. This safety check will guarantee that the algorithm maintains the polynomial complexity $O(\sqrt{n}\log(R/\epsilon))$ and achieves the order 2 of the convergence of the complementary gap to zero, since eventually no corrector (or centering) step is needed anymore in the rest of the iterations, according to the theorem.

Thus, we have shown that after the complementary gap becomes smaller than a fixed positive number, the pure primal-dual Newton method with the step-size choice in Variant 2 generates an iteration sequence which not only polynomially converges to an optimal solution pair, but one whose convergence is actually quadratic.

In practice, the step size, θ^k, in the predictor step can be simply chosen as the bound given in Lemma 7.7. Thus, no quartic equation solver is needed to guarantee our theoretical results. Also we see that the step size in Variant 2 converges to 1 superlinearly while the solution sequence remains "centered," i.e., $(x^k, s^k) \in \mathcal{N}_2(2\eta)$, without any explicit centering. This may partially explain why the large step strategy does not hurt the convergence of the algorithm in practice.

7.5 Notes

The issue of the asymptotic convergence of interior-point algorithms was first raised in Iri and Imai [195]. They showed that their (product) barrier function method with an exact line search procedure possesses quadratic convergence for nondegenerate LP. Then, Yamashita [463] showed that a variant of this method possesses both polynomial $O(nL)$ complexity and quadratic convergence for nondegenerate LP, and Tsuchiya and Tanabe [430] showed that Iri and Imai's method possesses quadratic convergence under a weaker nondegeneracy assumption.

Zhang, Tapia and Dennis [484, 483] first showed that a primal-dual algorithm exhibits $O(nL)$ complexity, with superlinear convergence under the assumption of the convergence of the iteration sequence, and quadratic convergence under the assumption of nondegeneracy. Kojima, Megiddo and Mizuno [225], Ji, Potra and Huang [205], and Zhang, Tapia and Potra [485] also showed quadratic convergence of a path-following algorithm for linear complementarity problems under the nondegeneracy assumption. McShane [268] showed that a primal-dual algorithm exhibits $O(\sqrt{n}L)$ complexity, with superlinear convergence under the assumption of the convergence of the iteration sequence. Other algorithms, interior or exterior, with quadratic convergence for nondegenerate LP include Coleman and Li's [89]. Some negative results on the asymptotic convergence of Karmarkar's original algorithm and a potential reduction method (with separate primal and dual updates) were given by Bayer and Lagarias [47], and Gonzaga and Todd [166], respectively.

Quadratic convergence for general LP, assuming neither the convergence of the iteration sequence nor nondegeneracy, was first established by Ye, Güler, Tapia and Zhang [476], and independently by Mehrotra [275] and Tsuchiya [427]. The algorithm of Mehrotra, and Ye et al., is based on the predictor-corrector algorithm of Mizuno et al.; also see Barnes, Chopra and Jensen.[44]. As we mentioned before, if one counts each iteration as solving one system of linear equations, as is usually done in the analysis of interior-point algorithms, the (average) order of convergence of the algorithm is only $\sqrt{2}$. Tsuchiya's result is based on Iri and Imai's $O(nL)$ method, which requires knowledge of the exact optimal objective value in advance. A standard way of dealing with this difficulty is to integrate the primal and dual problems into a single LP problem, whose size is twice that of the original problem. Thus, the (average) order of convergence would actually be below $\sqrt{2}$. The convergence order 2 algorithm for general LP, counting each iteration as solving one system of linear equations of the size of the original problem, was first given in Ye [471].

Quadratic convergence for the monotone LCP, described in Section 7.3, is based on Ye and Anstreicher [475]. They also give an example to show that the predictor step cannot achieve superlinear convergence if the LCP has no a strictly complementary solution. Monteiro and Wright [306] further show that any algorithm that behaves like Newton's method near the solution set cannot converge superlinearly when applied to an LCP that does not have a strictly complementary solution.

Recently, Mizuno [290] proposed a superlinearly convergent infeasible-interior-point algorithm for geometrical LCPs without the strictly complementary condition.

Most recently, Gonzaga and Tapia [165, 164] proved that the iteration sequence (x^k, y^k, s^k) generated by the predictor-corrector algorithm converges to an optimal solution on the interior of the optimal face. Consequently, Luo et al. [253] announced a genuine quadratically convergent algorithm. Bonnans and Gonzaga [69] developed a simplified predictor-corrector where the same Jacobian matrix is used in both the predictor and corrector steps within one iteration. The convergence order of the complementary gap to zero is $T + 1$, where T is the number of predictor steps in each iteration. Tsuchiya and Monteiro [428] showed that a variant of the long-step affine scaling algorithm is two-step superlinearly convergent. Luo, Sturm and Zhang [250] and later Potra and Sheng [349] and Ji, Potra and Sheng [206] analyzed the superlinear convergence behavior of the predictor-corrector algorithm for positive semi-definite programming without any assumption.

In the analysis of interior-point algorithms, the error function is chosen as the primal-dual gap or complementary $x^T s$ which should be zero at an optimal solution pair. For an optimization problem that possesses a strict complementarity solution, this error bound will lead to the same convergence rate for distances from iterates to the solution set, see Hoffman [186], Mangasarian [260, 261], and Luo and Tseng [252], and references therein.

7.6 Exercises

7.1 *Prove Proposition 7.1.*

7.2 *Prove that the sequence with $r^k = (1/k)^k$ is of order unity and is superlinearly convergent.*

7.3 *Let $(P, Z) = (P^*, Z^*)$ be the strict complementarity partition of the LP problem and (x^k, s^k) be generated from the predictor-corrector algorithm. Prove*
$$\xi \le x_j^k \le 1/\xi \quad \text{for} \quad j \in P$$
$$\xi \le s_j^k \le 1/\xi \quad \text{for} \quad j \in Z,$$
where $\xi < 1$ is a fixed positive quantity independent of k.

7.4 *Consider the predictor-corrector monotone LCP algorithm. Prove:*

1.
$$(x^k, s^k) \in \mathcal{N}_2(\eta)$$
as long as $0 < \eta \le 1/4$ and $0 < \tau \le \eta$.

2.
$$(x')^T s' = (1-\theta^k)(x^k)^T s^k + (\theta^k)^2 (d_x)^T d_s$$
$$(x^{k+1})^T s^{k+1} = (x')^T s' + (d'_x)^T d'_s.$$

3.
$$(d_x)^T d_s \le (x^k)^T s^k / 4$$
$$(d'_x)^T d'_s \le (x')^T s' / (8n).$$

4. Let $\delta^k = D_x d_s / \mu^k$ in the kth predictor step. Then,
$$\|\delta^k\| \le \sqrt{2}n/4.$$

7.5 Prove Lemma 7.7 using the preceding exercise.

7.6 Why does Variant 1 of the predictor-corrector algorithm have a higher order of convergence than the algorithm in Section 7.2, even though it uses more predictor steps in each iteration?

7.7 Prove the safety check described at the end of Section 7.4.2, i.e., to see if
$$|\log((x^{k+1})^T s^{k+1})|/|\log((x^k)^T s^k)| \ge 1.5$$
$$(x^{k+1})^T s^{k+1} / (x^k)^T s^k \le 1 - \Omega(1/\sqrt{n}),$$

will guarantee that Variant 2 of the algorithm maintains the polynomial complexity $O(\sqrt{n}\log(R/\epsilon))$ and achieves the order 2 of the convergence of the complementary gap to zero.

Chapter 8

Convex Optimization

In this chapter, we discuss interior-point algorithms for solving non-smooth and/or nonlinear convex optimization problems. These algorithms illustrate how widely applicable potential functions and interior-point algorithms could be in solving broader convex optimization problems.

8.1 Analytic Centers of Nested Polytopes

The problem is to find the "analytic" centers of *all* nested polytopes $\Omega^k \subset \mathcal{R}^m$, $m \le k \le n$, where for given (b^k, P^k, a^k)

$$\Omega^k := \{y \in \mathcal{R}^m : b^k \le (P^k)^T y \le a^k\}.$$

The data (b^k, P^k, c^k) are recursively related. Initially, P^m is an $m \times m$ nonsingular matrix and vectors $b^m, a^m \in \mathcal{R}^m$. Then, for $k \ge m$,

$$b^{k+1} = \begin{pmatrix} b^k \\ b_{k+1} \end{pmatrix} \in \mathcal{R}^{k+1}, \ a^{k+1} = \begin{pmatrix} a^k \\ a_{k+1} \end{pmatrix} \in \mathcal{R}^{k+1}$$

and

$$P^{k+1} = (P^k, \ p_{k+1}) \in \mathcal{R}^{m \times (k+1)}.$$

Here p_{k+1} is an $m \times 1$ vector, and b_{k+1} and a_{k+1} are two scalars. Clearly $\Omega^{k+1} \subset \Omega^k$ for $k \ge m$ so we call them "nested." This problem has applications in dynamic system identification and parameter estimation.

We assume that the interior of Ω^n is non-empty and in fact there is a point y such that

$$b^k + \epsilon e \le (P^n)^T y \le a^k - \epsilon e, \tag{8.1}$$

where $\epsilon > 0$ is a fixed positive number and e is the vector of all ones. For $k = 1, \ldots, n$, Let

$$\bar{w}^k = \max_{j=1,\ldots,k} \frac{a_j - b_j}{2}.$$

For each k we may directly apply the state of the art interior-point linear programming algorithm to find an approximate analytic center $\hat{y}^k$ of Ω^k. The number of Newton iterations to obtain $\hat{y}^k$ will be bounded by $O(\sqrt{2k}\log(\bar{w}^k/\epsilon))$, as we discussed earlier. Thus, to generate the sequence of approximate centers for all $k = m, \ldots, n$ we would need total $O(n^{1.5}\log(\bar{w}^n/\epsilon))$ Newton iterations.

In this section, we present a recursive interior-point algorithm where the number of total Newton iterations is bounded by $O(n\log(\bar{w}^n/\epsilon))$. This is reduced by a factor of $n^{.5}$ from the above bound. Note that the "average cost" (cost per center) to generate all $n - m + 1$ centers (from $k = m, \ldots, n$) is $O(\frac{n}{n-m+1}\log(\bar{w}^n/\epsilon))$. As $n \geq 2m$, $\frac{n}{n-m+1} \leq 2$, and the average cost becomes independent of m and n.

The basic idea is as follows. Starting from $k = m$, we generate an approximate analytic center $\hat{y}^k$ of Ω^k. Then, using the computation work to generate $\hat{y}^k$ we proceed to compute an approximate analytic center $\hat{y}^{k+1}$ of Ω^{k+1}. In other words, all earlier computation work would not be wasted in computing the current approximate center.

8.1.1 Recursive potential reduction algorithm

We now consider the primal potential algorithm of Section 3.3.2 to compute a $(3/4)$-approximate center of Ω^k, $k = 1, \ldots, n$. Let us rewrite Ω^k as

$$\Omega^k = \{y \in \mathcal{R}^m : (A^k)^T y \leq c^k\},$$

where

$$A^m = (p_1, -p_1, p_2, -p_2, \ldots, p_m, -p_m) \in \mathcal{R}^{m \times 2m}, \quad c^m = \begin{pmatrix} a_1 \\ -b_1 \\ a_2 \\ -b_2 \\ \ldots \\ a_m \\ -b_m \end{pmatrix} \in \mathcal{R}^{2m},$$

and for $k \geq m$

$$A^{k+1} = (A^k, p_{k+1}, -p_{k+1}) \in \mathcal{R}^{m \times 2(k+1)}, \quad c^{k+1} = \begin{pmatrix} c^k \\ a_{k+1} \\ -b_{k+1} \end{pmatrix} \in \mathcal{R}^{2(k+1)}.$$

8.1. ANALYTIC CENTERS OF NESTED POLYTOPES

For any k, $m \leq k \leq n$, consider the primal (homogeneous) potential function

$$\mathcal{P}(x, \Omega^k) = 2k \log(c^k)^T x - \sum_{j=1}^{2k} \log x_j,$$

where $x > 0$ and $A^k x = 0$. For $k = 1, ..., n$, let

$$x^k(0) = \begin{pmatrix} 2/(a_1 - b_1) \\ 2/(a_1 - b_1) \\ 2/(a_2 - b_2) \\ 2/(a_2 - b_2) \\ \cdots \\ 2/(a_k - b_k) \\ 2/(a_k - b_k) \end{pmatrix} \in \mathcal{R}^{2k}.$$

Then, for $k = m, \ldots, n$ we have $A^k x^k(0) = 0$ and $(c^k)^T x^k(0) = 2k$, and

$$\begin{aligned}\mathcal{P}(x^k(0), \Omega^k) &= 2k \log(2k) + 2\sum_{j=1}^{k} \log((a_j - b_j)/2) \\ &\leq 2k \log(2k) + 2k \log \bar{w}. \end{aligned} \quad (8.2)$$

Starting at $k = m$, we apply the primal potential reduction algorithm to generate an $(3/4)$-approximate analytic center of Ω^m from $x^m(0)$. The algorithm reduces the primal potential function $\mathcal{P}(x, \Omega^m)$ by a constant $1/6$ per iteration. In each iteration, we have normalized the computed iterates $x^m(1), x^m(2), \ldots$ such that $(c^m)^T x^m(\cdot) = 2m$. Let T^m be the number of total potential reduction iterations to generate a $(3/4)$-approximate analytic center $\hat{y}^m$ of Ω^m. We must have

$$\mathcal{P}(x^m(T^m), \Omega^m) - \mathcal{P}(x^m(0), \Omega^m) \leq -(1/6)T^m.$$

Now consider $k = m + 1$. Let

$$x^{m+1}(T^m) = \begin{pmatrix} x^m(T^m) \\ 2/(a_{m+1} - b_{m+1}) \\ 2/(a_{m+1} - b_{m+1}) \end{pmatrix} \in \mathcal{R}^{2(m+1)}.$$

Note that we have $A^{m+1} x^{m+1}(T^m) = 0$, $x^{m+1}(0)$ and $x^{m+1}(T^m)$ share the same last two components, and $(c^k)^T x^k(T^m) = 2k$, $k = m, m + 1$. Thus, we must have

$$\mathcal{P}(x^{m+1}(T^m), \Omega^{m+1}) - \mathcal{P}(x^{m+1}(0), \Omega^{m+1})$$

$$\begin{aligned}
&= 2(m+1)\log(2m+2) - \sum_{j=1}^{2(m+1)} \log x^{m+1}(T^m)_j \\
&\quad -2(m+1)\log(2m+2) + \sum_{j=1}^{2(m+1)} \log x^{m+1}(0)_j \\
&= -\sum_{j=1}^{2m} \log x^{m+1}(T^m)_j + \sum_{j=1}^{2m} \log x^{m+1}(0)_j \\
&= -\sum_{j=1}^{2m} \log x^m(T^m)_j + \sum_{j=1}^{2m} \log x^m(0)_j \\
&= 2m\log(2m) - \sum_{j=1}^{2m} \log x^m(T^m)_j + 2m\log(2m) + \sum_{j=1}^{2m} \log x^m(0)_j \\
&= \mathcal{P}(x^m(T^m), \Omega^m) - \mathcal{P}(x^m(0), \Omega^m) \\
&\leq -(1/6)T^m.
\end{aligned}$$

This inequality implies that when reduce $\mathcal{P}(x, \Omega^m)$ we also simultaneously reduce $\mathcal{P}(x, \Omega^{m+1})$ by a same amount.

Starting from $x^{m+1}(T^m)$ we reduce the (new) primal potential function $\mathcal{P}(x, \Omega^{m+1})$ by a constant $1/6$ per iteration and normalize the generated iterates $x^{m+1}(T^m+1), x^{m+1}(T^m+2), \ldots$ such that $(c^m)^T x^m(.) = 2(m+1)$. We stop the procedure as soon as a $(3/4)$-approximate analytic center, $\hat{y}^{m+1}$, of Ω^{m+1} is generated. Assume that $x^{m+1}(T^{m+1})$, $T^{m+1} \geq T^m$, is the last iterate. Then, we have

$$\mathcal{P}(x^{m+1}(T^{m+1}), \Omega^{m+1}) - \mathcal{P}(x^{m+1}(T^m), \Omega^{m+1}) \leq -(1/6)(T^{m+1} - T^m).$$

Hence,

$$\begin{aligned}
&\mathcal{P}(x^{m+1}(T^{m+1}), \Omega^{m+1}) - \mathcal{P}(x^{m+1}(0), \Omega^{m+1}) \\
&= \mathcal{P}(x^{m+1}(T^{m+1}), \Omega^{m+1}) - \mathcal{P}(x^{m+1}(T^m), \Omega^{m+1}) \\
&\quad + \mathcal{P}(x^{m+1}(T^m), \Omega^{m+1}) - \mathcal{P}(x^{m+1}(0), \Omega^{m+1}) \\
&\leq -(1/6)(T^{m+1} - T^m) - (1/6)T^m \\
&= -(1/6)T^{m+1}.
\end{aligned}$$

Therefore, we can continue this process for $k = m+2, m+3, \ldots, n$ to generate the sequence of $(3/4)$-approximate analytic centers $\hat{y}^k$ of Ω^k, $k = m, \ldots, n$. Immediately after we generate $\hat{y}^k$ with the primal iterate $x^k(T^k)$, we have $(c^k)^T x^k(T^k) = 2k$ and

$$\mathcal{P}(x^k(T^k), \Omega^k) - \mathcal{P}(x^k(0), \Omega^k) \leq -(1/6)T^k. \tag{8.3}$$

8.1. ANALYTIC CENTERS OF NESTED POLYTOPES

We let
$$x^{k+1}(T^k) = \begin{pmatrix} x^k(T^k) \\ 2/(a_{k+1} - b_{k+1}) \\ 2/(a_{k+1} - b_{k+1}) \end{pmatrix} \in \mathcal{R}^{2(k+1)}.$$

We always have $A^{k+1}x^{k+1}(T^k) = 0$, $x^{k+1}(0)$ and $x^{k+1}(T^k)$ share the same last two components, and $(c^k)^T x^k(T^k) = 2k$ and $(c^{k+1})^T x^{k+1}(T^k) = 2(k+1)$. Thus, we have

$$\begin{aligned} & \mathcal{P}(x^{k+1}(T^k), \Omega^{k+1}) - \mathcal{P}(x^{k+1}(0), \Omega^{k+1}) \\ & = \mathcal{P}(x^k(T^k), \Omega^k) - \mathcal{P}(x^k(0), \Omega^k) \\ & \leq -(1/6)T^k. \end{aligned}$$

Starting from $x^{k+1}(T^k)$ we reduce the (new) primal potential function $\mathcal{P}(x, \Omega^{k+1})$ by a constant $1/6$ per iteration and normalize the generated iterates $x^{k+1}(T^k+1), x^{k+1}(T^k+2), \ldots$ such that $(c^k)^T x^k(.) = 2(k+1)$. We stop the procedure as soon as a $(3/4)$-approximate analytic center, $\hat{y}^{k+1}$, of Ω^{k+1} is generated. Assume that $x^{k+1}(T^{k+1})$, $T^{k+1} \geq T^k$, is the last iterate. Then, we have

$$\mathcal{P}(x^{k+1}(T^{k+1}), \Omega^{k+1}) - \mathcal{P}(x^{k+1}(T^k), \Omega^{k+1}) \leq -(1/6)(T^{k+1} - T^k),$$

which implies that

$$\begin{aligned} & \mathcal{P}(x^{k+1}(T^{k+1}), \Omega^{k+1}) - \mathcal{P}(x^{k+1}(0), \Omega^{k+1}) \\ & = \mathcal{P}(x^k(T^k), \Omega^k) - \mathcal{P}(x^k(0), \Omega^k) - (1/6)(T^{k+1} - T^k) \\ & \leq -(1/6)T^{k+1}. \end{aligned}$$

According to the recursive nature of the procedure, T^k, $k = m, \ldots, n$, represents the number of total potential reduction iterations to compute the sequence of $(3/4)$-approximate analytic centers, $\hat{y}^t$, of Ω^t for all $t = m, \ldots, k$.

8.1.2 Complexity analysis

First, we have the following lemma:

Lemma 8.1 *For all $k = m, \ldots, n$,*

$$\mathcal{P}(x, \Omega^k) \geq \log 2k \log(\epsilon) + 2k \log(2k).$$

Proof. From (8.1), there is a y such that

$$c^k - (A^k)^T y \geq \epsilon e.$$

Thus, from the inequality in Section 2.3.2,

$$\begin{aligned}\mathcal{P}(x,\Omega^k) &\geq \mathcal{B}(y,\Omega^k) + 2k\log(2k) \\ &= \sum_{j=1}^{2k}\log(c^k - (A^k)^T y)_j + 2k\log(2k) \\ &\geq 2k\log(\epsilon) + 2k\log(2k).\end{aligned}$$

□

From the lemma and inequalities (8.2) and (8.3), we derive Theorem 8.2.

Theorem 8.2 *For any $k = m, \ldots, n$, the number of total potential reduction iterations to compute the sequence of $(3/4)$-approximate analytic centers, $\{\hat{y}^t\}$, of all Ω^t, $t = m, \ldots, k$, is bounded by $O(k \log(\bar{w}^k/\epsilon))$.*

Proof. From inequalities (8.2), (8.3) and Lemma 8.1, we have for $k = m, \ldots, n$,

$$\begin{aligned}(1/6)T^k &\leq \mathcal{P}(x^k(0),\Omega^k) - \mathcal{P}(x^k(T^k),\Omega^k) \\ &\leq 2k\log(2k) + 2k\log(\bar{w}^k) - 2k\log(\epsilon) - 2k\log(2k) \\ &= 2k\log(\bar{w}^k/\epsilon).\end{aligned}$$

□

8.2 Convex (Non-Smooth) Feasibility

The problem studied in this section is that of finding an interior point in a general convex set Γ, where $\Gamma \subset \mathcal{R}^m$ has a nonempty interior and is contained in the cube $\Omega^0 = \{y \in \mathcal{R}^m : 0 \leq y \leq e\} = [0,1]^m$. Since any bounded region can be scaled to fit in the cube, this is not much of a restriction. The algorithm starts by representing the cube with $2m$ linear inequalities and testing whether its analytic center, y^0, is an element of $\overset{\circ}{\Gamma}$. If yes, the algorithm stops; but if not, it uses a separating hyperplane, $a^T y = a^T y^0$, that passes through the center and divide the polytope into two parts—one of which, say $\{y \in \mathcal{R}^m : a^T y \leq a^T y^0\}$, contains Γ. Without loss of generality, we assume that a is normalized so that $\|a\| = 1$. The inequality $a^T y \leq a^T y^0$ is then added to the list of inequalities and an approximate analytic center, y^1, of the new, smaller, polytope is computed. The new point is tested and the procedure repeats until a point in $\overset{\circ}{\Gamma}$ is found; see Figure 8.1. This algorithm is an example of central-section or

8.2. CONVEX (NON-SMOOTH) FEASIBILITY

cutting plane algorithms. Since adding a new inequality adds a variable to the primal problem and adds a column to its constraint matrix, these kinds of algorithms are also called column generation algorithms.

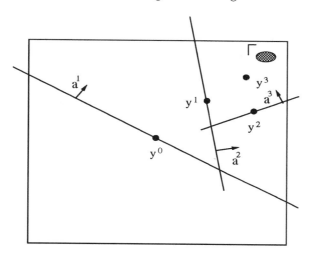

Figure 8.1. Illustration of the central-section algorithm for finding an interior-point in a convex set Γ; the sizes of a sequence of containing polytopes decreases as cuts added at the analytic centers.

The hyperplane used in each iteration is generated by a *separating oracle*. For example, consider the problem finding an interior point in a convex set defined by a system (finite or infinite) of convex inequalities

$$\Gamma = \{y \in \mathcal{R}^m : f_i(y) \leq 0, \quad i = 1, 2, ...\},$$

where each $f_i : \mathcal{R}^m \mapsto \mathcal{R}$ is convex, can be cast in this manner. In particular, the separating oracle just needs to select a to be $g_i/\|g_i\|$, where g_i is an arbitrary subgradient of any function f_i satisfying $f_i(\bar{y}) \geq 0$, i.e. $g_i \in \partial f_i(\bar{y})$ (the subdifferential of f_i). Note that for any $g_i \in \partial f_i(\bar{y})$, $f_i(y) \leq f_i(\bar{y})$ implies $g_i^T(y - \bar{y}) \leq 0$. Thus, if $f_i(\bar{y}) \geq 0$ and $f_i(y) \leq 0$ then $g_i^T(y - \bar{y}) \leq 0$. (In fact, the requirement that $f_i(\bar{y})$ be computed exactly and $g_i \in \partial f_i(\bar{y})$ can be significantly relaxed.)

In general, the algorithm in this section can use any separating oracle that answers the following query: is $y^0 \in \overset{\circ}{\Gamma}$; and if not, what is a separating hyperplane such that $\Gamma \subset \{y \in \mathcal{R}^m : a^T y \leq a^T y^0\}$, where $\|a\| = 1$?

The problem under investigation may also be cast as that of finding the solution to an infinite system of linear inequalities, which is defined implicitly by the oracle $\Gamma = \{y \in \mathcal{R}^m : G^T y \leq g\}$ for some $G \in \mathcal{R}^{m \times d}$ and $g \in \mathcal{R}^d$ and d is infinite. The classical centering methods that have been suggested for the above convex feasibility problem include the center of gravity method, the max-volume sphere method, the ellipsoid method, the max-volume ellipsoid method, and the volumetric center method.

The analytic center column generation or cutting plane algorithm computes $\bar{y}$ as the analytic center of the system of inequalities generated so far. In this section, we show that for any given convex feasibility problem with a nonempty interior, the algorithm is a fully polynomial-time approximation scheme that uses only linear inequalities to approximate the solution set. A fully polynomial-time approximation scheme means that for every ϵ, the accuracy at termination, the running time is a polynomial in the dimension m and $1/\epsilon$.

8.2.1 Max-potential reduction

Now, we use an approximate center y^k to generate a cut, where (y^k, s^k) is an interior point in $\Omega = \{y \in \mathcal{R}^m : A^T y \leq c\}$ and an $x^k > 0$ is known such that $Ax^k = 0$ and
$$\|X^k s^k - e\| \leq \eta \tag{8.4}$$
for some $0 < \eta < 1$. Let us place a cut exactly at y^k. That is, we add a new inequality $a^T y \leq a^T y^k$ to Ω, and consider the new set
$$\Omega^+ = \{y : A^T y \leq c, \quad a^T y \leq a^T y^k\}.$$

We now prove a lemma resembling Theorem 2.10.

Lemma 8.3 *Denote by $(\bar{y}, \bar{s})$ the analytic center of Ω and let*
$$\bar{r} = \sqrt{a^T (A\bar{S}^{-2} A^T)^{-1} a} \ .$$

Then the max-potential of Ω^+
$$\mathcal{B}(\Omega^+) \leq \mathcal{B}(\Omega) + \log(\bar{r}) - \delta$$
for some constant δ depending only on η. Moreover, if $0 < \eta < 1/100$, then we have $\delta > 0$.

Proof. Denote by $\bar{y}^+$ the analytic center for Ω^+. Let $\bar{s}^+ = c - A^T \bar{y}^+ > 0$ and $\bar{s}^+_{n+1} = a^T y^k - a^T \bar{y}^+$. Then we have
$$\begin{aligned}\bar{s}^+_{n+1} &= a^T (y^k - \bar{y}^+) \\ &= a^T (A\bar{S}^{-2} A^T)^{-1} (A\bar{S}^{-2} A^T)(y^k - \bar{y}^+)\end{aligned}$$

8.2. CONVEX (NON-SMOOTH) FEASIBILITY

$$
\begin{aligned}
&= a^T(A\bar{S}^{-2}A^T)^{-1}A\bar{S}^{-2}(A^Ty^k - A^T\bar{y}^+) \\
&= a^T(A\bar{S}^{-2}A^T)^{-1}A\bar{S}^{-2}(-c + A^Ty^k + c - A^T\bar{y}^+) \\
&= a^T(A\bar{S}^{-2}A^T)^{-1}A\bar{S}^{-2}(\bar{s}^+ - s^k) \\
&= a^T(A\bar{S}^{-2}A^T)^{-1}A\bar{S}^{-1}(\bar{S}^{-1}\bar{s}^+ - \bar{S}^{-1}s^k) \\
&\leq \|a^T(A\bar{S}^{-2}A^T)^{-1}A\bar{S}^{-1}\|\|\bar{S}^{-1}\bar{s}^+ - \bar{S}^{-1}s^k\| \\
&= \bar{r}\|\bar{S}^{-1}\bar{s}^+ - e + e - \bar{S}^{-1}s^k\| \\
&\leq \bar{r}(\|\bar{S}^{-1}\bar{s}^+ - e\| + \|e - \bar{S}^{-1}s^k\|) \\
&\leq \bar{r}(\|\bar{S}^{-1}\bar{s}^+ - e\| + \frac{\eta}{(1-\eta)^2}) \quad \text{(from Theorem 3.2)}.
\end{aligned}
$$

Using techniques similar to the proof of Theorem 2.10, we get

$$
\left(\|\bar{S}^{-1}\bar{s}^+ - e\| + \frac{\eta}{(1-\eta)^2}\right) \prod_{j=1}^n \frac{\bar{s}_j^+}{\bar{s}_j} \leq 4\exp\left(\frac{8\eta - 3}{2(1-\eta)^2}\right).
$$

Thus,

$$
\mathcal{B}(\Omega^+) - \mathcal{B}(\Omega) \leq \log(\bar{r}) + \log(4) + \frac{8\eta - 3}{2(1-\eta)^2}.
$$

Let

$$
\delta = -\log(4) - \frac{8\eta - 3}{2(1-\eta)^2}.
$$

Let $\eta = 1/100$. Then we have $\delta > 0$ and the desired result.

□

8.2.2 Compute a new approximate center

In this section, we show how to construct a pair (x, y, s) from (x^k, y^k, s^k) such that $(A, a)x = 0$ with $x > 0$, (y, s) is in the interior of Ω^+, and

$$
\|Xs - e\| < \eta < 1.
$$

Suppose a pair (x^k, y^k, s^k) is given which satisfies (8.4), we use the *dual* scaling for the construction of (x, y, s). Let

$$
\begin{aligned}
r^k &= \sqrt{a^T(A(S^k)^{-2}A^T)^{-1}a}, \\
\Delta y &= -(\beta/r^k)(A(S^k)^{-2}A^T)^{-1}a, \\
\Delta s &= (\beta/r^k)A^T(A(S^k)^{-2}A^T)^{-1}a, \\
\Delta x &= -(\beta/r^k)(S^k)^{-2}A^T(A(S^k)^{-2}A^T)^{-1}a.
\end{aligned}
$$

Then we set
$$y = y^k + \Delta y$$
and
$$x = \begin{pmatrix} x^k + \Delta x \\ \beta/r^k \end{pmatrix} = \begin{pmatrix} x^k - (\beta/r^k)(S^k)^{-2}A^T(A(S^k)^{-2}A^T)^{-1}a \\ \beta/r^k \end{pmatrix}.$$

It can be readily verified that
$$s = \begin{pmatrix} c - A^T(y^k + \Delta y) \\ a^T y^k - a^T(y^k + \Delta y) \end{pmatrix} = \begin{pmatrix} s^k + \Delta s \\ \beta r^k \end{pmatrix}$$
$$= \begin{pmatrix} s^k + (\beta/r^k)A^T(A(S^k)^{-2}A^T)^{-1}a \\ \beta r^k \end{pmatrix}.$$

First, we have
$$(A, a)x = Ax^k - (\beta/r^k)a + (\beta/r^k)a = 0.$$

Second, we have
$$s^k + (\beta/r^k)A^T(A(S^k)^{-2}A^T)^{-1}a = (S^k)(e + p^k)$$
and
$$x^k - (\beta/r^k)(S^k)^{-2}A^T(A(S^k)^{-2}A^T)^{-1}a = (S^k)^{-1}(X^k S^k e - p^k),$$
where
$$p^k = (\beta/r^k)(S^k)^{-1}A^T(A(S^k)^{-2}A^T)^{-1}a.$$

Note that we have
$$\|p^k\| = \beta. \tag{8.5}$$

On the other hand, we have
$$x_j^k s_j^k \geq 1 - \eta.$$

Thus, if we select η and β such that
$$1 - \eta - \beta > 0, \tag{8.6}$$
then both
$$s = \begin{pmatrix} (S^k)(e + p^k) \\ \beta r^k \end{pmatrix} > 0$$
and
$$x = \begin{pmatrix} (S^k)^{-1}(X^k S^k e - p^k) \\ \beta/r^k \end{pmatrix} > 0.$$

8.2. CONVEX (NON-SMOOTH) FEASIBILITY

A simple calculation yields

$$Xs - e = \begin{pmatrix} X^k s^k - e \\ 0 \end{pmatrix} - \begin{pmatrix} (p^k)^2 \\ 1 - \beta^2 \end{pmatrix} + \begin{pmatrix} (X^k S^k - I)p^k \\ 0 \end{pmatrix}$$

where the vector

$$(p^k)^2 = ((p_1^k)^2, (p_2^k)^2, \ldots (p_n^k)^2)^T.$$

Therefore, we have

$$\begin{aligned} \|Xs - e\| &\leq \|X^k s^k - e\| + \sqrt{\|(p^k)^2\|^2 + (1-\beta^2)^2} + \|X^k s^k - e\|\|p^k\| \\ &\leq \eta + \sqrt{\beta^4 + (1-\beta^2)^2} + \eta\beta, \end{aligned}$$

where the last step follows from (8.4) and (8.5). Let $\beta = 1/\sqrt{2}$ and $\eta = 0.15$. Then,

$$\|Xs - e\| \leq \gamma = 0.15 + 1/\sqrt{2} + 0.15/\sqrt{2} < 1.$$

Or, we can let

$$\eta = 1/100, \quad \text{and have} \quad \gamma := 1/100 + 1.01/\sqrt{2} < 1.$$

Furthermore, it can also be easily verified that (8.6) holds.

Hence, using this (y, s) as a starting pair, we can apply the *dual* Newton procedure of Chapter 3 to generate a pair (y^{k+1}, s^{k+1}) and $x^{k+1} = x(y^{k+1})$ such that

$$\begin{aligned} (A, a)x^{k+1} &= 0, \quad x^{k+1} > 0, \\ s^{k+1} &= (c^T, a^T y^k)^T - (A, a)^T y^{k+1} > 0, \end{aligned}$$

and

$$\|X^{k+1} s^{k+1} - e\| \leq \eta.$$

By Theorem 3.3 in Section 3.2.1 and the above given values of γ and η, this can be accomplished in 4 *dual* Newton steps due to the fact $\gamma^{16} \leq \eta$. This column generation process can be repeated, and from Lemma 8.3 the nested sequence of polyhedral sets Ω^k generated by the algorithm satisfies

$$\mathcal{B}(\Omega^{k+1}) \leq \mathcal{B}(\Omega^k) + \log(\bar{r}^k) - \delta \tag{8.7}$$

where δ is some constant,

$$\bar{r}^k = \sqrt{a_{k+1}^T (A(\bar{S}^k)^{-2} A^T)^{-1} a_{k+1}},$$

$(\bar{y}^k, \bar{s}^k)$ is the analytic center of Ω^k, and a_{k+1} is the cut generated at the kth iteration. Note that $(\bar{y}^k, \bar{s}^k)$ is solely used for analysis, and the algorithm does not need any knowledge of $(\bar{y}^k, \bar{s}^k)$.

8.2.3 Convergence and complexity

Let the solution set Γ be contained in $\Omega^0 = \{y \in \mathcal{R}^m : 0 \leq y \leq e\}$, and $\overset{\circ}{\Gamma}$ contain a full dimensional closed ball with $\epsilon < \frac{1}{2}$ radius. We also assume that there exists an oracle which for every $\bar{y} \in \Omega^0$ either returns that $\bar{y} \in \overset{\circ}{\Gamma}$ or generates a separating hyperplane $\{y : a^T y \leq a^T \bar{y}\} \supset \Gamma$, with $\|a\| = 1$ being assumed.

The column-generation, or cutting plane, algorithm from approximate analytic centers is as follows:

Algorithm 8.1 *Let*

$$A^0 = (I, -I) \in \mathcal{R}^{m \times 2m}, \quad c^0 = \begin{pmatrix} e \\ 0 \end{pmatrix} \in \mathcal{R}^{2m}, \tag{8.8}$$

$$y^0 = \frac{1}{2} e \in \mathcal{R}^m, \quad s^0 = c^0 - (A^0)^T y^0 = \frac{1}{2} e \in \mathcal{R}^{2m}, \quad x^0 = 2e \in \mathcal{R}^{2m}.$$

Set
$$k := 0.$$

While $y^k \notin \overset{\circ}{\Gamma}$ **do**

1. *Query the oracle to generate a hyperplane $\{y : a_{k+1}^T y \leq a_{k+1}^T y^k\} \supset \Gamma$ with $\|a_{k+1}\| = 1$, and let*

$$\Omega^{k+1} = \{y \in \mathcal{R}^m : c^{k+1} - (A^{k+1})^T y \geq 0\},$$

   where

$$A^{k+1} = (A^k, a_{k+1}) \quad \text{and} \quad c^{k+1} = \begin{pmatrix} c^k \\ a_{k+1}^T y^k \end{pmatrix}.$$

2. *Compute $(y^{k+1}, s^{k+1}, x^{k+1})$ such that y^{k+1} is an η-approximate analytic center of Ω^{k+1}, using the Newton method with the updating scheme of Section 8.2.2 and starting from (y^k, s^k, x^k), an η-approximate of Ω^k.*

3. *Let $k := k + 1$ and return to Step 1.*

Let the potential function computed at the *exact* analytic center $\bar{y}^k$ be

$$\mathcal{B}(\Omega^k) = \sum_{j=1}^{2m+k} \log(c^k - (A^k)^T \bar{y}^k)_j.$$

Clearly the following relations hold, provided that termination has not occurred:

8.2. CONVEX (NON-SMOOTH) FEASIBILITY

$$\Gamma \subset \Omega^k \quad \forall k, \tag{8.9}$$

and

$$\mathcal{B}(\Omega^{k+1}) \leq \mathcal{B}(\Omega^k) + \frac{1}{2}\log(\bar{r}^k)^2 - \delta \quad \text{(by Lemma 8.3)}, \tag{8.10}$$

where

$$(\bar{r}^k)^2 = a_{k+1}^T(A^k(\bar{S}^k)^{-2}(A^k)^T)^{-1}a_{k+1} \quad \text{and} \quad \bar{s}^k = c^k - (A^k)^T\bar{y}^k.$$

Lemma 8.4 *For all $k \geq 0$,*

$$\mathcal{B}(\Omega^k) \geq (2m+k)\log \epsilon.$$

Proof. From (8.9), $\Gamma \subset \Omega^k$. Thus Ω^k contains a full dimensional ball with ϵ radius. Let the center of this ball be $\bar{y}$. Then $c^k - (A^k)^T\bar{y} \geq \epsilon e$; thus

$$\mathcal{B}(\Omega^k) = \sum_{j=1}^{2m+k} \log(c^k - (A^k)^T\bar{y}^k)_j \geq \sum_{j=1}^{2m+k} \log(c^k - (A^k)^T\bar{y})_j \geq \sum_{j=1}^{2m+k} \log \epsilon,$$

where $\bar{y}^k$ denotes the analytic center of Ω^k.

□

Lemma 8.5 *Let $s = c^k - (A^k)^T y$ for any $y \in \Omega^k$. Then*

i) $0 \leq s_j \leq 1, \quad j = 1, \ldots, 2m$

ii) $0 \leq s_j \leq \sqrt{m}, \quad j = 2m+1, \ldots, 2m+k.$

Proof. For $j = 1, \ldots, m$, $s_j = 1 - y_j$; since $0 \leq y_j \leq 1$, $0 \leq s_j \leq 1$.
For $j = m+1, \ldots, 2m$, $s_j = y_{j-m}$; since $0 \leq y_{j-m} \leq 1$, $0 \leq s_j \leq 1$.
For $j = 2m+1, \ldots, 2m+k$,

$$s_j = a_{j-2m}^T y^{j-2m} - a_{j-2m}^T y \leq \|a_{j-2m}\| \, \|y^{j-2m} - y\| = \|y^{j-2m} - y\| \leq \sqrt{m}.$$

The last inequality is due to the fact that $0 \leq y^{j-2m} \leq e$ and $0 \leq y \leq e$ or $y^{j-2m} \in \Omega^0$ and $y \in \Omega^0$.

□

Lemma 8.4 indicates that, in order to prove finite convergence, one needs to show that $\mathcal{B}(\Omega^k)$ grows more slowly than $2m + k$. By Lemma 8.3, this means finding upper bounds on $\bar{r}^k$. In the following Lemma this is achieved by using a construction that bounds $A^k(\bar{S}^k)^{-2}(A^k)^T$ from below by using a certain matrix B^k, which is simple enough to handle.

Lemma 8.6 Let $s = c^k - (A^k)^T y$ for any $y \in \Omega^k$ and $B^0 = 8I$, $B^{k+1} = B^k + \frac{1}{m} a_{k+1} a_{k+1}^T$. Then

$$A^k S^{-2} (A^k)^T \succeq B^k;$$

that is,

$$A^k S^{-2} (A^k)^T - B^k$$

is positive semi-definite.

Proof. Let $Y = \text{diag}(y)$. Then

$$\begin{aligned} A^k S^{-2} (A^k)^T &= Y^{-2} + (I - Y)^{-2} + \sum_{j=1}^k \frac{a_j a_j^T}{(s_{2m+j})^2} \\ &\succeq Y^{-2} + (I - Y)^{-2} + \frac{1}{m} \sum_{j=1}^k a_j a_j^T \quad \text{(by Lemma 8.5)} \\ &\succeq 8I + \frac{1}{m} \sum_{j=1}^k a_j a_j^T \quad \text{(as } 0 \leq y \leq e\text{)} \\ &= B^k. \end{aligned}$$

$\square$

Lemma 8.7 Let $\bar{s}^k = c^k - (A^k)^T \bar{y}^k$ be the slack vector at the analytic center $\bar{y}^k$ of Ω^k and $(\omega^k)^2 = a_{k+1}^T (B^k)^{-1} a_{k+1}$. Then

$$(\omega^k)^2 \geq a_{k+1}^T (A^k (\bar{S}^k)^{-2} (A^k)^T)^{-1} a_{k+1} = (\bar{r}^k)^2.$$

This lemma implies that upper bounds on the series of $(\omega^k)^2$ will lead to upper bounds on the series $(\bar{r}^k)^2$.

Lemma 8.8

$$\sum_{j=0}^k (\omega^j)^2 \leq 2m^2 \log \left(1 + \frac{k+1}{8m^2} \right).$$

Proof. Note that

$$\det B^{k+1} = \det \left(B^k + \frac{1}{m} a_{k+1} a_{k+1}^T \right) = \left(1 + \frac{(\omega^k)^2}{m} \right) \det B^k.$$

Thus

$$\log \det B^{k+1} = \log \det B^k + \log \left(1 + \frac{(\omega^k)^2}{m} \right).$$

8.2. CONVEX (NON-SMOOTH) FEASIBILITY

But

$$\frac{(\omega^k)^2}{m} \leq \frac{1}{8} a_{k+1}^T a_{k+1} = \frac{1}{8};$$

hence

$$\log(1 + \frac{(\omega^k)^2}{m}) \geq \frac{(\omega^k)^2}{m} - \frac{(\frac{(\omega^k)^2}{m})^2}{2(1 - \frac{(\omega^k)^2}{m})}$$

$$= \frac{(\omega^k)^2}{m}(1 - \frac{\frac{(\omega^k)^2}{m}}{2(1 - \frac{(\omega^k)^2}{m})})$$

$$\geq \frac{(\omega^k)^2}{2m}.$$

Thus we have

$$\log \det B^{k+1} \geq \log \det B^0 + \sum_{j=0}^{k} \frac{(\omega^j)^2}{2m} = m \log 8 + \sum_{j=0}^{k} \frac{(\omega^j)^2}{2m}.$$

Moreover,

$$\frac{1}{m} \log \det B^{k+1} \leq \log \frac{\text{trace } B^{k+1}}{m} = \log\left(8 + \frac{k+1}{m^2}\right).$$

Therefore,

$$\sum_{j=0}^{k} \frac{(\omega^j)^2}{2m} \leq m \log\left(8 + \frac{k+1}{m^2}\right) - m \log 8$$

or

$$\sum_{j=0}^{k} (\omega^j)^2 \leq 2m^2 \log\left(1 + \frac{k+1}{8m^2}\right).$$

□

Theorem 8.9 *The cutting plane algorithm stops with a feasible solution as soon as k satisfies:*

$$\frac{\epsilon^2}{m} \geq \frac{\frac{1}{2} + 2m \log(1 + \frac{k+1}{8m^2})}{2m + k + 1} \exp\left(-2\delta \frac{k+1}{k+1+2m}\right).$$

Proof. From relation (8.10) and Lemma 8.4,

$$(2m + k + 1) \log \epsilon \leq \mathcal{B}(\Omega^{k+1})$$

$$\leq \mathcal{B}(\Omega^0) + \frac{1}{2}\sum_{j=0}^{k}\log(\bar{r}^j)^2 - (k+1)\delta$$

$$= 2m\log\frac{1}{2} + \frac{1}{2}\sum_{j=0}^{k}\log(\bar{r}^j)^2 - (k+1)\delta.$$

Thus

$$\log\epsilon + \frac{k+1}{2m+k+1}\delta$$

$$\leq \frac{1}{2(2m+k+1)}\left(2m\log\frac{1}{4} + \sum_{j=0}^{k}\log(\bar{r}^j)^2\right)$$

$$\leq \frac{1}{2}\log\frac{2m\frac{1}{4} + \sum_{j=0}^{k}(\bar{r}^j)^2}{2m+k+1} \quad \text{(from the concavity of log)}$$

$$\leq \frac{1}{2}\log\frac{\frac{m}{2} + \sum_{j=0}^{k}(\omega^j)^2}{2m+k+1} \quad \text{(from Lemma 8.7)}$$

$$\leq \frac{1}{2}\log\frac{\frac{m}{2} + 2m^2\log(1+\frac{k+1}{8m^2})}{2m+k+1} \quad \text{(from Lemma 8.8)}$$

or

$$\frac{\epsilon^2}{m} \leq \frac{\frac{1}{2} + 2m\log(1+\frac{k+1}{8m^2})}{2m+k+1}\exp\left(-2\delta\frac{k+1}{k+1+2m}\right).$$

□

Theorem 8.9 implies that the complexity of the column generation scheme, counted by the calls to the oracle, is $O^*(\frac{m^2}{\epsilon^2})$; the notation O^* means that logarithmic terms are ignored. The largest value of η that guarantees $\gamma^{16} \leq \eta$ (so that four dual Newton steps are enough to recenter) is about $\eta = .09$ with $\beta = .691$. In this case constant δ may be negative; nonetheless the algorithm will still terminate after $O^*(\frac{m^2}{\epsilon^2})$ iterations.

Theorem 8.10 *The approximate analytic center algorithm, which uses the updating scheme of Section 8.2.2 and the Newton method, is, for appropriate values of η and β which depend on the exact mix of recentering and updating steps, a fully polynomial-time approximation scheme.*

8.3 Positive Semi-Definite Programming

Recall that $\mathcal{M}^n$ denotes the set of symmetric matrices in $\mathcal{R}^{n\times n}$. Let $\mathcal{M}^n_+$ denote the set of positive semi-definite matrices and $\overset{\circ}{\mathcal{M}}^n_+$ the set of positive

8.3. POSITIVE SEMI-DEFINITE PROGRAMMING

definite matrices in $\mathcal{M}^n$. The goal of this section is to extend interior-point algorithms to solving the positive semi-definite programming problem (PSP) and (PSD) presented in Section 1.3.8.

(PSP) and (PSD) are analogues to linear programming (LP) and (LD). In fact, as the notation suggest, (LP) and (LD) can be expressed as a positive semi-definite program by defining

$$C = \text{diag}(c), \quad A_i = \text{diag}(a_{i\cdot}), \quad b = b,$$

where $a_{i\cdot}$ is the ith row of matrix A. Many of the theorems and algorithms used in LP have analogues in PSP. However, while interior-point algorithms for LP are generally considered competitive with the simplex method in practice and outperform it as problems become large, interior-point methods for PSP outperform other methods on even small problems.

Denote the primal feasible set by $\mathcal{F}_p$ and the dual by $\mathcal{F}_d$. We assume that both $\overset{\circ}{\mathcal{F}}_p$ and $\overset{\circ}{\mathcal{F}}_d$ are nonempty. Thus, the optimal solution sets for both (PSP) and (PSD) are bounded and the central path exists, see Section 2.5 Let z^* denote the optimal value and $\mathcal{F} = \mathcal{F}_p \times \mathcal{F}_d$. In this section, we are interested in finding an ϵ approximate solution for the PSP problem:

$$C \bullet X - b^T y = S \bullet X \le \epsilon.$$

For simplicity, we assume that a central path pair (X^0, y^0, S^0), which satisfies

$$(X^0)^{.5} S^0 (X^0)^{.5} = \mu^0 I \quad \text{and} \quad \mu^0 = X^0 \bullet S^0 / n,$$

is known. We will use it as our initial point throughout this section.

Let $X \in \overset{\circ}{\mathcal{F}}_p$, $(y, S) \in \overset{\circ}{\mathcal{F}}_d$, and $z \le z^*$. Then consider the primal potential function

$$\mathcal{P}(X, z) = (n + \rho) \log(C \bullet X - z) - \log \det X,$$

and the primal-dual potential function

$$\psi(X, S) = (n + \rho) \log(S \bullet X) - \log \det XS,$$

where $\rho = \sqrt{n}$. Let $z = b^T y$. Then $S \bullet X = C \bullet X - z$, and we have

$$\psi(x, s) = \mathcal{P}(x, z) - \log \det S.$$

Like in Chapter 4, these functions will be used to solve PSP problems.

Define the "∞-norm," which is the traditional l_2 operator norm for matrices, of $\mathcal{M}^n$ by

$$\|X\|_\infty := \max_{j \in \{1, \ldots, n\}} \{|\lambda_j(X)|\},$$

where $\lambda_j(X)$ is the jth eigenvalue of X, and the "Euclidean" or l_2 norm, which is the traditional Frobenius norm, by

$$\|X\| := \|X\|_f = \sqrt{X \bullet X} = \sqrt{\sum_{j=1}^{n}(\lambda_j(X))^2} \ .$$

We rename these norms because they are perfect analogues to the norms of vectors used in LP. Furthermore, note that, for $X \in \mathcal{M}^n$,

$$\text{tr}(X) = \sum_{j=1}^{n} \lambda_j(X) \quad \text{and} \quad \det(I + X) = \prod_{j=1}^{n}(1 + \lambda_j(X)).$$

Then, we have the following lemma which resembles Lemma 3.1.

Lemma 8.11 *Let $X \in \mathcal{M}^n$ and $\|X\|_\infty < 1$. Then,*

$$tr(X) \geq \log\det(I + X) \geq tr(X) - \frac{\|X\|^2}{2(1 - \|X\|_\infty)} \ .$$

8.3.1 Potential reduction algorithm

Consider a pair of $(X^k, y^k, S^k) \in \mathcal{\overset{\circ}{F}}$. Fix $z^k = b^T y^k$, then the gradient matrix of the primal potential function at X^k is

$$\nabla P(X^k, z^k) = \frac{n+\rho}{S^k \bullet X^k} C - (X^k)^{-1}.$$

The following corollary is an analog to inequality (3.14).

Corollary 8.12 *Let $X^k \in \overset{\circ}{\mathcal{M}}^n_+$ and $\|(X^k)^{-.5}(X - X^k)(X^k)^{-.5}\|_\infty < 1$. Then, $X \in \overset{\circ}{\mathcal{M}}^n_+$ and*

$$\mathcal{P}(X, z^k) - \mathcal{P}(X^k, z^k) \leq \nabla \mathcal{P}(X^k, z^k) \bullet (X - X^k)$$
$$+ \frac{\|(X^k)^{-.5}(X - X^k)(X^k)^{-.5}\|^2}{2(1 - \|(X^k)^{-.5}(X - X^k)(X^k)^{-.5}\|_\infty)} \ .$$

Let

$$\mathcal{A} = \begin{pmatrix} A_1 \\ A_2 \\ \dots \\ A_m \end{pmatrix}.$$

8.3. POSITIVE SEMI-DEFINITE PROGRAMMING

Then, define
$$\mathcal{A}X = \begin{pmatrix} A_1 \bullet X \\ A_2 \bullet X \\ ... \\ A_m \bullet X \end{pmatrix} = b,$$

and
$$\mathcal{A}^T y = \sum_{i=1}^m y_i A_i.$$

Then, we directly solve the following "ball-constrained" problem:

$$\begin{array}{ll} \text{minimize} & \nabla \mathcal{P}(X^k, z^k) \bullet (X - X^k) \\ \text{s.t.} & \mathcal{A}(X - X^k) = 0, \\ & \|(X^k)^{-.5}(X - X^k)(X^k)^{-.5}\| \le \alpha < 1. \end{array}$$

Let $X' = (X^k)^{-.5} X (X^k)^{-.5}$. Note that for any symmetric matrices $Q, T \in \mathcal{M}^n$ and $X \in \overset{\circ}{\mathcal{M}}{}^n_+$,

$$Q \bullet X^{.5} T X^{.5} = X^{.5} Q X^{.5} \bullet T \text{ and } \|XQ\|_. = \|QX\|_. = \|X^{.5} Q X^{.5}\|_..$$

Then we transform the above problem into

$$\begin{array}{ll} \text{minimize} & (X^k)^{.5} \nabla \mathcal{P}(X^k, z^k)(X^k)^{.5} \bullet (X' - I) \\ \text{s.t.} & \mathcal{A}'(X' - I) = 0, \ i = 1, 2, ..., i, \\ & \|X' - I\| \le \alpha, \end{array}$$

where
$$\mathcal{A}' = \begin{pmatrix} A'_1 \\ A'_2 \\ ... \\ A'_m \end{pmatrix} := \begin{pmatrix} (X^k)^{.5} A_1 (X^k)^{.5} \\ (X^k)^{.5} A_2 (X^k)^{.5} \\ ... \\ (X^k)^{.5} A_m (X^k)^{.5} \end{pmatrix}.$$

Let the minimizer be X' and let $X^{k+1} = (X^k)^{.5} X' (X^k)^{.5}$. Then

$$X' - I = -\alpha \frac{P^k}{\|P^k\|},$$

$$X^{k+1} - X^k = -\alpha \frac{(X^k)^{.5} P^k (X^k)^{.5}}{\|P^k\|}, \qquad (8.11)$$

where
$$\begin{aligned} P^k &= \mathcal{P}_{\mathcal{A}'}(X^k)^{.5} \nabla \mathcal{P}(X^k, z^k)(X^k)^{.5} \\ &= (X^k)^{.5} \nabla \mathcal{P}(X^k, z^k)(X^k)^{.5} - \mathcal{A}'^T y^k \end{aligned}$$

or
$$P^k = \frac{n+\rho}{S^k \bullet X^k}(X^k)^{.5}(C - \mathcal{A}^T y^k)(X^k)^{.5} - I,$$
and
$$y^k = \frac{S^k \bullet X^k}{n+\rho}(\mathcal{A}'\mathcal{A}'^T)^{-1}\mathcal{A}'(X^k)^{.5}\nabla \mathcal{P}(X^k, z^k)(X^k)^{.5}.$$

Here, $\mathcal{P}_{\mathcal{A}'}$ is the projection operator onto the null space of $\mathcal{A}'$, and

$$\mathcal{A}'\mathcal{A}'^T := \begin{pmatrix} A_1' \bullet A_1' & A_1' \bullet A_2' & \dots & A_1' \bullet A_m' \\ A_2' \bullet A_1' & A_2' \bullet A_2' & \dots & A_2' \bullet A_m' \\ \dots & \dots & \dots & \dots \\ A_m' \bullet A_1' & A_m' \bullet A_2' & \dots & A_m' \bullet A_m' \end{pmatrix} \in \mathcal{M}^m.$$

In view of Corollary 8.12 and

$$\begin{aligned}\nabla \mathcal{P}(X^k, z^k) \bullet (X^{k+1} - X^k) &= -\alpha \frac{\nabla \mathcal{P}(X^k, z^k) \bullet (X^k)^{.5} P^k (X^k)^{.5}}{\|P^k\|} \\ &= -\alpha \frac{(X^k)^{.5} \nabla \mathcal{P}(X^k, z^k)(X^k)^{.5} \bullet P^k}{\|P^k\|} \\ &= -\alpha \frac{\|P^k\|^2}{\|P^k\|} = -\alpha \|P^k\|,\end{aligned}$$

we have

$$\mathcal{P}(X^{k+1}, z^k) - \mathcal{P}(X^k, z^k) \le -\alpha \|P^k\| + \frac{\alpha^2}{2(1-\alpha)}.$$

Thus, as long as $\|P^k\| \ge \beta > 0$, we may choose an appropriate α such that

$$\mathcal{P}(X^{k+1}, z^k) - \mathcal{P}(X^k, z^k) \le -\delta$$

for some positive constant δ.

Now, we focus on the expression of P^k, which can be rewritten as

$$P(z^k) := P^k = \frac{n+\rho}{S^k \bullet X^k}(X^k)^{.5} S(z^k)(X^k)^{.5} - I \quad (8.12)$$

with
$$S(z^k) = C - \mathcal{A}^T y(z^k) \quad (8.13)$$
and
$$y(z^k) := y^k = y_2 - \frac{S^k \bullet X^k}{n+\rho} y_1 = y_2 - \frac{C \bullet X^k - z^k}{n+\rho} y_1, \quad (8.14)$$

8.3. POSITIVE SEMI-DEFINITE PROGRAMMING

where y_1 and y_2 are given by

$$\begin{aligned} y_1 &= (\mathcal{A}'\mathcal{A}'^T)^{-1}\mathcal{A}'I = (\mathcal{A}'\mathcal{A}'^T)^{-1}b, \\ y_2 &= (\mathcal{A}'\mathcal{A}'^T)^{-1}\mathcal{A}'(X^k)^{.5}C(X^k)^{.5}. \end{aligned} \qquad (8.15)$$

Regarding $\|P^k\| = \|P(z^k)\|$, we have the following lemma resembling Lemma 4.8.

Lemma 8.13 *Let*

$$\mu^k = \frac{S^k \bullet X^k}{n} = \frac{C \bullet X^k - z^k}{n} \quad \text{and} \quad \mu = \frac{S(z^k) \bullet X^k}{n}.$$

If

$$\|P(z^k)\| < \min\left(\beta\sqrt{\frac{n}{n+\beta^2}}, 1-\beta\right), \qquad (8.16)$$

then the following three inequalities hold:

$$S(z^k) \succ 0, \quad \|(X^k)^{.5}S(z^k)(X^k)^{.5} - \mu e\| < \beta\mu, \quad \text{and} \quad \mu < (1-.5\beta/\sqrt{n})\mu^k. \qquad (8.17)$$

Proof. The proof is by contradiction. For example, if the first inequality of (8.17) is not true, then $(X^k)^{.5}S(z^k)(X^k)^{.5}$ has at least one eigenvalue less than or equal to zero, and

$$\|P(z^k)\| \geq 1.$$

The proof of the second and third inequalities are similar to that of Lemma 4.8.

□

Based on this lemma, we have the following potential reduction theorem.

Theorem 8.14 *Given $X^k \in \overset{\circ}{\mathcal{F}}_p$ and $(y^k, S^k) \in \overset{\circ}{\mathcal{F}}_d$, let $\rho = \sqrt{n}$, $z^k = b^T y^k$, X^{k+1} be given by (8.11), and $y^{k+1} = y(z^k)$ in (8.14) and $S^{k+1} = S(z^k)$ in (8.13). Then, either*

$$\psi(X^{k+1}, S^k) \leq \psi(X^k, S^k) - \delta$$

or

$$\psi(X^k, S^{k+1}) \leq \psi(X^k, S^k) - \delta,$$

where $\delta > 1/20$.

Proof. If (8.16) does not hold, i.e.,

$$\|P(z^k)\| \geq \min\left(\beta\sqrt{\frac{n}{n+\beta^2}}, 1-\beta\right),$$

then, since $\psi(X^{k+1}, S^k) - \psi(X^k, S^k) = \mathcal{P}(X^{k+1}, z^k) - \mathcal{P}(X^k, z^k)$,

$$\psi(X^{k+1}, S^k) - \psi(X^k, S^k) \leq -\alpha \min\left(\beta\sqrt{\frac{n}{n+\beta^2}}, 1-\beta\right) + \frac{\alpha^2}{2(1-\alpha)}.$$

Otherwise, from Lemma 8.13 the inequalities of (8.17) hold:

i) The first of (8.17) indicates that y^{k+1} and S^{k+1} are in $\overset{\circ}{\mathcal{F}}_d$.

ii) Using the second of (8.17) and applying Lemma 8.11 to matrix $(X^k)^{.5} S^{k+1}(X^k)^{.5}/\mu$, we have

$$n \log S^{k+1} \bullet X^k - \log \det S^{k+1} X^k$$
$$= n \log S^{k+1} \bullet X^k/\mu - \log \det (X^k)^{.5} S^{k+1}(X^k)^{.5}/\mu$$
$$= n \log n - \log \det (X^k)^{.5} S^{k+1}(X^k)^{.5}/\mu$$
$$\leq n \log n + \frac{\|(X^k)^{.5} S^{k+1}(X^k)^{.5}/\mu - I\|^2}{2(1 - \|(X^k)^{.5} S^{k+1}(X^k)^{.5}/\mu - I\|_\infty)}$$
$$\leq n \log n + \frac{\beta^2}{2(1-\beta)}$$
$$\leq n \log S^k \bullet X^k - \log \det S^k X^k + \frac{\beta^2}{2(1-\beta)}.$$

iii) According to the third of (8.17), we have

$$\sqrt{n}(\log S^{k+1} \bullet X^k - \log S^k \bullet X^k) = \sqrt{n} \log \frac{\mu}{\mu^k} \leq -\frac{\beta}{2}.$$

Adding the two inequalities in ii) and iii), we have

$$\psi(X^k, S^{k+1}) \leq \psi(X^k, S^k) - \frac{\beta}{2} + \frac{\beta^2}{2(1-\beta)}.$$

Thus, by choosing $\beta = .43$ and $\alpha = .3$ we have the desired result.

□

Theorem 8.14 establishes an important fact: the *primal-dual* potential function can be reduced by a constant no matter where X^k and y^k are. In practice, one can perform the line search to minimize the primal-dual potential function. This results in the following potential reduction algorithm.

8.3. POSITIVE SEMI-DEFINITE PROGRAMMING 253

Algorithm 8.2 *Given $x^0 \in \overset{\circ}{\mathcal{F}}_p$ and $(y^0, s^0) \in \overset{\circ}{\mathcal{F}}_d$. Let $z^0 = b^T y^0$. Set $k := 0$.*
  While $S^k \bullet X^k \geq \epsilon$ **do**

  1. *Compute y_1 and y_2 from (8.15).*

  2. *Set $y^{k+1} = y(\bar{z})$, $S^{k+1} = S(\bar{z})$, $z^{k+1} = b^T y^{k+1}$ with*

  $$\bar{z} = \arg\min_{z \geq z^k} \psi(X^k, S(z)).$$

  If $\psi(X^k, S^{k+1}) > \psi(X^k, S^k)$ then $y^{k+1} = y^k$, $S^{k+1} = S^k$, $z^{k+1} = z^k$.

  3. *Let $X^{k+1} = X^k - \bar{\alpha}(X^k)^{.5} P(z^{k+1})(X^k)^{.5}$ with*

  $$\bar{\alpha} = \arg\min_{\alpha \geq 0} \psi(X^k - \alpha(X^k)^{.5} P(z^{k+1})(X^k)^{.5}, S^{k+1}).$$

  4. *Let $k := k+1$ and return to Step 1.*

The performance of the algorithm results from the following corollary:

Corollary 8.15 *Let $\rho = \sqrt{n}$. Then, Algorithm 8.2 terminates in at most $O(\sqrt{n}\log(C \bullet X^0 - b^T y^0)/\epsilon)$ iterations with*

$$C \bullet X^k - b^T y^k \leq \epsilon.$$

Proof. In $O(\sqrt{n}\log(S^0 \bullet X^0/\epsilon))$ iterations

$$\begin{aligned}-\sqrt{n}\log(S^0 \bullet X^0/\epsilon) &= \psi(X^k, S^k) - \psi(X^0, S^0) \\ &\geq \sqrt{n}\log S^k \bullet X^k + n\log n - \psi(X^0, S^0) \\ &= \sqrt{n}\log(S^k \bullet X^k / S^0 \bullet X^0).\end{aligned}$$

Thus,
$$\sqrt{n}\log(C \bullet X^k - b^T y^k) = \sqrt{n}\log S^k \bullet X^k \leq \sqrt{n}\log \epsilon,$$

i.e.,
$$C \bullet X^k - b^T y^k = S^k \bullet X^k \leq \epsilon.$$

$\square$

8.3.2 Primal-dual algorithm

Once we have a pair $(X, y, S) \in \overset{\circ}{\mathcal{F}}$ with $\mu = S \bullet X/n$, we can apply the primal-dual Newton method to generate a new iterate X^+ and (y^+, S^+) as follows: Solve for d_X, d_y and d_S from the system of linear equations:

$$\begin{aligned} D^{-1}d_X D^{-1} + d_S &= R := \gamma\mu X^{-1} - S, \\ \mathcal{A} d_X &= 0, \\ -\mathcal{A}^T d_y - d_S &= 0, \end{aligned} \quad (8.18)$$

where

$$D = X^{.5}(X^{.5}SX^{.5})^{-.5}X^{.5}.$$

Note that $d_S \bullet d_X = 0$.

This system can be written as

$$\begin{aligned} d_{X'} + d_{S'} &= R', \\ \mathcal{A}' d_{X'} &= 0, \\ -{\mathcal{A}'}^T d_y - d_{S'} &= 0, \end{aligned} \quad (8.19)$$

where

$$d_{X'} = D^{-.5}d_X D^{-.5}, \quad d_{S'} = D^{.5}d_S D^{.5}, \quad R' = D^{.5}(\gamma\mu X^{-1} - S)D^{.5},$$

and

$$\mathcal{A}' = \begin{pmatrix} A'_1 \\ A'_2 \\ \cdots \\ A'_m \end{pmatrix} := \begin{pmatrix} D^{.5}A_1 D^{.5} \\ D^{.5}A_2 D^{.5} \\ \cdots \\ D^{.5}A_m D^{.5} \end{pmatrix}.$$

Again, we have $d_{S'} \bullet d_{X'} = 0$, and

$$d_y = (\mathcal{A}'{\mathcal{A}'}^T)^{-1}\mathcal{A}' R', \quad d_{S'} = -{\mathcal{A}'}^T d_y, \text{ and } d_{X'} = R' - d_{S'}.$$

Then, assign

$$d_S = \mathcal{A}^T d_y \quad \text{and} \quad d_X = D(R - d_S)D.$$

Let

$$V^{1/2} = D^{-.5}XD^{-.5} = D^{.5}SD^{.5} \in \overset{\circ}{\mathcal{M}}^n_+.$$

Then, we can verify that $S \bullet X = I \bullet V$. We now present the following lemma, whose proof is very similar to that for Lemmas 3.12 and 4.11 and will be omitted.

8.3. POSITIVE SEMI-DEFINITE PROGRAMMING

Lemma 8.16 *Let the direction d_X, d_y and d_S be generated by equation (8.18) with $\gamma = n/(n+\rho)$, and let*

$$\theta = \frac{\alpha}{\|V^{-1/2}\|_\infty \|\frac{I \bullet V}{n+\rho} V^{-1/2} - V^{1/2}\|}, \qquad (8.20)$$

where α is a positive constant less than 1. Let

$$X^+ = X + \theta d_X, \quad y^+ = y + \theta d_y, \quad \text{and} \quad S^+ = S + \theta d_S.$$

Then, we have $(X^+, y^+, S^+) \in \overset{\circ}{\mathcal{F}}$ and

$$\psi(X^+, S^+) - \psi(X, S)$$

$$\leq -\alpha \frac{\|V^{-1/2} - \frac{n+\rho}{I \bullet V} V^{1/2}\|}{\|V^{-1/2}\|_\infty} + \frac{\alpha^2}{2(1-\alpha)}.$$

Applying Lemma 4.12 to $v \in \mathcal{R}^n$ as the vector of the n eigenvalues of V, we can prove the following lemma:

Lemma 8.17 *Let $V \in \overset{\circ}{\mathcal{M}}^n_+$ and $\rho \geq \sqrt{n}$. Then,*

$$\frac{\|V^{-1/2} - \frac{n+\rho}{I \bullet V} V^{1/2}\|}{\|V^{-1/2}\|_\infty} \geq \sqrt{3/4}.$$

From these two lemmas we have

$$\psi(X^+, S^+) - \psi(X, S)$$

$$\leq -\alpha \sqrt{3/4} + \frac{\alpha^2}{2(1-\alpha)} = -\delta$$

for a constant δ. This leads to Algorithm 8.3.

Algorithm 8.3 *Given $(X^0, y^0, S^0) \in \overset{\circ}{\mathcal{F}}$. Set $\rho = \sqrt{n}$ and $k := 0$.*
While $S^k \bullet X^k \geq \epsilon$ do

1. Set $(X, S) = (X^k, S^k)$ and $\gamma = n/(n+\rho)$ and compute (d_X, d_y, d_S) from (8.18).

2. Let $X^{k+1} = X^k + \bar{\alpha} d_X$, $y^{k+1} = y^k + \bar{\alpha} d_y$, and $S^{k+1} = S^k + \bar{\alpha} d_S$, where

$$\bar{\alpha} = \arg\min_{\alpha \geq 0} \psi(X^k + \alpha d_X, S^k + \alpha d_S).$$

3. Let $k := k + 1$ and return to Step 1.

Theorem 8.18 *Let $\rho = \sqrt{n}$. Then, Algorithm 8.3 terminates in at most $O(\sqrt{n} \log(S^0 \bullet X^0/\epsilon))$ iterations with*

$$C \bullet X^k - b^T y^k \leq \epsilon.$$

Primal-dual adaptive path-following algorithms, the predictor-corrector algorithms and the wide-neighborhood algorithms similar to those in Section 4.5 can also be developed for solving (PSP).

8.4 Monotone Complementarity Problem

In this section we present a generalization of the homogeneous self-dual linear programming (LP) algorithm to solving the nonlinear monotone complementarity problem (MCP) of Section 1.3.10, which includes finding a KKT point for convex optimization.

Here we may let the domain of $f(x)$ be an open set, e.g., the interior of the positive orthant. Then, (MCP) is said to be (asymptotically) feasible if and only if there is a *bounded* sequence $(x^t > 0, s^t > 0)$, $t = 1, 2, ...$, such that
$$\lim_{t \to \infty} s^t - f(x^t) \to 0,$$
where any limit point $(\hat{x}, \hat{s})$ of the sequence is called an *asymptotically feasible point* for (MCP). (MCP) has an interior feasible point if it has an (asymptotically) feasible point $(\hat{x} > 0, \hat{s} > 0)$. (MCP) is said to be (asymptotically) solvable if there is an (asymptotically) feasible $(\hat{x}, \hat{s})$, such that $\hat{x}^T \hat{s} = 0$, where $(\hat{x}, \hat{s})$ is called the "optimal" or "complementary" solution for (MCP). (MCP) is (strongly) infeasible if and only if there is no sequence $(x^t > 0, s^t > 0)$, $t = 1, 2, ...$, such that
$$\lim_{t \to \infty} s^t - f(x^t) \to 0.$$

Consider a class of (MCP), where f satisfies the following condition: Let
$$\upsilon : (0, 1) \to (1, \infty)$$
be a monotone increasing function, such that
$$\|X(f(x + d_x) - f(x) - \nabla f(x) d_x)\|_1 \leq \upsilon(\alpha) d_x^T \nabla f(x) d_x \quad (8.21)$$
whenever
$$d_x \in \mathcal{R}^n, \quad x \in \mathcal{R}_{++}^n, \quad \|X^{-1} d_x\|_\infty \leq \alpha < 1.$$
Then, f is said to be scaled Lipschitz in $\overset{\circ}{\mathcal{R}}_+^n$.

8.4. MONOTONE COMPLEMENTARITY PROBLEM

Given a central path point $x^0 > 0$, $s^0 = f(x^0) > 0$ and $X^0 s^0 = \mu^0 e$ one can develop an interior-point algorithm that generates a maximal complementary solution of the scaled Lipschitz (MCP) in $O(\sqrt{n}\log(\mu^0/\epsilon))$ interior-point iterations, where ϵ is the complementarity error.

However, the initial point x^0 is generally unknown. In fact, we don't even know whether such a point exists or not, that is, (MCP) might be infeasible or feasible but have no positive feasible point. To overcome this difficulty, in Section 5.3 we developed a homogeneous linear programming (LP) algorithm based on the construction of a homogeneous and self-dual LP model. In this section, we present a homogeneous model for solving the monotone complementarity problem. The algorithm again possesses the following desired features:

- It achieves $O(\sqrt{n}\log(1/\epsilon))$-iteration complexity if f satisfies the scaled Lipschitz condition.

- If (MCP) has a solution, the algorithm generates a sequence that approaches feasibility and optimality *simultaneously*; if the problem is (strongly) infeasible, the algorithm generates a sequence that converges to a certificate proving infeasibility.

8.4.1 A convex property

Let $f(x)$ be a continuous monotone mapping from $\overset{\circ}{\mathcal{R}}{}^n_+$ to $\mathcal{R}^n$. Consider the set of residuals

$$R_{++} = \{s - f(x) \in \mathcal{R}^n : \quad (x,s) > 0\},$$

and for a $r \in \mathcal{R}^n$ let

$$S_{++}(r) = \{(x,s) \in \mathcal{R}^{2n}_{++} : s = f(x) + r\}.$$

Since f is continuous in $\overset{\circ}{\mathcal{R}}{}^n_+$, we can easily verify that R_{++} is an open set. Furthermore, we have

Theorem 8.19 *Consider the mapping $F = (Xy, s - f(x)) \in \mathcal{R}^{2n}$ from $(x,s) \in \mathcal{R}^{2n}$. Then F maps $\mathcal{R}^{2n}_{++}$ onto $\overset{\circ}{\mathcal{R}}{}^n_+ \times R_{++}$ homeomorphically, that is, F is one-to-one on $\mathcal{R}^{2n}_{++}$, F maps $\mathcal{R}^{2n}_{++}$ onto $\overset{\circ}{\mathcal{R}}{}^n_+ \times R_{++}$, and the inverse mapping F^{-1} is continuous on $\overset{\circ}{\mathcal{R}}{}^n_+ \times R_{++}$.*

Simply using the monotone of f, we also have the following lemma:

Lemma 8.20 *Let* $r \in \mathcal{R}^n$. *Assume that* $(x^1, s^1) \in S_{++}(\theta^1 r)$ *and* $(x^2, s^2) \in S_{++}(\theta^2 r)$ *where* θ^1 *and* θ^2 *are two real numbers. Then*

$$(\theta^2 - \theta^1) r^T (x^2 - x^1) \leq (x^2 - x^1)^T (s^2 - s^1).$$

This lemma leads to the next lemma:

Lemma 8.21 *Let* $r \in \mathcal{R}^n$ *and* $\theta^0 \leq \theta^1$. *Assume* $S_{++}(\theta^0 r) \neq \emptyset$ *and* $S_{++}(\theta^1 r) \neq \emptyset$. *Then, for every* $\delta > 0$, *the union of*

$$C_{++}(\theta r, \delta) = \{(x, y) \in S_{++}(\theta r) : x^T y \leq \delta\}, \quad \theta \in [\theta^0, \theta^1]$$

is bounded.

Proof. Let $(x^0, s^0) \in S_{++}(\theta^0 r)$ and $(x^1, s^1) \in S_{++}(\theta^1 r)$, and

$$\max((s^0)^T x^0, (s^1)^T x^1) \leq \delta.$$

Let $\theta \in [\theta^0, \theta^1]$ and $(x, s) \in C_{++}(\theta r, \delta)$. Then we have by Lemma 8.20 that

$$(s^1)^T x + (x^1)^T s \leq (\theta^1 - \theta) r^T x + c^1$$

and

$$(s^0)^T x + (x^0)^T s \leq (\theta^0 - \theta) r^T x + c^0,$$

where

$$c^1 = (\theta^1 - \theta^0) |r^T x^1| + 2\delta \text{ and } c^0 = (\theta^1 - \theta^0) |r^T x^0| + 2\delta.$$

Thus, if $\theta^0 < \theta^1$ then we have

$$(\theta - \theta^0)((s^1)^T x + (x^1)^T s) + (\theta^1 - \theta)((s^0)^T x + (x^0)^T s)$$

$$\leq (\theta - \theta^0) c^1 + (\theta^1 - \theta) c^0.$$

Thus, we have

$$e^T x + e^T s \leq \frac{\max\{c^1, c^0\}}{\min\{(x^0, s^0, x^1, s^1)\}},$$

which implies that (x, s) is bounded. The lemma is obviously true if $\theta^0 = \theta^1$. □

To prove the convexity of R_{++}, it suffices to prove that if the system

$$s = f(x) + r^0 + \theta r, \quad (x, s) > 0$$

has a solution at $\theta = 0$ and $\theta = 1$, then it has a solution for every $\theta \in [0, 1]$. Without loss of generality, we may assume $r^0 = 0$. Let $(x^0, s^0) \in S_{++}(0)$,

8.4. MONOTONE COMPLEMENTARITY PROBLEM

$(x^1, s^1) \in S_{++}(r)$, and $\max((s^0)^T x^0, (s^1)^T x^1) \leq \delta^*$. Now consider the system

$$Xs = (1-\theta)X^0 s^0 + \theta X^1 s^1 \quad \text{and} \quad s - f(x) = \theta r, \ (x,s) > 0. \tag{8.22}$$

Let
$$\Theta = \{\theta \in \mathcal{R} : \text{system (8.22) has a solution}\}.$$

Then, from the openness of R_{++} and Theorem 8.19 we derive Lemma 8.22.

Lemma 8.22 Θ *is an open set and system (8.22) has a unique solution $(x(\theta), s(\theta))$ for every $\theta \in \Theta$. Moreover, $(x(\theta), s(\theta))$ is continuous in $\theta \in \Theta$.*

We now ready to prove the following theorem.

Theorem 8.23 R_{++} *is an open convex subset of $\mathcal{R}^n$.*

Proof. The openness has been discussed earlier. Let
$$\theta^* = \inf\{\theta : [\theta, 1] \subset \Theta\}.$$

Since $1 \in \Theta$, we know by Lemma 8.22 that $\theta^* < 1$ and $\theta^* \notin \Theta$. If $\theta^* < 0$, $\theta r \in R_{++}$ for every $\theta \in [0,1]$; hence the theorem follows. Suppose on the contrary that $\theta^* > 0$. Let $\{\theta^k \in (\theta^*, 1]\}$ be a sequence converging to θ^*. Then, for $k = 1, 2, ...$, we have

$$X(\theta^k)s(\theta^k) = (1-\theta^k)X^0 s^0 + \theta^k X^1 s^1,$$

$$s(\theta^k) - f(x(\theta^k)) = \theta^k r, \quad \text{and} \quad (x(\theta^k), s(\theta^k)) > 0,$$

which implies that

$$(x(\theta^k))^T s(\theta^k) = (1-\theta^k)(x^0)^T s^0 + \theta^k (x^1)^T s^1 \leq \delta^*.$$

Thus, $(x(\theta^k), s(\theta^k))$ is in the union of $C_{++}(\theta r, \delta^*)$, $\theta \in [0,1]$. Since the union is a bounded subset by Lemma 8.21, we may assume without loss of generality that the sequence $\{(x(\theta^k), s(\theta^k))\}$ converges to some $(\bar{x}, \bar{s}) \in \mathcal{R}^{2n}_+$. By the continuity of the mapping $Xy : \mathcal{R}^{2n}_+ \to \mathcal{R}^n_+$, the point $(\bar{x}, \bar{s})$ satisfies

$$\bar{X}\bar{s} = (1-\theta^*)X^0 s^0 + \theta^* X^1 s^1 \in \mathcal{R}^n_{++}.$$

By the continuity of $f : \mathcal{R}^n_{++} \to \mathcal{R}^n$, we then see that

$$\bar{s} - f(\bar{x}) = (1-\theta^*)0 + \theta^* r = \theta^* r.$$

This implies that $(\bar{x}, \bar{s}) \in S_{++}(\theta^* r)$ and $\theta^* \in \Theta$, the contradiction of which is $\theta^* \notin \Theta$. Thus we have shown that R_{++} is a convex set. □

8.4.2 A homogeneous MCP model

Consider an augmented homogeneous model related to (MCP):

$(HMCP)$ minimize $x^T s + \tau \kappa$

$$\text{s.t.} \quad \begin{pmatrix} s \\ \kappa \end{pmatrix} = \begin{pmatrix} \tau f(x/\tau) \\ -x^T f(x/\tau) \end{pmatrix}, \ (x, \tau, s, \kappa) \geq 0.$$

Let

$$\psi(x, \tau) = \begin{pmatrix} \tau f(x/\tau) \\ -x^T f(x/\tau) \end{pmatrix} : \overset{\circ}{\mathcal{R}}_+^{n+1} \to \mathcal{R}^{n+1}. \tag{8.23}$$

Then, it is easy to verify that $\nabla \psi$ is positive semi-definite as shown in the following lemma.

Lemma 8.24 *Let ∇f be positive semi-definite in $\mathcal{R}_+^n$. Then $\nabla \psi$ is positive semi-definite in $\overset{\circ}{\mathcal{R}}_+^{n+1}$, i.e. given $(x; \tau) > 0$,*

$$(d_x; d_\tau)^T \nabla \psi(x, \tau)(d_x; d_\tau) \geq 0$$

for any $(d_x; d_\tau) \in \mathcal{R}^{n+1}$, where

$$\nabla \psi(x, \tau) = \begin{pmatrix} \nabla f(x/\tau) & f(x/\tau) - \nabla f(x/\tau)(x/\tau) \\ -f(x/\tau)^T - (x/\tau)^T \nabla f(x/\tau)^T & (x/\tau)^T \nabla f(x/\tau)(x/\tau) \end{pmatrix}. \tag{8.24}$$

Proof.

$$\begin{aligned} & (d_x; d_\tau)^T \nabla \psi(x, \tau)(d_x; d_\tau) \\ = \ & d_x^T \nabla f(x/\tau) d_x - d_x^T \nabla f(x/\tau) x (d_\tau/\tau) \\ & -(d_\tau/tau) x^T \nabla f(x/\tau)^T d_x + d_\tau^2 x^T \nabla f(x/\tau) x / \tau^2 \\ = \ & (d_x - d_\tau x/\tau)^T \nabla f(x/\tau)(d_x - d_\tau x/\tau) \end{aligned} \tag{8.25}$$

□

Furthermore, we have the following theorem, part of which is related to Exercise 8.6.

Theorem 8.25 *Let ψ be given by (8.23). Then,*

i) *ψ is a continuous homogeneous function in $\overset{\circ}{\mathcal{R}}_+^{n+1}$ with degree 1 and for any $(x; \tau) \in \overset{\circ}{\mathcal{R}}_+^{n+1}$*

$$(x; \tau)^T \psi(x, \tau) = 0$$

and

$$(x; \tau)^T \nabla \psi(x, \tau) = -\psi(x, \tau)^T.$$

8.4. MONOTONE COMPLEMENTARITY PROBLEM

ii) If f is a continuous monotone mapping from $\mathcal{R}_+^n$ to $\mathcal{R}^n$, then ψ is a continuous monotone mapping from $\overset{\circ}{\mathcal{R}}_+^{n+1}$ to $\mathcal{R}^{n+1}$.

iii) If f is scaled Lipschitz with $\upsilon = \upsilon_f$, then ψ is scaled Lipschitz, that is, it satisfies condition (8.21) with

$$\upsilon = \upsilon_\psi(\alpha) = \left(1 + \frac{2\upsilon_f(2\alpha/(1+\alpha))}{1-\alpha}\right)\left(\frac{1}{1-\alpha}\right).$$

iv) (HMCP) is (asymptotically) feasible and every (asymptotically) feasible point is an (asymptotically) complementary solution.

Now, let $(x^*, \tau^*, s^*, \kappa^*)$ be a maximal complementary solution for (HMCP). Then

v) (MCP) has a solution if and only if $\tau^* > 0$. In this case, $(x^*/\tau^*, s^*/\tau^*)$ is a complementary solution for (MCP).

vi) (MCP) is (strongly) infeasible if and only if $\kappa^* > 0$. In this case, $(x^*/\kappa^*, s^*/\kappa^*)$ is a certificate to prove (strong) infeasibility.

Proof. The proof of (i) is straightforward.

We leave the proof of (ii) as an exercise.

We now prove (iii). Assume $(x; \tau) \in R_{++}^{n+1}$ and let $(d_x; d_\tau)$ be given such that $\|(X^{-1}d_x; \tau^{-1}d_\tau)\|_\infty \leq \alpha < 1$. To prove ψ is scaled Lipschitz we must bound

$$\left\|\begin{pmatrix} X & 0 \\ 0 & \tau \end{pmatrix}\left(\psi(x+d_x, \tau+d_\tau) - \psi(x,\tau) - \nabla\psi(x,\tau)\begin{pmatrix} d_x \\ d_\tau \end{pmatrix}\right)\right\|_1. \tag{8.26}$$

From (8.23) and (8.24), the upper part in (8.26) is identical to

$$\begin{aligned} & X\left(f(y+d_y)(\tau+d_\tau) - f(y)\tau - (\nabla f(y)d_x + f(y)d_\tau - \nabla f(y)xd_\tau/\tau)\right) \\ = & (\tau+d_\tau)X\left(f(y+d_y) - f(y) - \nabla f(y)d_y\right) \\ = & \tau(\tau+d_\tau)Y\left(f(y+d_y) - f(y) - \nabla f(y)d_y\right), \end{aligned} \tag{8.27}$$

where

$$y = x/\tau \quad \text{and} \quad y + d_y = \frac{x+d_x}{\tau+d_\tau}, \tag{8.28}$$

that is,

$$d_y = \frac{\tau d_x - x d_\tau}{\tau(\tau+d_\tau)} = \frac{d_x - (d_\tau/\tau)x}{\tau+d_\tau}. \tag{8.29}$$

Note
$$\begin{aligned}\|Y^{-1}d_y\|_\infty &= \|\tau X^{-1}(\tau d_x - d_\tau x)/(\tau(\tau+d_\tau))\|_\infty\\ &= \|(\tau X^{-1}d_x - d_\tau e)/(\tau+d_\tau)\|_\infty\\ &\le (\|X^{-1}d_x\|_\infty + \alpha)/(1-\alpha)\\ &\le 2\alpha/(1-\alpha).\end{aligned} \qquad (8.30)$$

Per the assumption that f is scaled Lipschitz with $v = v_f$, it follows for $\alpha \in [0,1)$ that

$$\begin{aligned}&\|\tau(\tau+d_\tau)Y(f(y+d_y) - f(y) - \nabla f(y)d_y)\|_1\\ \le\;& \tau(\tau+d_\tau)v_f(2\alpha/(1-\alpha))d_y^T \nabla f(y)d_y\\ =\;& \tfrac{\tau v_f(2\alpha/(1-\alpha))}{\tau+d_\tau}(d_x - xd_\tau/\tau)^T \nabla f(y)(d_x - xd_\tau/\tau)\\ =\;& \tfrac{v_f(2\alpha/(1-\alpha))}{1+d_\tau/\tau}(d_x;d_\tau)^T \nabla\psi(x,\tau)(d_x;d_\tau)\\ \le\;& \tfrac{v_f(2\alpha/(1-\alpha))}{1-\alpha}(d_x;d_\tau)^T \nabla\psi(x,\tau)(d_x;d_\tau).\end{aligned} \qquad (8.31)$$

Next we bound the lower part of (8.26). This part is equal to

$$\begin{aligned}&\tau\left(-f(y+d_y)^T(x+d_x) - (-f(y)^T x)\right.\\ &\left.\quad -[-f(y)^T d_x - x^T \nabla f(y)^T d_x/\tau + x^T f(y)xd_\tau/\tau^2]\right)\\ =\;& \tau\bigl((x+d_x)^T(-f(y+d_y) + f(y) + \nabla f(y)d_y)\\ &\quad -(x+d_x)^T \nabla f(y)d_y + (x/\tau)^T \nabla f(y)d_y(\tau+d_\tau)\bigr)\\ =\;& \tau\bigl((x+d_x)^T(-f(y+d_y) + f(y) + \nabla f(y)d_y)\\ &\quad -(d_x - d_\tau x/\tau)^T \nabla f(y)d_y\bigr)\\ =\;& \tau^2(e + X^{-1}d_x)^T Y\bigl(-f(y+d_y) + f(y) + \nabla f(y)d_y\bigr)\\ &\quad -\tau(\tau+d_\tau)d_y^T \nabla f(y)d_y.\end{aligned}$$

Thus, using (8.25) and (8.30)

$$\begin{aligned}&|\tau\bigl(-f(y+d_y)^T(x+d_x) - (-f(y)^T x)\\ &\quad -[-f(y)^T d_x - x^T \nabla f(y)^T d_x/\tau + x^T f(y)xd_\tau/\tau^2]\bigr)|\\ \le\;& \tau^2 \|e + X^{-1}d_x\|_\infty \|Y(-f(y+d_y) + (-f(y) - \nabla f(y)d_y))\|_1\\ &\quad +|\tau(\tau+d_\tau)|d_y^T \nabla f(y)d_y\\ \le\;& \bigl(\tau^2(1+\alpha)v_f(2\alpha/(1-\alpha)) + \tau(\tau+d_\tau)\bigr)d_y^T \nabla f(y)d_y\\ =\;& \tfrac{\tau^2(1+\alpha)v_f(2\alpha/(1-\alpha))+\tau(\tau+d_\tau)}{(\tau+d_\tau)^2}(d_x^T - d_\tau x/\tau)^T \nabla f(y)(d_x - d_\tau x/\tau)\\ =\;& \tfrac{(1+\alpha)v_f(2\alpha/(1-\alpha))+(1+d_\tau/\tau)}{(1+d_\tau/\tau)^2}(d_x;d_\tau)^T \nabla\psi(x,\tau)(d_x;d_\tau)\\ \le\;& \left(\tfrac{(1+\alpha)v_f(2\alpha/(1-\alpha))}{(1-\alpha)^2} + \tfrac{1}{1-\alpha}\right)(d_x;d_\tau)^T \nabla\psi(x,\tau)(d_x;d_\tau).\end{aligned}$$
$$(8.32)$$

The sum of (8.31) and (8.32) is equal to

$$v_\psi(\alpha)(d_x;d_\tau)^T \nabla\psi(x,\tau)(d_x;d_\tau)$$

8.4. MONOTONE COMPLEMENTARITY PROBLEM

and it bounds the term in (8.26) leading to the desired result.

We leave the proof of (iv) as an exercise too.

We now prove (v). If $(x^*, \tau^*, s^*, \kappa^*)$ is a solution for $(HMCP)$ and $\tau^* > 0$, then we have

$$s^*/\tau^* = f(x^*/\tau^*) \quad \text{and} \quad (x^*)^T s^*/(\tau^*)^2 = 0,$$

that is, $(x^*/\tau^*, s^*/\tau^*)$ is a solution for (MCP). Let $(\hat{x}, \hat{s})$ be a solution to (MCP). Then $\tau = 1$, $x = \hat{x}$, $s = \hat{s}$, and $\kappa = 0$ is a solution for $(HMCP)$. Thus, every maximal solution of $(HMCP)$ must have $\tau^* > 0$.

Finally, we prove (vi). Consider the set

$$R_{++} = \{s - f(x) \in \mathcal{R}^n : \quad (x, s) > 0\}.$$

As proved in Theorem 8.23, R_{++} is an open convex set. If (MCP) is strongly infeasible, then we must have $0 \notin \hat{R}_{++}$ where $\hat{R}_{++}$ represents the closure of R_{++}. Thus, there is a hyperplane that separates 0 and $\hat{R}_{++}$, that is, there is a vector $a \in \mathcal{R}^n$ with $||a|| = 1$ and a positive number ξ such that

$$a^T(s - f(x)) \geq \xi > 0 \quad \forall \, x \geq 0, \, s \geq 0. \tag{8.33}$$

For $j = 1, 2, ..., n$, set s_j sufficiently large, but fix x and the rest of s, $a_j \geq 0$ must be true. Thus,

$$a \geq 0, \quad \text{or} \quad a \in \mathcal{R}^n_+.$$

On the other hand, for any fixed x, we set $s = 0$ and see that

$$-a^T f(x) \geq \xi > 0 \quad \forall \, x \geq 0. \tag{8.34}$$

In particular,

$$-a^T f(ta) \geq \xi > 0 \quad \forall \, t \geq 0. \tag{8.35}$$

From the monotone of f, for every $x \in \mathcal{R}^n_+$ and any $t \geq 0$ we have

$$(tx - x)^T (f(tx) - f(x)) \geq 0.$$

Thus,

$$x^T f(tx) \geq x^T f(x) \tag{8.36}$$

and

$$\lim_{t \to \infty} x^T f(tx)/t \geq 0. \tag{8.37}$$

Thus, from (8.35) and (8.37),

$$\lim_{t \to \infty} a^T f(ta)/t = 0.$$

For an $x \in \mathcal{R}_+^n$, denote

$$f^\infty(x) := \lim_{t \to \infty} f(tx)/t,$$

where $f^\infty(x)$ represents the limit of any subsequence and its values may include ∞ or $-\infty$.

We now prove $f^\infty(a) \geq 0$. Suppose that $f^\infty(a)_j < -\delta$. Then consider the vector $x = a + \epsilon e_j$ where e_j is the vector with the jth component being 1 and zeros everywhere else. Then, for ϵ sufficiently small and t sufficiently large we have

$$\begin{aligned}
x^T f(tx)/t &= (a + \epsilon e_j)^T f(t(a + \epsilon e_j))/t \\
&= a^T f(t(a + \epsilon e_j))/t + \epsilon e_j^T f(t(a + \epsilon e_j))/t \\
&< \epsilon e_j^T f(t(a + \epsilon e_j))/t \quad \text{(from (8.34))} \\
&= \epsilon \frac{f(t(a + \epsilon e_j))_j - f(ta)_j}{t} + \epsilon \frac{f(ta)_j}{t} \\
&\leq \epsilon \left(O(\epsilon) + \frac{f(ta)_j}{t} \right) \quad \text{(from continuity of } f\text{)} \\
&= \epsilon(O(\epsilon) - \delta/2) \\
&\leq -\epsilon \delta/4.
\end{aligned}$$

But this contradicts relation (8.37). Thus, we must have

$$f^\infty(a) \geq 0.$$

We now further prove that $f^\infty(a)$ is bounded. Consider

$$\begin{aligned}
0 &\leq (ta - e)^T (f(ta) - f(e))/t \\
&= a^T f(ta) - e^T f(ta)/t - a^T f(e) + e^T f(e)/t \\
&< -e^T f(ta)/t - a^T f(e) + e^T f(e)/t.
\end{aligned}$$

Taking as a limit $t \to \infty$ from both sides, we have

$$e^T f^\infty(a) \leq -a^T f(e).$$

Thus, $f^\infty(a) \geq 0$ is bounded. Again, we have $a^T f(ta) \leq -\xi$ from (8.35) and $a^T f(ta) \geq a^T f(a)$ from (8.36). Thus, $\lim a^T f(ta)$ is bounded. To summarize, ($\bar{H}MCP$) has an asymptotical solution ($x^* = a, \tau^* = 0, s^* = f^\infty(a), \kappa^* = \lim -a^T f(ta) \geq \xi$).

Conversely, if there is a bounded sequence ($x^k > 0, \tau^k > 0, s^k > 0, \kappa^k > 0$), then

$$\lim s^k = \lim \tau^k f(x^k/\tau^k) \geq 0, \quad \lim \kappa^k = \lim -(x^k)^T f(x^k/\tau^k) \geq \xi > 0.$$

8.4. MONOTONE COMPLEMENTARITY PROBLEM

Then, we claim that there is no feasible point $(x \geq 0, s \geq 0)$ such that $s - f(x) = 0$. We prove this fact by contradiction. If there is one, then

$$\begin{aligned} 0 &\leq ((x^k; \tau^k) - (x; 1))^T (\psi(x^k, \tau^k) - \psi(x, 1)) \\ &= (x^k - x)^T (\tau^k f(x^k/\tau^k) - f(x)) + (\tau^k - 1)^T (xf(x) - (x^k)^T f(x^k/\tau^k)). \end{aligned}$$

Therefore,

$$(x^k)^T f(x^k/\tau^k) \geq (x^k)^T f(x) + \tau^k x^T f(x^k/\tau^k) - \tau^k x^T f(x).$$

Since the first two terms at the right-hand side are positive and $\lim \tau^k = 0$, we must have

$$\lim (x^k)^T f(x^k/\tau^k) \geq 0,$$

which is a contradiction to $\kappa^k = -(x^k)^T f(x^k/\tau^k) \geq \xi > 0$. Also, any limit of x^k is a separating hyperplane, i.e., a certificate proving infeasibility.

□

8.4.3 The central path

Due to Theorem 8.25, we can solve (MCP) by finding a maximal complementary solution of $(HMCP)$. Select $x^0 > 0$, $s^0 > 0$, $\tau^0 > 0$ and $\kappa^0 > 0$ and let the residual vectors

$$r^0 = s^0 - \tau^0 f(x^0/\tau^0), \quad z^0 = \kappa^0 + (x^0)^T f(x^0/\tau^0).$$

Also let

$$\bar{n} = (r^0)^T x^0 + z^0 \tau^0 = (x^0)^T s^0 + \tau^0 \kappa^0.$$

For simplicity, we set

$$x^0 = e, \ \tau^0 = 1, \ s^0 = e, \ \kappa^0 = 1, \ \theta^0 = 1,$$

with

$$X^0 s^0 = e \quad \text{and} \quad \tau^0 \kappa^0 = 1.$$

Note that $\bar{n} = n + 1$ in this setting.

We present the next theorem.

Theorem 8.26 *Consider* $(HMCP)$.

i) *For any $0 < \theta \leq 1$, there exists a strictly positive point $(x > 0, \tau > 0, s > 0, \kappa > 0)$ such that*

$$\begin{pmatrix} s \\ \kappa \end{pmatrix} - \psi(x, \tau) = \begin{pmatrix} s - \tau f(x/\tau) \\ \kappa + x^T f(x/\tau) \end{pmatrix} = \theta \begin{pmatrix} r^0 \\ z^0 \end{pmatrix}. \tag{8.38}$$

ii) *Starting from* $(x^0 = e, \tau^0 = 1, s^0 = e, \kappa^0 = 1)$, *for any* $0 < \theta \leq 1$ *there is a unique strictly positive point* $(x(\theta), \tau(\theta), s(\theta), \kappa(\theta))$ *that satisfies equation (8.38) and*

$$\begin{pmatrix} Xz \\ \tau\kappa \end{pmatrix} = \theta e. \tag{8.39}$$

iii) *For any* $0 < \theta \leq 1$, *the solution* $(x(\theta), \tau(\theta), s(\theta), \kappa(\theta))$ *in [ii] is bounded. Thus,*

$$\mathcal{C}(\theta) := \left\{ (x, \tau, s, \kappa) : \begin{pmatrix} s \\ \kappa \end{pmatrix} - \psi(x, \tau) = \theta \begin{pmatrix} r^0 \\ z^0 \end{pmatrix}, \begin{pmatrix} Xz \\ \tau\kappa \end{pmatrix} = \theta e \right\} \tag{8.40}$$

for $0 < \theta \leq 1$ *is a continuous bounded trajectory.*

iv) *The limit point* $(x(0), \tau(0), s(0), \kappa(0))$ *is a maximal complementary solution for* $(HMCP)$.

Proof. We prove (i). Again, the set

$$H_{++} := \left\{ \begin{pmatrix} s \\ \kappa \end{pmatrix} - \psi(x, \tau) : \quad (x, \tau, s, \kappa) > 0 \right\}$$

is open and convex. We have $(r^0; z^0) \in H_{++}$ by construction. On the other hand, $0 \in \bar{H}_{++}$ from Theorem 8.25. Thus,

$$\theta \begin{pmatrix} r^0 \\ z^0 \end{pmatrix} \in H_{++}.$$

The proof of (ii) is due to Theorem 8.19.

We now prove (iii). Again, the existence is due to Theorem 8.19. We prove the boundedness. Assume $(x, \tau, s, \kappa) \in \mathcal{C}(\theta)$ then

$$\begin{aligned}
&(x; \tau)^T (r^0; z^0) \\
&= (x; \tau)^T (s^0; \kappa^0) - (x; \tau)^T \psi(x^0; \tau^0) \\
&= (x; \tau)^T (s^0; \kappa^0) + (s; \kappa)^T (x^0; \tau^0) - (s; \kappa)^T (x^0; \tau^0) - (x; \tau)^T \psi(x^0; \tau^0) \\
&= (x; \tau)^T (s^0; \kappa^0) + (s; \kappa)^T (x^0; \tau^0) \\
&\quad - (x^0; \tau^0)^T (\theta(r^0; z^0) + \psi(x, \tau)) - (x; \tau)^T \psi(x^0; \tau^0) \\
&= (x; \tau)^T (s^0; \kappa^0) + (s; \kappa)^T (x^0; \tau^0) \\
&\quad - \theta(x^0; \tau^0)^T (r^0; z^0) - (x^0; \tau^0)^T \psi(x, \tau) - (x; \tau)^T \psi(x^0; \tau^0) \\
&\geq (x; \tau)^T (s^0; \kappa^0) + (s; \kappa)^T (x^0; \tau^0) \\
&\quad - \theta(x^0; \tau^0)^T (r^0; z^0) - (x; \tau)^T \psi(x, \tau) - (x^0; \tau^0)^T \psi(x^0; \tau^0) \\
&= (x; \tau)^T (s^0; \kappa^0) + (s; \kappa)^T (x^0; \tau^0) - \theta(x^0; \tau^0)^T (r^0; z^0) \\
&= (x; \tau)^T (s^0; \kappa^0) + (s; \kappa)^T (x^0; \tau^0) - \theta(x^0; \tau^0)^T ((s^0; \kappa^0) - \psi(x^0, \tau^0)) \\
&= (x; \tau)^T (s^0; \kappa^0) + (s; \kappa)^T (x^0; \tau^0) - \theta(x^0; \tau^0)^T (s^0; \kappa^0).
\end{aligned}$$

8.4. MONOTONE COMPLEMENTARITY PROBLEM

Also for $0 < \theta \leq 1$,

$$\theta(x;\tau)^T(r^0;z^0) = (x;\tau)^T((s;\kappa) - \psi(x,\tau)) = (x;\tau)^T(s;\kappa)$$
$$= \theta(n+1) = \theta(x^0;\tau^0)^T(s^0;\kappa^0).$$

From the above two relations, we have

$$(x;\tau)^T(s^0;\kappa^0) + (s;\kappa)^T(x^0;\tau^0) \leq (1+\theta)(x^0;\tau^0)^T(s^0;\kappa^0).$$

Thus, $(x;\tau;s;\kappa)$ is bounded.

Finally, we prove (iv). Let $(x^*, \tau^*, s^*, \kappa^*)$ be any maximal complementarity solution for $(HMCP)$ such that

$$(s^*;\kappa^*) = \psi(x^*;\tau^*) \quad \text{and} \quad (x^*)^T s^* + \tau^* \kappa^* = 0,$$

and it is normalized by

$$(r^0;z^0)^T(x^*;\tau^*) = (r^0;z^0)^T(x^0;\tau^0) = (s^0;\kappa^0)^T(x^0;\tau^0) = (n+1).$$

For any $0 < \theta \leq 1$, let (x,τ,s,κ) be the solution on the path. Then, we have

$$((x;\tau) - (x^*;\tau^*))^T((s;\kappa) - (s^*;\kappa^*))$$
$$= ((x;\tau) - (x^*;\tau^*))^T(\psi(x;\tau) - \psi(x^*;\tau^*)) + \theta(r^0;z^0)^T((x;\tau) - (x^*;\tau^*))$$
$$\geq \theta(r^0;z^0)^T((x;\tau) - (x^*;\tau^*)).$$

Therefore,

$$(x;\tau)^T(s^*;\kappa^*) + (s;\kappa)^T(x^*;\tau^*)$$
$$\leq (x;\tau)^T(s;\kappa) - \theta(r^0;z^0)^T((x;\tau) - (x^*;\tau^*))$$
$$= (x;\tau)^T(s;\kappa) - (x;\tau)^T(s;\kappa) + \theta(r^0;z^0)^T(x^*;\tau^*)$$
$$= \theta(r^0;z^0)^T(x^*;\tau^*)$$
$$= \theta(n+1).$$

Using $x_j s_j = \theta$ we obtain,

$$(x;\tau)^T(s^*;\kappa^*) + (s;\kappa)^T(x^*;\tau^*)$$
$$= \theta \sum \frac{s_j^*}{s_j} + \frac{\kappa^*}{\tau} + \sum \frac{x_j^*}{x_j} + \frac{\tau^*}{\kappa}$$
$$\leq \theta(n+1).$$

Thus, we have

$$\frac{s_j^*}{s_j} \leq (n+1), \quad \text{and} \quad \frac{\kappa^*}{\kappa} \leq (n+1)$$

and
$$\frac{x_j^*}{x_j} \leq (n+1), \quad \text{and} \quad \frac{\tau^*}{\tau} \leq (n+1).$$

Thus, the limit point, $(x(0), \tau(0), s(0), \kappa(0))$, is a maximal complementarity solution for $(HMCP)$.

□

We now present an interior-point algorithm that generates iterates within a neighborhood of $\mathcal{C}(\theta)$. For simplicity, in what follows we let $x := (x; \tau) \in \mathcal{R}^{n+1}$, $s := (s; \kappa) \in \mathcal{R}^{n+1}$, and $r^0 := (r^0; z^0)$. Recall that, for any $x, s > 0$,

$$x^T \psi(x) = 0 \quad \text{and} \quad x^T \nabla \psi(x) = -\psi(x)^T. \tag{8.41}$$

Furthermore, ψ is monotone and satisfies the scaled Lipschitz. We will use these facts frequently in our analyses.

8.4.4 An interior-point algorithm

At iteration k with iterate $(x^k, s^k) > 0$, the algorithm solves a system of linear equations for direction (d_x, d_s) from

$$d_s - \nabla \psi(x^k) d_x = -\eta r^k \tag{8.42}$$

and

$$X^k d_s + S^k d_x = \gamma \mu^k e - X^k s^k, \tag{8.43}$$

where η and γ are proper given parameters between 0 and 1, and

$$r^k = s^k - \psi(x^k) \quad \text{and} \quad \mu^k = \frac{(x^k)^T s^k}{n+1}.$$

First we prove the following lemma:

Lemma 8.27 . *The direction (d_x, d_s) satisfies*

$$d_x^T d_s = d_x^T \nabla \psi(x^k) d_x + \eta(1 - \eta - \gamma)(n+1)\mu^k.$$

Proof. Premultiplying each side of (8.42) by d_x^T gives

$$d_x^T d_s - d_x^T \nabla \psi(x^k) d_x = -\eta d_x^T (s^k - \psi(x^k)). \tag{8.44}$$

8.4. MONOTONE COMPLEMENTARITY PROBLEM

Multiplying each side of (8.42) by x^k and using (8.41) give

$$\begin{aligned}(x^k)^T d_s + \psi(x^k) d_x &= -\eta(x^k)^T r^k \\ &= -\eta(x^k)^T (s^k - \psi(x^k)) \\ &= -\eta(x^k)^T s^k \\ &= -\eta(n+1)\mu^k.\end{aligned} \quad (8.45)$$

These two equalities in combination with (8.43) imply

$$\begin{aligned}d_x^T d_s &= d_x^T \nabla\psi(x^k) d_x - \eta(d_x^T s^k + d_s^T x^k + \eta(n+1)\mu^k) \\ &= d_x^T \nabla\psi(x^k) d_x - \eta(-(n+1)\mu^k + \gamma(n+1)\mu^k + \eta(n+1)\mu^k) \\ &= d_x^T \nabla\psi(x^k) d_x + \eta(1 - \gamma - \eta)(n+1)\mu^k.\end{aligned}$$

$\square$

For a step-size $\alpha > 0$, let the new iterate

$$x^+ := x^k + \alpha d_x > 0, \quad (8.46)$$

and

$$\begin{aligned}s^+ &:= s^k + \alpha d_s + \psi(x^+) - \psi(x^k) - \alpha \nabla\psi(x^k) d_x \\ &= \psi(x^+) + (s^k - \psi(x^k)) + \alpha(d_s - \nabla\psi(x^k) d_x) \\ &= \psi(x^+) + (s^k - \psi(x^k)) - \alpha\eta(s^k - \psi(x^k)) \\ &= \psi(x^+) + (1 - \alpha\eta) r^k.\end{aligned} \quad (8.47)$$

The last two equalities come from (8.42) and the definition of r^k. Also let

$$r^+ = s^+ - \psi(x^+).$$

Then, we have

Lemma 8.28. *Consider the new iterate (x^+, s^+) given by (8.46) and (8.47).*

i) $r^+ = (1 - \alpha\eta) r^k$

ii) $(x^+) s^+ = (x^k)^T s^k (1 - \alpha(1-\gamma)) + \alpha^2 \eta(1 - \eta - \gamma)(n+1)\mu^k.$

Proof. From (8.47)

$$\begin{aligned}r^+ &= s^+ - \psi(x^+) \\ &= (1 - \alpha\eta) r^k.\end{aligned}$$

Next we prove (ii) Using (8.41), (8.43), and Lemma 8.27, we have

$$\begin{aligned}
(x^+)^T s^+ &= (x^+)^T(s^k + \alpha d_s + \psi(x^+) - \psi(x^k) - \alpha \nabla\psi(x^k)d_x) \\
&= (x^+)^T(s^k + \alpha d_s) - (x^+)^T(\psi(x^k) + \alpha \nabla\psi(x^k)d_x) \\
&= (x^+)^T(s^k + \alpha d_s) - (x^k + \alpha d_x)^T(\psi(x^k) + \alpha \nabla\psi(x^k)d_x) \\
&= (x^+)^T(s^k + \alpha d_s) - \alpha(x^k)^T \nabla\psi(x^k)d_x - \alpha d_x^T \psi(x^k) - \alpha^2 d_x^T \nabla\psi(x^k)d_x \\
&= (x^+)^T(s^k + \alpha d_s) - \alpha^2 d_x^T \nabla\psi(x^k)d_x \\
&= (x^k + \alpha d_x)^T(s^k + \alpha d_s) - \alpha^2 d_x^T \nabla\psi(x^k)d_x \\
&= (x^k)^T s^k + \alpha(d_x^T s^k + d_s^T x^k) + \alpha^2(d_x^T d_s - d_x^T \nabla\psi(x^k)d_x) \\
&= (x^k)^T s^k + \alpha(d_x^T s^k + d_s^T x^k) + \alpha^2 \eta(1 - \eta - \gamma)(n+1)\mu^k \\
&= (1 - \alpha(1-\gamma))(x^k)^T s^k + \alpha^2 \eta(1-\eta-\gamma)(n+1)\mu^k.
\end{aligned}$$

$\square$

This lemma shows that, for setting $\eta = 1 - \gamma$, the infeasibility residual and the complementarity gap are reduced at exactly the same rate, as in the homogeneous linear programming algorithm. Now we prove the following:

Theorem 8.29 *Assume that ψ is scaled Lipschitz with $\upsilon = \upsilon_\psi$ and at iteration k*

$$\|X^k s^k - \mu^k e\| \le \beta \mu^k, \quad \mu^k = \frac{(x^k)^T s^k}{n+1},$$

where

$$\beta = \frac{1}{3 + 4\upsilon_\psi(\sqrt{2}/2)} \le 1/3.$$

Furthermore, let $\eta = \beta/\sqrt{n+1}$, $\gamma = 1 - \eta$, and $\alpha = 1$ in the algorithm. Then, the new iterate

$$x^+ > 0, \quad s^+ = \psi(x^+) + (1-\eta)r^k > 0,$$

and

$$\|X^+ s^+ - \mu^+ e\| \le \beta \mu^+, \quad \mu^+ = \frac{(x^+)^T s^+}{n+1}.$$

Proof. It follows from Lemma 8.28 that $\mu^+ = \gamma \mu^k$. From (8.43) we further have

$$S^k d_x + X^k d_s = -X^k s^k + \mu^+ e.$$

Hence,

$$D^{-1} d_x + D d_s = -(X^k S^k)^{-1/2}(X^k s^k - \mu^+ e),$$

where $D = (X^k)^{1/2}(S^k)^{-1/2}$. Note

$$d_x^T d_s = d_x^T \nabla\psi(x^k) d_x \ge 0$$

8.4. MONOTONE COMPLEMENTARITY PROBLEM

from Lemma 8.27 and $\gamma = 1 - \eta$. This together with the assumption of the theorem imply

$$\|D^{-1}d_x\|^2 + \|Dd_s\|^2 \le \|(X^k S^k)^{-1/2}(X^k s^k - \mu^+ e)\|^2 \le \frac{\|X^k s^k - \mu^+ e\|^2}{(1-\beta)\mu^k}.$$

Also note

$$\begin{aligned}
\|X^k s^k - \mu^+ e\|^2 &= \|X^k s^k - \mu^k e + (1-\gamma)\mu^k e\|^2 \\
&= \|X^k s - \mu^k e\|^2 + ((1-\gamma)\mu^k)^2 \|e\|^2 \\
&\le (\beta^2 + \eta^2(n+1))(\mu^k)^2 \\
&= 2\beta^2 (\mu^k)^2.
\end{aligned}$$

Thus,

$$\|(X^k)^{-1} d_x\| = \|(X^k S^k)^{-1/2} D^{-1} d_x\| \le \frac{\|D^{-1} d_x\|}{\sqrt{(1-\beta)\mu^k}}$$

$$\le \frac{\sqrt{2}\beta\mu^k}{(1-\beta)\mu^k} = \frac{\sqrt{2}\beta}{1-\beta} \le \frac{\sqrt{2}}{2},$$

since $\beta \le 1/3$. This implies that $x^+ = x^k + d_x > 0$. Furthermore, we have

$$\begin{aligned}
\|D_x d_s\| &= \|D^{-1} D_x D d_s\| \\
&\le \|D^{-1} d_x\| \|D d_s\| \\
&\le (\|D^{-1} d_x\|^2 + \|Dd_s\|^2)/2 \\
&\le \|(X^k S^k)^{-1/2}(X^k s^k - \mu^+)\|^2/2 \\
&\le \frac{\|X^k s^k - \mu^+ e\|^2}{2(1-\beta)\mu^k} \\
&\le \frac{2\beta^2(\mu^k)^2}{2(1-\beta)\mu^k} \\
&= \frac{\beta^2 \mu^k}{1-\beta},
\end{aligned}$$

and

$$d_x^T d_s = d_x^T D^{-1} D d_s \le \|D^{-1} d_x\| \|Dd_s\| \le \frac{\beta^2 \mu^k}{1-\beta}.$$

Consider

$$\begin{aligned}
& X^+ s^+ - \mu^+ e \\
&= X^+(s^k + d_s + \psi(x^+) - \psi(x^k) - \nabla\psi(x^k)d_x) - \mu^+ e \\
&= (X^k + D_x)(s^k + d_s) - \mu^+ e + X^+(\psi(x^+) - \psi(x^k) - \nabla\psi(x^k)d_x) \\
&= D_x d_s + X^+(\psi(x^+) - \psi(x^k) - \nabla\psi(x^k)d_x).
\end{aligned}$$

Using that ψ is the scaled Lipschitz, $d_x^T \nabla\psi(x^k)d_x = d_x^T d_s$ and the above four relations we obtain

$$\begin{aligned}
&\|X^+s^+ - \mu^+ e\| \\
&= \|D_x d_s + X^+(\psi(x^+) - \psi(x^k) - \nabla\psi(x^k)d_x)\| \\
&= \|D_x d_s + (X^k)^{-1}X^+ X^k(\psi(x^+) - \psi(x^k) - \nabla\psi(x^k)d_x)\| \\
&\leq \|D_x d_s\| + \|(X^k)^{-1}X^+\|_\infty \|X^k(\psi(x^+) - \psi(x^k) - \nabla\psi(x^k)d_x)\|_1 \\
&\leq \|D_x d_s\| + 2 \|X^k(\psi(x^+) - \psi(x^k) - \nabla\psi(x^k)d_x)\|_1 \\
&\leq \|D_x d_s\| + 2v_\psi(\sqrt{2}/2) d_x^T \nabla\psi(x^k)d_x \\
&= \|D_x d_s\| + 2v_\psi(\sqrt{2}/2) d_x^T d_s \\
&\leq \frac{\beta^2 \mu^k}{1-\beta} + 2v_\psi(\sqrt{2}/2) \frac{\beta^2 \mu^k}{1-\beta} \\
&\leq \frac{(1 + 2v_\psi(\sqrt{2}/2))\beta^2 \mu^k}{1-\beta}.
\end{aligned}$$

Finally, $\beta = 1/(3 + 4v_\psi(\sqrt{2}/2))$ implies that

$$\frac{(1 + 2v_\psi(\sqrt{2}/2))\beta^2 \mu^k}{1-\beta} \leq \frac{\beta}{2}$$

and

$$\|X^+ s^+ - \mu^+ e\| \leq \beta \mu^k / 2 < \beta\gamma\mu^k = \beta\mu^+.$$

It is easy to verify that $x^+ > 0$ and $\|X^+ s^+ - \mu^+ e\| < \beta\mu^+$ implies $s^+ > 0$.

□

The above theorem shows that the homogeneous algorithm will generate a sequence $(x^k, s^k) > 0$ with $(x^{k+1}, s^{k+1}) := (x^+, s^+)$ such that $s^k = \psi(x^k) + r^k$ and $\|X^k s^k - \mu^k\| \leq \beta\mu^k$, where both $\|r^k\|$ and $(x^k)^T s^k$ converge to zero at a global rate $\gamma = 1 - \beta/\sqrt{n+1}$. We see that if $v_\psi(\sqrt{2}/2)$ is a constant, or $v_f(2/(1+\sqrt{2}))$ is a constant in (MCP) due to (iii) of Theorem 8.25, then it results in an $O(\sqrt{n}\log(1/\epsilon))$ iteration algorithm with error ϵ. It generates a maximal solution for $(HMCP)$, which is either a solution or a certificate proving infeasibility for (MCP), due to (v) and (vi) of Theorem 8.25.

One more comment is that our results should hold for the case where $f(x)$ is a continuous *monotone* mapping from $\overset{\circ}{\mathcal{R}}^n_+$ to $\mathcal{R}^n$. In other words, $f(x)$ may not exist at the boundary of $\mathcal{R}^n_+$.

8.5 Notes

The result in Section 8.1 is based on the work of Bai and Ye [41], where they try to find an estimate of a unknown parameter vector for a single input-single output discrete-time dynamic system. The result can be also used to find the frontier of multiple objectives.

Various centers were considered for the central-section method as we mentioned earlier. Goffin, Haurie and Vial [143], Sonnevend [383], and Ye [470] were among the first to propose the analytic central-section or cutting plane method. Its complexity issues were addressed by Atkinson and Vaidya [38], Goffin, Luo and Ye [144, 145], and Nesterov [325]. In particular, Atkinson and Vaidya developed a scheme to delete unnecessary inequalities and managed to prove a polynomial analytic central-section algorithm. The analytic central-section method was used and tested for a variety of large scale problems, where they performed quite well; see, for example, Bahn, Goffin, Vial and Merle [39], Bahn, Merle, Goffin and Vial [40], and Mitchell [284, 285].

The primal potential reduction algorithm for positive semi-definite programming is due to Alizadeh [10, 9], in which Ye has "suggested studying the primal-dual potential function for this problem" and "looking at symmetric preserving scalings of the form $X_0^{-1/2} X X_0^{-1/2}$," and to Nesterov and Nemirovskii [327], and the primal-dual algorithm described here is due to Nesterov and Todd [329, 330]. One can also develop a dual potential reduction algorithm. In general, consider

$$(PSP) \quad \inf \quad C \bullet X$$
$$\text{s.t.} \quad A \bullet X = b, \ X \in K,$$

and its dual

$$(PSD) \quad \sup \quad b^T y$$
$$\text{s.t.} \quad A^* \bullet Y + S = C, \ S \in K,$$

where K is a convex homogeneous cone.

Interior-point algorithms compute a search direction (d_X, d_Y, d_S) and a new strictly feasible primal-dual pair X^+ and $(Y^+; S^+)$ is generated from

$$X^+ = X + \alpha d_X, \ Y^+ = Y + \beta d_Y, \ S^+ = S + \beta d_S,$$

for some step-sizes α and β.

The search direction (d_X, d_Y, d_S) is determined by the following equations:

$$A \bullet d_X = 0, \quad d_S = -A^* \bullet d_Y \quad \text{(feasibility)} \tag{8.48}$$

and

$$d_X + F''(S)d_S = -\frac{n+\rho}{X \bullet S}X - F'(S) \quad \text{(dual scaling)}, \qquad (8.49)$$

or

$$d_S + F''(X)d_X = -\frac{n+\rho}{X \bullet S}S - F'(X) \quad \text{(primal scaling)}, \qquad (8.50)$$

or

$$d_S + F''(Z)d_X = -\frac{n+\rho}{X \bullet S}S - F'(X) \quad \text{(joint scaling)}, \qquad (8.51)$$

where Z is chosen to satisfy

$$S = F''(Z)X. \qquad (8.52)$$

The differences among the three algorithms are the computation of the search direction and their theoretical close-form step-sizes. All three generate an ϵ-optimal solution (X, Y, S), i.e.,

$$X \bullet S \leq \epsilon$$

in a guaranteed polynomial time.

Other primal-dual algorithms for positive semi-definite programming are in Alizadeh, Haeberly and Overton [11, 12], Boyd, Ghaoui, Feron and Balakrishnan [72], Helmberg, Rendl, Vanderbei and Wolkowicz [181], Jarre [203], de Klerk, Roos and Terlaky.[224], Kojima, Shindoh and Hara [233], Monteiro and Zhang [307], Nesterov, Todd and Ye [331], Potra and Sheng [348], Shida, Shindoh and Kojima [379], Sturm and Zhang [392], Tseng [423], Vandenberghe and Boyd [440, 441], and references therein. Efficient interior-point algorithms are also developed for optimization over the second-order cone; see Andersen and Christiansen [21], Lobo, Vandenberghe and Boyd [246], and Xue and Ye [462]. These algorithms have established the best approximation complexity results for some combinatorial problems.

The scaled Lipschitz condition used in Section 8.4 was proposed by Kortanek and Zhu [237] for linearly constrained convex minimization, related to the work of Monteiro and Adler [300], and later extended by Potra and Ye [351] for the monotone complementary problem. This condition is included in a more general condition analyzed by Nesterov and Nemirovskii [327], den Hertog [182], den Hertog, Jarre, Roos and Terlaky [183], and Jarre [204].

Results in Section 8.4.1 are based on Kojima, Megiddo and Mizuno [225]. A similar augmented transformation in Section 8.4.2 has been discussed in Ye and Tse [480] and it is closely related to the recession function in convex analyses of Rockafellar [364]. All other results in Section 8.4 are based on Andersen and Ye [18]. Interior-point algorithms for convex programming include: Abhyankar, Morin and Trafalis [1] for multiple objective optimization, Anstreicher, den Hertog, Roos and Terlaky [29], Ben–Daya and Shetty [50], Bonnans and Bouhtou [68], Carpenter, Lustig, Mulvey and Shanno [78], Goldfarb and Liu [149], Jarre [202], Kapoor and Vaidya [216], Mehrotra and Sun [279], Pardalos, Ye and Han [340], Ponceleon [344], Ye [467], Ye and Tse [480], etc. for quadratic programming; Ding and Li [100], Güler [174], Harker and Xiao [179] Ji, Potra and Huang [205], Polak, Higgins and Mayne [341], Shanno and Simantiraki [376], Sun and Zhao [398], Tseng [422], Zhao [487], for the monotone complementarity problem; Ben–Tal and Nemirovskii [51], Faybusovich [112, 113], Goldfarb and Scheinberg [150], Güler [175], Güler and Tuncel [176], Luo and Sun [249], Luo, Sturm and Zhang [250, 251], Monteiro and Pang [303], Muramatsu [312], Ramana [354] Ramana, Tuncel and Wolkowicz [355], Saigal and Lin [371], Todd, Toh and Tutuncu [414], Vandenberghe, Boyd, and Wu [442], for nonpolyhedral optimization; Anstreicher and Vial [32], Byrd, Gilbert and Nocedal [76], Coleman and Li [89] Güler [175], den Hertog, Roos and Terlaky [185], Kortanek, Potra and Ye [234], Mehrotra and Sun [280], Monteiro [297], Nash and Sofer [314], Potra and Ye [350], Sun and Qi [397], Tanabe [192], Wang, Monteiro, and Pang [455], Zhang [481], for nonlinear programming; Asic and Kovacevic-Vujcic [35], Ferris and Philpott [114], Todd [407], Tunçel and Todd [433], for semi-infinite programming; Birge and Holmes [56], Birge and Qi [57], for stochastic programming.

A new homotopy method, the smoothing method, for solving complementarity problems and its relation to interior-point methods have been developed by Burke and Xu [75], B. Chen and Harker [81], C. Chen and Mangasarian [82], X. Chen [83], Gabriel and Moré [129], Hotta and Yoshise [188], Kanzow [215], and Qi and Chen [353].

Applications, decompositions, inexact iteration, and special data structures of interior-point algorithms were described in Bixby, Gregory, Lustig, Marsten and Shanno [59], Choi and Goldfarb [84], Christiansen and Kortanek [87], Gondzio [155], Han, Pardalos and Ye [178], Ito, Kelley and Sachs [196], Kaliski [212], Pardalos, Ye and Han [340], Ponnambalam, Vannelli and Woo [345], Resende and Veiga [361], Tone [418], Wallacher and Zimmermann [454].

8.6 Exercises

8.1 *Prove Lemma 8.7*

8.2 Prove the following convex quadratic inequality
$$(Ay+b)^T(Ay+b) - c^T y - d \leq 0$$
is equivalent to a matrix inequality
$$\begin{pmatrix} I & Ay+b \\ (Ay+b)^T & c^T y + d \end{pmatrix} \succeq 0.$$
Using this relation to formulate a convex quadratic minimization problem with convex quadratic inequalities as a (PSD) problem.

8.3 Prove Corollary 8.12.

8.4 Prove Lemma 8.16.

8.5 Describe and analyze a dual potential algorithm for positive semidefinite programming in the standard form.

8.6 Let $f(x) : \mathcal{R}^n \to \mathcal{R}^n$ be a monotone function over a open convex set $\Omega = \{x : f(x) \geq 0, \ x \geq 0\}$, i.e.,
$$(y-x)^T(f(y) - f(x)) \geq 0$$
for all x, y in Ω. Prove for any convergent sequence $(x^k \geq 0, \tau^k > 0)$ such that
$$\lim \tau^k f(x^k/\tau^k) \geq 0 \quad \text{and} \quad \lim \tau^k = 0,$$
we must have
$$\lim (x^k)^T f(x^k/\tau^k) \geq 0.$$

8.7 If (MCP) has a solution, then the solution set is convex and it contains a maximal solution (x^*, s^*) where the number positive components in (x^*, s^*) is maximal. Moreover, the indices of those positive components are invariant among all maximal solutions for (MCP).

8.8 Let $f(x) : \mathcal{R}^n \to \mathcal{R}^n$ be a monotone function in $\mathcal{R}^n_+$. Then for any $(x' \geq 0; \tau' > 0)$ and $(x'' \geq 0; \tau'' > 0)$, prove
$$(x' - x'')^T (\tau' f(x'/\tau') - \tau'' f(x''/\tau'')) +$$
$$(\tau' - \tau'')^T ((x'')^T f(x''/\tau'') - (x')^T f(x'/\tau')) \geq 0.$$

8.9 Prove Theorem 8.19.

8.10 Prove Lemma 8.20.

8.11 Prove Lemma 8.22.

8.12 Prove (ii) and (iv) of Theorem 8.25.

Chapter 9

Nonconvex Optimization

The aim of this chapter is to describe some results in interior-point algorithms for solving "hard" problems, such as the fractional programming problem, the non-monotone linear complementarity problem (LCP), and the general quadratic programming (QP) problem, and to suggest some directions in which future progress might be made. These problems play important roles in optimization theory. In one sense they are continuous optimization and fundamental sub-problems for general nonlinear programming, but they are also considered the most challenging combinatorial optimization problems.

9.1 von Neumann Economic Growth Problem

Consider the von Neumann economic growth (NEG) problem:

$$\gamma^* := \max\{\gamma \mid \exists\, y \neq 0 : \quad y \geq 0; \quad (B - \gamma A)y \geq 0\},$$

where $A = \{a_{ij} \geq 0\}$ and $B = \{b_{ij} \geq 0\}$ are two given nonnegative matrices in $\mathcal{R}^{m \times n}$. Each row-index i stands for a "good," while each column index j stands for a "process." Process j can convert a_{ij} units of good i, in one time period, into b_{ij} units of good i. So a process uses goods as materials or inputs, and gives goods as products or outputs. Matrix B is referred as output matrix, and A is the input matrix. Component y_j of y denotes the "intensity" by which we let process j work. Vector By gives the amounts of outputs produced, and Ay gives the amounts of inputs consumed, during one time period. Then, γ represents the growth factor at intensity y. So

$By \geq \gamma Ay$ requires that, for each good i, the amount of good i produced in period t is at least the amount of good i required in period $(t+1)$ with the growth factor γ. The NEG problem is to find the largest growth factor using an optimal intensity vector.

The NEG problem is an example of a fractional programming problem, and the results developed in this section can be extended to problems of the form:
$$\max_y \left\{ \min_j \frac{g_j(y)}{f_j(y)} \right\},$$
where $f, g : \mathcal{R}^n \to \mathcal{R}_+^m$.

To solve the NEG problem, we will make a few assumptions. Since they are basic assumptions for a meaningful economic growth model, they pose minimal restriction.

Assumption 9.1 *A has no all-zero columns.*

Note that γ^* is bounded above based on this assumption. In fact,
$$\gamma^* \leq \max_j \frac{\sum_{i=1}^m b_{ij}}{\sum_{i=1}^m a_{ij}}.$$

There is a related dual NEG problem,
$$\eta^* := \min\{\eta \mid \exists\, x \neq 0 : \quad x \geq 0; \quad (\eta A - B)^T x \geq 0\}.$$

We further assume:

Assumption 9.2 *B has no all-zero rows.*

Then, η^* is also bounded below. But a duality overlap may exist, i.e., it is possible $\eta^* < \gamma^*$. However, under the following irreducibility assumption the model is well behaved: $\eta^* = \gamma^*$.

Assumption 9.3 *There is no (proper) subset S of the rows and no subset T of the columns such that $A_{ij} = 0$ for all $i \in S$ and all $j \in T$, and such that for all $i \in \{1, 2, ..., m\} \setminus S$, $B_{ij} > 0$ for some $j \in T$.*

Moreover, the γ-level set,
$$\Gamma(\gamma) := \{y \in R^n : e^T y = 1, \quad y \geq 0; \quad (B - \gamma A)y \geq 0\}, \qquad (9.1)$$
has a nonempty interior for $\gamma < \gamma^*$, meaning in this paper that
$$\overset{\circ}{\Gamma}(\gamma) = \{y \in R^n : e^T y = 1, \quad y > 0; \quad (B - \gamma A)y > 0\}$$

9.1. VON NEUMANN ECONOMIC GROWTH PROBLEM

is nonempty for $\gamma < \gamma^*$. Note that we have replaced $y \neq 0$ with $e^T y = 1$ in the NEG problem, where e is the vector of all ones. This is without loss of generality since the system is homogeneous in y.

Obviously, the underlying decision problem related to the NEG problem can be solved in polynomial time: Given matrices A and B, and a number γ, does the linear system

$$\{e^T y = 1, \quad y \geq 0; \quad (B - \gamma A) y \geq 0\}$$

has a feasible point? Let $0 \leq \gamma^* \leq R$ for some positive R. Then, one can use the *bisection* method to generate a $\bar{\gamma}$ such that $\gamma^* - \epsilon \leq \bar{\gamma} \leq \gamma^*$ in $O(\log(R/\epsilon))$ bisection steps where each step solves a linear feasibility problem with data A, B and γ. Therefore, the NEG problem is polynomially solvable.

In this section, we directly solve the NEG problem using an interior-point algorithm that is in the spirit of earlier central-section algorithms. That is, it reduces the size of the γ-level set $\Gamma(\gamma)$ and its associated max-potential by by increasing γ. In each iteration, the algorithm increases γ and applies the primal-dual Newton procedure to compute an approximate analytic center of $\Gamma(\gamma)$. We show that the algorithm generates an ϵ approximate solution in $O((m+n)(\log(R/\epsilon) + \log(m+n)))$ iterations where each iteration solves a system of $(m+n)$ linear equations.

9.1.1 Max-potential of $\Gamma(\gamma)$

We apply the analytic center and the max-potential theory to the inequality system $\Gamma(\gamma)$ of (9.1) for a fixed $\gamma < \gamma^*$. Recall that the max-potential of $\Gamma(\gamma)$, if it has a nonempty interior, is defined as

$$\mathcal{B}(\gamma) := \mathcal{B}(\Gamma(\gamma)) = \max_{y \in \Gamma(\gamma)} \left(\sum_{i=1}^{m} \log(By - \gamma Ay)_i + \sum_{j=1}^{n} \log y_j \right).$$

In the following, we frequently use the slack vector $s := By - \gamma Ay$.

Clearly, since B is a nonnegative matrix, we must have $\gamma^* \geq 0$. Without loss of generality, we further assume that $\Gamma(0)$ of (9.1) has a nonempty interior. This fact automatically holds if Assumption 9.2 holds. We also need the system $\Gamma(\gamma)$ to have a bounded and nonempty interior for all $\gamma < \gamma^*$, so that the analytic center and the max-potential are well defined for all $\gamma < \gamma^*$. As we discussed earlier, this is true under Assumptions 9.1 and 9.3. In what follows, we replace Assumption 9.3 by a weaker assumption:

Assumption 9.4 *There exists an optimal intensity vector $y^* \in \Gamma(\gamma^*)$ such that*
$$\bar{\pi} := (B + A)y^* > 0.$$

The following lemma, similar to Proposition 4.2, will help determine how small $\mathcal{B}(\gamma)$ must be to ensure that γ and any $y \in \Gamma(\gamma)$ is an ϵ-approximate solution.

Lemma 9.1 *Let $0 \leq \gamma < \gamma^*$. Then, the system $\Gamma(\gamma)$ under Assumptions 9.1, 9.2, and 9.4 has a bounded and nonempty interior. Moreover, the max-potential*
$$\mathcal{B}(\gamma) \geq (m+n)\log(\frac{\gamma^* - \gamma}{\gamma^* + 1}) + \sum_{i=1}^{m}\log(\bar{\pi}_i/2) + n\log(\bar{\xi}/n),$$
where
$$\bar{\xi} = \max\left\{z : (\frac{1}{\gamma^*+1}\bar{\pi} + Ae/n)z \leq \frac{1}{2(\gamma^*+1)}\bar{\pi}\right\} > 0.$$

Proof. $\Gamma(\gamma)$ being bounded is obvious since $e^T y = 1$ and $y \geq 0$. Let y^* be the one in $\Gamma(\gamma^*)$ such that
$$e^T y^* = 1; \quad y^* \geq 0; \quad (B - \gamma^* A)y^* \geq 0, \tag{9.2}$$
and it satisfies
$$(B + A)y^* = \bar{\pi} > 0.$$
Let
$$\delta = \frac{\gamma(1+\gamma^*)}{1+\gamma}.$$
Then, for $0 \leq \gamma < \gamma^*$ we must have
$$\gamma < \delta < \gamma^*.$$
The left inequality follows from $1 + \gamma^* > 1 + \gamma$, and the right inequality follows from $\gamma < \gamma^*$, which implies $\gamma(1+\gamma^*) < \gamma^*(1+\gamma)$, which further implies
$$\gamma^* > \frac{\gamma(1+\gamma^*)}{1+\gamma} = \delta.$$
Thus, we have
$$(\gamma^* - \delta)(B+A)y^* = (\gamma^* - \delta)\bar{\pi} > 0. \tag{9.3}$$
Adding two inequalities (9.2) and (9.3), we have
$$((1+\gamma^* - \delta)B - \delta A)y^* \geq (\gamma^* - \delta)\bar{\pi} > 0,$$

9.1. VON NEUMANN ECONOMIC GROWTH PROBLEM 281

or
$$\left(B - \frac{\delta}{1+\gamma^*-\delta}A\right)y^* \geq \frac{\gamma^*-\delta}{1+\gamma^*-\delta}\bar{\pi} > 0,$$

which implies
$$(B - \gamma A)y^* \geq \frac{\gamma^*-\gamma}{1+\gamma^*}\bar{\pi} > 0,$$

since
$$\gamma = \frac{\delta}{1+\gamma^*-\delta}.$$

Therefore, there is an $0 < \omega < 1$ such that
$$\bar{y} = (1-\omega)y^* + \omega e/n > 0$$

and
$$(B - \gamma A)\bar{y} > 0.$$

That is, $\bar{y}$ is in the interior of $\Gamma(\gamma)$.

Specifically, let $\omega = \bar{\xi}(\gamma^* - \gamma)/\gamma^*$. Then, we have

$$\begin{aligned}
&(B - \gamma A)\bar{y} \\
&= (B - \gamma A)((1-\omega)y^* + \omega e/n) \\
&= (B - \gamma A)\left(\left(1 - \frac{\bar{\xi}(\gamma^*-\gamma)}{\gamma^*}\right)y^* + \frac{\bar{\xi}(\gamma^*-\gamma)}{\gamma^*}e/n\right) \\
&= \left(1 - \frac{\bar{\xi}(\gamma^*-\gamma)}{\gamma^*}\right)(B - \gamma A)y^* + \frac{\bar{\xi}(\gamma^*-\gamma)}{\gamma^*}(B - \gamma A)e/n \\
&\geq \left(1 - \frac{\bar{\xi}(\gamma^*-\gamma)}{\gamma^*}\right)(B - \gamma A)y^* - \frac{\bar{\xi}(\gamma^*-\gamma)}{\gamma^*}\gamma A e/n \\
&\geq \left(1 - \frac{\bar{\xi}(\gamma^*-\gamma)}{\gamma^*}\right)(B - \gamma A)y^* - \bar{\xi}(\gamma^*-\gamma)Ae/n \\
&\geq \left(1 - \frac{\bar{\xi}(\gamma^*-\gamma)}{\gamma^*}\right)\frac{\gamma^*-\gamma}{1+\gamma^*}\bar{\pi} - \bar{\xi}(\gamma^*-\gamma)Ae/n \\
&\geq (1-\bar{\xi})\frac{\gamma^*-\gamma}{1+\gamma^*}\bar{\pi} - \bar{\xi}(\gamma^*-\gamma)Ae/n \\
&= \frac{\gamma^*-\gamma}{2(1+\gamma^*)}\bar{\pi} + \frac{\gamma^*-\gamma}{2(1+\gamma^*)}\bar{\pi} - \bar{\xi}\frac{\gamma^*-\gamma}{1+\gamma^*}\bar{\pi} - \bar{\xi}(\gamma^*-\gamma)Ae/n \\
&= \frac{\gamma^*-\gamma}{2(1+\gamma^*)}\bar{\pi} + (\gamma^*-\gamma)\left(\frac{1}{2(1+\gamma^*)}\bar{\pi} - \bar{\xi}\left(\frac{1}{1+\gamma^*}\bar{\pi} + Ae/n\right)\right) \\
&\geq \frac{\gamma^*-\gamma}{2(1+\gamma^*)}\bar{\pi}.
\end{aligned}$$

Furthermore, we have

$$\bar{y} \geq we/n = \frac{\bar{\xi}(\gamma^* - \gamma)}{\gamma^*}e/n > \frac{\bar{\xi}(\gamma^* - \gamma)}{1+\gamma^*}e/n.$$

Note that the max-potential

$$\mathcal{B}(\gamma) \geq \sum_{i=1}^m \log(B\bar{y} - \gamma A\bar{y})_i + \sum_{j=1}^n \log \bar{y}_j,$$

which together with the above two bounds give the desired result.

□

Note that the above lemma may not hold in general.

Example 9.1 *Let*

$$B = \begin{pmatrix} 2 & 1 \\ 0 & 1 \end{pmatrix} \quad \text{and} \quad A = \begin{pmatrix} 1 & 0 \\ 0 & 1 \end{pmatrix}.$$

Then, for this problem $\gamma^* = 2$. However, for any $1 < \gamma < 2$, it must be true that $(y_1 = 1, y_2 = 0)$ is on the boundary of $\Gamma(0)$.

It can be shown that Assumption 9.3 implies Assumption 9.4. However, Assumption (9.4) is weaker than Assumption 9.3. Consider

$$B = \begin{pmatrix} 1 & 0 \\ 0 & 1 \end{pmatrix} \quad \text{and} \quad A = \begin{pmatrix} 1 & 0 \\ 0 & 1 \end{pmatrix}.$$

This system is reducible but $\Gamma(\gamma)$ has a nonempty interior for all $\gamma < \gamma^* = 1$, and it satisfies Assumption 9.4.

There is also an economic interpretation for Assumption 9.4: if $\Gamma(\gamma)$ has an empty interior for some $\gamma < \gamma^*$, then at every optimal intensity vector y^*

$$(B + A)y^* \not> 0,$$

which implies that for some good i

$$b_i y^* = a_i y^* = 0$$

at every optimal intensity vector y^*, where b_i and a_i are the ith row of the matrices B and A, respectively. Thus, the ith good is irrelevant, as it is neither produced nor consumed, and so can be removed from further

9.1. VON NEUMANN ECONOMIC GROWTH PROBLEM

consideration. Therefore, we can set up a reduced NEG problem (both row and column dimensions may be reduced) such as

$$\max\{\gamma \mid \exists\, y \neq 0 : \quad y \geq 0; \quad (B_2 - \gamma A_2)y \geq 0; \quad (b_i + a_i)y = 0\},$$

where B_2 and A_2 are the remaining matrices of B and A after deleting b_i and a_i, respectively. Now the reduced γ-level set will have a nonempty interior for any $\gamma < \gamma^*$.

We now prove two more lemmas to indicate that the max-potential of $\Gamma(\gamma)$ is an effective *measure* for the γ-level set of the NEG problem.

Lemma 9.2 *Let $\gamma^0 < \gamma^1 < \gamma^*$. Then,*

$$\mathcal{B}(\gamma^0) > \mathcal{B}(\gamma^1).$$

Proof. Let y^1 be the analytic center of $\Gamma(\gamma^1)$. Then, since

$$(B - \gamma^0 A)y^1 = (B - \gamma^1 A)y^1 + (\gamma^1 - \gamma^0)Ay^1$$

and $(\gamma^1 - \gamma^0)Ay^1 \geq 0$, we must have

$$(B - \gamma^0 A)y^1 \geq (B - \gamma^1 A)y^1 > 0.$$

This shows that y^1 is in the interior of $\Gamma(\gamma^0)$. Moreover, $(\gamma^1 - \gamma^0)Ay^1 \neq 0$, since A has no all-zero columns and $y^1 > 0$. Thus,

$$\begin{aligned}
\mathcal{B}(\gamma^0) &\geq \sum_{i=1}^{m} \log(By^1 - \gamma^0 Ay^1)_i + \sum_{j=1}^{n} \log y_j^1 \\
&= \sum_{i=1}^{m} \log(By^1 - \gamma^1 Ay^1 + (\gamma^1 - \gamma^0)Ay^1)_i + \sum_{j=1}^{n} \log y_j^1 \\
&> \sum_{i=1}^{m} \log(By^1 - \gamma^1 Ay^1)_i + \sum_{j=1}^{n} \log y_j^1 \\
&= \mathcal{B}(\gamma^1).
\end{aligned}$$

□

Directly from Lemma 9.1, we have Lemma 9.3.

Lemma 9.3 *Let $\min(\bar{\pi}) \geq 1/R$ and $\bar{\xi} \geq 1/R$ for some positive R, and let γ satisfy $(\gamma^* - \gamma)/(\gamma^* + 1) \geq \epsilon$. Then, the max-potential*

$$\mathcal{B}(\gamma) \geq (m+n)\log(\epsilon/R) - m\log 2 - n\log n.$$

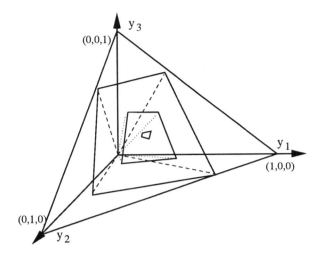

Figure 9.1. Illustration of the level set $\Gamma(\gamma)$ on the simplex polytope; the size of $\Gamma(\gamma)$ decreases as γ increases.

Lemma 9.3 indicates that if we reduce

$$\mathcal{B}(\gamma) \leq -O((m+n)(\log(R/\epsilon) + \log n)),$$

then it must be true that $(\gamma^* - \gamma)/(1 + \gamma^*) < \epsilon$ and any $y \in \Gamma(\gamma)$ is an ϵ approximate solution. The algorithm actually returns an approximate (analytic) center of $\Gamma(\gamma)$.

9.1.2 Approximate analytic centers of $\Gamma(\gamma)$

Most of technical results presented in this section are related to those discussed earlier in Section 2.2. Let $0 \leq \gamma^0 < \gamma^*$. Note that for any fixed γ, the $m+n$ inequalities defining $\Gamma(\gamma)$ have corresponding "primal" variables x and z. Thus, the analytic center y^0 of $\Gamma(\gamma^0)$, or simply Γ^0, satisfies the following conditions:

$$\begin{pmatrix} X^0 s^0 \\ Z^0 y^0 \end{pmatrix} = \begin{pmatrix} e \\ e \end{pmatrix}, \qquad (9.4)$$

where (s^0, y^0) is feasible with $\gamma = \gamma^0$ for the system

$$s - (B - \gamma A)y = 0; \quad e^T y = 1; \quad (s, y) \geq 0, \qquad (9.5)$$

and (x^0, z^0) is feasible with $\gamma = \gamma^0$ for the system

$$(B - \gamma A)^T x + z = (m+n)e; \quad (x, z) \geq 0. \qquad (9.6)$$

9.1. VON NEUMANN ECONOMIC GROWTH PROBLEM

Now let y^1 be the analytic center of $\Gamma^1 = \Gamma(\gamma^1)$ with $\gamma^1 = \gamma^0 + \Delta\gamma$. For a suitable choice of $\Delta\gamma$, we can show that $\gamma^0 < \gamma^1 < \gamma^*$ and

$$\mathcal{B}(\gamma^1) \leq \mathcal{B}(\gamma^0) - \Omega(1).$$

We prove a lemma on how to select $\Delta\gamma$.

Lemma 9.4 *Let*

$$\Delta\gamma = \beta \min\left\{\frac{1}{\|X^0 A Y^0 e\|}, \frac{1}{\|Y^0 A^T X^0 e\|}\right\},$$

for some constant $0 < \beta < 1$ and let $\gamma^1 = \gamma^0 + \Delta\gamma$. Furthermore, update the slack variables, letting

$$\bar{s} = s^0 - \Delta\gamma A y^0 \quad \text{and} \quad \bar{y} = y^0,$$

and

$$\bar{x} = x^0 \quad \text{and} \quad \bar{z} = z^0 + \Delta\gamma A^T x^0.$$

Then, $(\bar{s}, \bar{y})$ and $(\bar{x}, \bar{z})$ are feasible for systems (9.5) and (9.6) with $\gamma = \gamma^1$, respectively. Moreover,

$$\left\|\begin{pmatrix} \bar{X}\bar{s} - e \\ \bar{Z}\bar{y} - e \end{pmatrix}\right\| \leq \sqrt{2}\beta,$$

and

$$(\bar{s}, \bar{y}) > 0 \quad \text{and} \quad (\bar{x}, \bar{z}) > 0.$$

Proof. The two equations in systems (9.5) and (9.6) for $\gamma = \gamma^1$ can easily be verified. The inequality for the norm can be proved from relation (9.4) and the choice of $\Delta\gamma$.

$$\begin{aligned}
\|\bar{X}\bar{s} - e\|^2 &= \|X^0(s^0 - \Delta\gamma A y^0) - e\|^2 \\
&= \|\Delta\gamma X^0 A Y^0 e\|^2 \\
&\leq \beta^2.
\end{aligned}$$

Similarly,

$$\begin{aligned}
\|\bar{Z}\bar{y} - e\|^2 &= \|Y^0(z^0 + \Delta\gamma A^T x^0) - e\|^2 \\
&= \|\Delta\gamma Y^0 A^T X^0 e\|^2 \\
&\leq \beta^2.
\end{aligned}$$

These relations also imply that

$$\bar{s} > 0 \quad \text{and} \quad \bar{z} > 0,$$

since
$$\bar{x} = x^0 > 0 \quad \text{and} \quad \bar{y} = y^0 > 0,$$
which concludes the proof. □

The above lemma establishes a fact that $(\bar{s}, \bar{y})$ and $(\bar{x}, \bar{z})$ are approximate centers for systems (9.5) and (9.6) with $\gamma = \gamma^1$, respectively. Thus, $(\bar{x}, \bar{z})$ or $(\bar{s}, \bar{y})$ can be used as an initial pair of Newton's method to generate the new center pair (x^1, z^1) and (s^1, y^1).

We now state another technical result that will be used in the next section in conjunction with Lemma 9.4.

Proposition 9.5 *Let $H = \{h_{ij}\}$ be a nonnegative $m \times n$-matrix. Then*
$$\|He\| \leq e^T H e,$$
and
$$\|H^T e\| \leq e^T H^T e = e^T H e.$$

9.1.3 Central-section algorithm

We now present a conceptual algorithm, which uses the perfect center, to illustrate the basic idea of our approach:

1. Let $\gamma^0 = 0$. Use the primal-dual Newton method, which is described below, to generate the analytic center y^0 of $\Gamma^0 = \Gamma(\gamma^0)$, and set $k := 0$.

2. Let
$$\Delta \gamma^k = \beta \min \left\{ \frac{1}{\|X^k A Y^k e\|}, \frac{1}{\|Y^k A^T X^k e\|} \right\}$$
for some constant $0 < \beta < 1$ and update variables γ, s and z like in Lemma 9.4. Then use the Newton procedure to generate the analytic center y^{k+1} of $\Gamma^{k+1} = \Gamma(\gamma^{k+1})$.

3. If $\mathcal{B}(\gamma^{k+1}) > -O((m+n)(\log(R/\epsilon) + \log n))$, then let $k := k+1$ and return to Step 2.

The primal-dual Newton procedure is applied to the systems of equations in (9.4), (9.5) and (9.6) to find the analytic center of $\Gamma(\gamma)$. Let $(\bar{x}, \bar{z}, \bar{s}, \bar{y})$ be defined in Lemma 9.4 and repeatedly solve for (d_s, d_y) and (d_x, d_z):

$$\begin{aligned}
\bar{S} d_x + \bar{X} d_s &= e - \bar{x}\bar{s}, \\
\bar{Y} d_z + \bar{Z} d_y &= e - \bar{y}\bar{z}, \\
(B - \gamma^1 A)^T d_x + d_z &= 0, \\
d_s - (B - \gamma^1 A) d_y &= 0.
\end{aligned} \quad (9.7)$$

9.1. VON NEUMANN ECONOMIC GROWTH PROBLEM

Then, let
$$\bar{x} := \bar{x} + d_x \quad \text{and} \quad \bar{z} := \bar{z} + d_z,$$
and
$$\bar{s} := \bar{s} + d_s \quad \text{and} \quad \bar{y} := \bar{y} + d_y.$$

We now analyze the algorithm using approximate center pairs (s^0, y^0) and (x^0, z^0) that are feasible for systems (9.5) and (9.6) with $\gamma = \gamma^0$, respectively, and
$$\left\| \begin{pmatrix} X^0 s^0 - e \\ Z^0 y^0 - e \end{pmatrix} \right\| \leq \delta. \tag{9.8}$$

As we proved in Theorem 3.2(iii),
$$\mathcal{B}(\gamma^0) \geq \mathcal{B}(s^0, y^0) := \sum_{i=1}^m \log s_i^0 + \sum_{j=1}^n \log y_j^0 \geq \mathcal{B}(\gamma^0) - \frac{\delta^2}{2(1-\delta)}.$$

Thus, $\mathcal{B}(s, y)$ is close to $\mathcal{B}(\gamma)$ when (s, y) is an approximate center of $\Gamma(\gamma)$, and it can be used to terminate the algorithm.

The following lemma is an analogue to Lemma 9.4.

Lemma 9.6 *Let positive constants δ and β satisfy $\delta + \sqrt{2}\beta < 1$. Let (x^0, z^0) and (s^0, y^0) be a δ-approximate center pair for $\Gamma(\gamma^0)$. Let $\Delta\gamma$, γ^1, $(\bar{s}, \bar{y})$ and $(\bar{x}, \bar{z})$ be selected as in Lemma 9.4. Then, $(\bar{s}, \bar{y})$ and $(\bar{x}, \bar{z})$ are feasible for systems (9.5) and (9.6) with $\gamma = \gamma^1$, respectively. Moreover,*
$$\left\| \begin{pmatrix} \bar{X}\bar{s} - e \\ \bar{Z}\bar{y} - e \end{pmatrix} \right\| \leq \delta + \sqrt{2}\beta < 1,$$
and
$$(\bar{s}, \bar{y}) > 0 \quad \text{and} \quad (\bar{x}, \bar{z}) > 0.$$

Now, using $(\bar{x}, \bar{z})$ and $(\bar{s}, \bar{y})$ as the initial pair, we apply the Newton procedure described by (9.7) to generate a new approximate center pair (s^1, y^1) and (x^1, z^1) for $\Gamma(\gamma^1)$. Note that we terminate the procedure in *one* step, then assign
$$(s^1, y^1) = (\bar{s} + d_s, \bar{y} + d_y) \quad \text{and} \quad (x^1, z^1) = (\bar{x} + d_x, \bar{z} + d_z).$$

Note that (iv) of Theorem 3.2 and (9.6) indicate that
$$\left\| \begin{pmatrix} X^1 s^1 - e \\ Z^1 y^1 - e \end{pmatrix} \right\| \leq \frac{\sqrt{2}}{4} \frac{(\delta + \sqrt{2}\beta)^2}{1 - \delta - \sqrt{2}\beta}, \tag{9.9}$$

and
$$\|(Y^0)^{-1}(Y^1 - Y^0)\| \leq \frac{\delta + \sqrt{2}\beta}{1 - \delta - \sqrt{2}\beta}. \tag{9.10}$$

Next, for suitable constants δ and β, for example,
$$\delta = 1/12 \quad \text{and} \quad \beta = 1/(4\sqrt{2}), \tag{9.11}$$
we prove that the potential value at (s^1, y^1) in $\Gamma^1 = \Gamma(\gamma^1)$ is reduced by a constant from the value at (s^0, y^0) in $\Gamma^0 = \Gamma(\gamma^0)$.

Theorem 9.7 *Let δ and β be chosen as in (9.11), and (x^0, y^0, s^0, z^0) satisfy (9.8). Let*
$$\Delta\gamma = \beta \min\left\{\frac{1}{\|X^0 A Y^0 e\|}, \frac{1}{\|Y^0 A^T X^0 e\|}\right\}$$
and $\gamma^1 = \gamma^0 + \Delta\gamma$. Let (s^1, y^1) and (x^1, y^1) be generated in one step of the Newton procedure. Then,
$$\gamma^0 < \gamma^1 < \gamma^*,$$
$$\left\|\begin{pmatrix} X^1 s^1 - e \\ Z^1 y^1 - e \end{pmatrix}\right\| \leq \delta,$$
and
$$\mathcal{B}(s^1, y^1) \leq \mathcal{B}(s^0, y^0) - \Omega(1).$$

Proof. $\gamma^1 > \gamma^0$ because $\Delta\gamma > 0$ and $\gamma^1 < \gamma^*$ because $\Gamma^1 = \Gamma(\gamma^1)$ has a nonempty interior. From inequality (9.9) we have
$$\left\|\begin{pmatrix} X^1 s^1 - e \\ Z^1 y^1 - e \end{pmatrix}\right\| \leq \frac{\sqrt{2}}{4}\frac{1}{6} < \delta.$$

Now we prove the potential reduction inequality. We have
$$\begin{aligned}
(x^0)^T s^1 + (z^0)^T y^1 &= (x^0)^T (B - \gamma^1 A) y^1 + (z^0)^T y^1 \\
&= (y^1)^T ((B - \gamma^0 A)^T x^0 + z^0) - \Delta\gamma (x^0)^T A y^1 \\
&= (m+n) e^T y^1 - \Delta\gamma (x^0)^T A y^1 \\
&= (m+n) - \Delta\gamma (x^0)^T A y^1.
\end{aligned}$$

Thus,
$$\begin{aligned}
\prod_{i=1}^m (x_i^0 s_i^1) \prod_{i=1}^n (z_i^0 y_i^1) &\leq \left(\frac{(x^0)^T s^1 + (z^0)^T y^1}{m+n}\right)^{m+n} \\
&= \left(1 - \frac{\Delta\gamma (x^0)^T A y^1}{m+n}\right)^{m+n} \\
&\leq \exp(-\Delta\gamma (x^0)^T A y^1).
\end{aligned}$$

9.1. VON NEUMANN ECONOMIC GROWTH PROBLEM

Moreover, using $(x^0)^T A y^1 > 0$, Proposition 9.5 and relation (9.10), we have

$$\begin{aligned}
\Delta\gamma(x^0)^T A y^1 &= \Delta\gamma e^T X^0 A Y^1 e \\
&= \beta \min\left\{\frac{1}{\|X^0 A y^0\|}, \frac{1}{\|Y^0 A^T x^0\|}\right\} e^T X^0 A Y^1 e \\
&\geq \beta \frac{e^T X^0 A Y^1 e}{e^T X^0 A Y^0 e} \\
&\geq \beta \left(1 - \frac{\delta + \sqrt{2}\beta}{1 - \delta - \sqrt{2}\beta}\right) \\
&= \beta \frac{1 - 2(\delta + \sqrt{2}\beta)}{1 - \delta - \sqrt{2}\beta} .
\end{aligned}$$

Finally, we have

$$\begin{aligned}
&\mathcal{B}(s^1, y^1) - \mathcal{B}(s^0, y^0) \\
&= \sum_{i=1}^m \log \frac{s_i^1}{s_i^0} + \sum_{i=1}^n \log \frac{y_i^1}{y_i^0} \\
&= \sum_{i=1}^m \log(x_i^0 s_i^1) + \sum_{i=1}^n \log(z_i^0 y_i^1) - \sum_{i=1}^m \log(x_i^0 s_i^0) - \sum_{i=1}^n \log(z_i^0 y_i^0) \\
&\leq -\beta \frac{1 - 2(\delta + \sqrt{2}\beta)}{1 - \delta - \sqrt{2}\beta} - \sum_{i=1}^m \log(x_i^0 s_i^0) - \sum_{i=1}^n \log(z_i^0 y_i^0) \\
&\leq -\beta \frac{1 - 2(\delta + \sqrt{2}\beta)}{1 - \delta - \sqrt{2}\beta} + \frac{\delta^2}{2(1 - \delta)} \quad \text{(from Lemma 3.1 and (9.8)).}
\end{aligned}$$

One can verify

$$\beta \frac{1 - 2(\delta + \sqrt{2}\beta)}{1 - \delta - \sqrt{2}\beta} - \frac{\delta^2}{2(1 - \delta)} = \frac{1}{8\sqrt{2}} - \frac{1}{264} > 0.$$

$\square$

We now formally state the algorithm.

Algorithm 9.1 *Let $\gamma^0 = 0$. Then generate a δ-approximate center y^0 of $\Gamma^0 = \Gamma(\gamma^0)$. Set $k := 0$.*
__While__ $\mathcal{B}(s^{k+1}, y^{k+1}) > -O((m+n)(\log(R/\epsilon) + \log n))$ __do__

1. Let

$$\Delta\gamma^k = \beta \min\left\{\frac{1}{\|X^k A Y^k e\|}, \frac{1}{\|Y^k A^T X^k e\|}\right\}$$

for some constant $0 < \beta < 1$ and let $\gamma^{k+1} = \gamma^k + \Delta\gamma^k$. Let

$$\bar{s} = s^k - \Delta\gamma Ay^k \quad \text{and} \quad \bar{y} = y^k,$$

and

$$\bar{x} = x^k \quad \text{and} \quad \bar{z} = z^k + \Delta\gamma A^T x^k.$$

2. Solve for (d_s, d_y) and (d_x, d_z) from (9.7) and let

$$x^{k+1} = \bar{x} + d_x \quad \text{and} \quad z^{k+1} = \bar{z} + d_z,$$

and

$$s^{k+1} = \bar{s} + d_s \quad \text{and} \quad y^{k+1} = \bar{y} + d_y.$$

3. Let $k := k + 1$ and return to Step 1.

It is well known that an initial δ-approximate center pair, (z^0, y^0) and (x^0, z^0), can be generated in no more than $O((m + n)(\log(R/\epsilon) + \log n))$ interior-point algorithm iterations. Thus, we conclude the following:

Theorem 9.8 *Algorithm 9.1, with a suitable choice of δ and β, terminates in $k = O((m + n)(\log(R/\epsilon) + \log n))$ iterations and each iteration solves a system of $(m+n)$ linear equations. The resulting $y^k \in \Gamma(\gamma^k)$ and γ^k satisfy*

$$0 < \frac{\gamma^* - \gamma}{1 + \gamma^*} < \epsilon.$$

The algorithm also generates the optimal dual vector. More precisely, we prove the following result.

Proposition 9.9 *Any limit point of*

$$\lim_{k \to \infty} \frac{x^k}{e^T x^k},$$

where x^k is generated by Algorithm 9.1, is a solution for the dual NEG problem with $\eta = \eta^*$ under Assumptions 9.1, 9.2, and 9.4.

Proof. For simplicity, we assume that (x^k, y^k, s^k, z^k) is exactly centered, i.e., it satisfies relations (9.4), (9.5), and (9.6). Since at least one component of s^k converges to zero, $e^T x^k = e^T (S^k)^{-1} e$ tends to $+\infty$. Moreover, from (9.6) we have

$$(B - \gamma^k A)^T \frac{x^k}{e^T x^k} + \frac{z^k}{e^T x^k} = (m + n) \frac{e}{e^T x^k}.$$

9.2. LINEAR COMPLEMENTARITY PROBLEM

Thus, the right-hand vector of the above equation converges to zero. Since $z^k/e^T x^k > 0$ for all k,

$$\lim_{k\to\infty} (B - \gamma^k A)^T \frac{x^k}{e^T x^k} = - \lim_{k\to\infty} \frac{z^k}{e^T x^k} \leq 0.$$

Furthermore, under Assumptions 9.1, 9.2 and 9.4 we have $\gamma^k \to \gamma^* = \eta^*$. Therefore, any limit point of the sequence of positive $x^k/e^T x^k$ is a solution for the dual.

□

Finally we turn our attention to the question raised earlier, that is, what happens if $\Gamma(\gamma)$ has an empty interior for some $\gamma < \gamma^*$? It turns out that there exists a nice duality theorem for the NEG problem, that is, under Assumptions 9.1 and 9.2, $\Gamma(\gamma)$ has a nonempty interior for all $\gamma < \eta^* \leq \gamma^*$ (see Kemeny et al. and Gale in Section 9.6). Thus, the algorithm discussed in this paper will precisely generate η^* under only Assumptions 9.1 and 9.2. Similarly, the η-level set of the dual has an nonempty interior for all $\eta > \gamma^* \geq \eta^*$. Thus, one can apply the algorithm for solving the dual to generate γ^* in the same manner. Thus, we can solve the NEG problem under only Assumptions 9.1 and 9.2, which are basic assumptions for a meaningful economic growth model.

9.2 Linear Complementarity Problem

This section analyzes the linear complementarity problem described in Section 1.3.7. We have used $\mathcal{F}$ to denote the "feasible region," i.e.,

$$\mathcal{F} = \{(x,s) : s = Mx + q,\ x \geq 0 \text{ and } s \geq 0\}.$$

We further assume, without loss of generality, that

$$\overset{\circ}{\mathcal{F}} = \{(x,s) : s = Mx + q,\ x > 0 \text{ and } s > 0\}$$

is nonempty. The problem was solved in Chapters 4 and 7 for the case when M is monotone.

We present a potential reduction algorithm in this section to solve the general case. Similar to solving the LP problem, we will describe a "condition-based" iteration complexity bound for solving the LCP. This condition number characterizes the degree of difficulty of the LCP solution when a potential reduction algorithm is used. We show how the condition number depends on the data (M, q).

9.2.1 Potential reduction algorithm

We use the familiar primal-dual potential function

$$\psi(x,s) = \psi_{n+\rho}(x,s) = (n+\rho)\log(x^T s) - \sum_{j=1}^{n}\log(x_j s_j)$$

for an interior feasible point (x,s). As described in Chapter 2, $\rho \geq 0$.
Starting from an interior point (x^0, s^0) with

$$\psi(x^0, s^0) =: \psi^0,$$

the potential reduction algorithm generates a sequence of interior feasible points $\{x^k, s^k\}$ terminating at a point such that $(x^k)^T s^k \leq \epsilon$. Such a point is found when

$$\psi(x^k, s^k) \leq \rho\log\epsilon + n\log n$$

since, from the arithmetic-geometric mean inequality,

$$n\log((x^k)^T s^k) - \sum_{j=1}^{n}\log(x_j^k s_j^k) \geq n\log(n) \geq 0.$$

Note that $\psi(x,s) \leq \psi^0$ implies that $x^T s \leq \psi^0/\rho$. Hence, the boundedness of $\{(x,s) \in \mathcal{F} : x^T s \leq \psi^0/\rho)\}$ guarantees the boundedness of $\{(x,s) \in \overset{\circ}{\mathcal{F}} : \psi(x,s) \leq \psi^0\}$.

To achieve a potential reduction, we again use the scaled gradient projection method. The gradient vector of the potential function with respect to x is

$$\nabla\psi_x = \frac{n+\rho}{\Delta}s - X^{-1}e$$

and with respect to s is

$$\nabla\psi_s = \frac{n+\rho}{\Delta}x - S^{-1}e,$$

where $\Delta = x^T s$. At the kth iteration, we solve the following linear program subject to an ellipsoid constraint:

$$\begin{aligned}
(EP) \quad \text{minimize} \quad & \nabla^T\psi_{x^k}d_x + \nabla^T\psi_{s^k}d_s \\
\text{s.t.} \quad & d_s = Md_x \\
& \|(X^k)^{-1}d_x\|^2 + \|(S^k)^{-1}d_s\|^2 \leq \alpha^2 < 1.
\end{aligned}$$

Denote by d_x and d_s the minimal solution for (EP). Then, we have

$$\begin{pmatrix} (X^k)^{-1}d_x \\ (S^k)^{-1}d_s \end{pmatrix} = -\alpha\frac{p^k}{\|p^k\|}, \tag{9.12}$$

9.2. LINEAR COMPLEMENTARITY PROBLEM

where
$$p^k = \begin{pmatrix} p_x^k \\ p_s^k \end{pmatrix} = \begin{pmatrix} \frac{n+\rho}{\Delta^k} X^k(s^k + M^T\pi) - e \\ \frac{n+\rho}{\Delta^k} S^k(x^k - \pi) - e \end{pmatrix}, \qquad (9.13)$$

and
$$\pi = ((S^k)^2 + M(X^k)^2 M^T)^{-1}(S^k - MX^k)\left(X^k s^k - \frac{\Delta^k}{n+\rho}e\right). \qquad (9.14)$$

From the concavity of log function and Lemma 3.1 (also see Exercise 9.4),

$$\psi(x^k + d_x, s^k + d_s) - \psi(x^k, s^k) \le -\alpha \|p^k\| + \frac{\alpha^2}{2}\left(n + \rho + \frac{1}{1-\alpha}\right). \qquad (9.15)$$

Letting
$$\alpha = \min\left\{\frac{\|p^k\|}{n+\rho+2}, \frac{1}{n+\rho+2}\right\} \le 1/2, \qquad (9.16)$$

We have
$$\psi(x^k + d_x, s^k + d_s) - \psi(x^k, s^k) \le -\min\left\{\frac{\|p^k\|^2}{2(n+\rho+2)}, \frac{1}{2(n+\rho+2)}\right\}. \qquad (9.17)$$

The algorithm can be described as follows:

Algorithm 9.2 *Given $x^0, s^0 > 0$ and $s^0 = Mx^0 + q$ and $k := 0$.*
While $(x^k)^T s^k \ge \epsilon$ do

1. *Compute π of (9.14) and p^k of (9.13), and select α of (9.16); construct d_x and d_s of (9.12).*

2. *Let $x^{k+1} = x^k + d_x$ and $s^{k+1} = s^k + d_s$.*

3. *Let $k := k+1$ and return to Step 1.*

Clearly from inequality (9.17), $\|p^k\|^2$ can be used to measure the potential reduction at the kth iteration of the potential reduction algorithm. For any $x, s \in \mathcal{R}_+^n$, let
$$g(x,s) = \frac{n+\rho}{\Delta} Xs - e$$

and
$$H(x,s) = 2I - (XM^T - S)(S^2 + MX^2 M^T)^{-1}(MX - S).$$

Note that $H(x,s)$ is positive semi-definite (PSD), and

$$\|p^k\|^2 = g^T(x^k, s^k) H(x^k, s^k) g(x^k, s^k).$$

Recall that $\|g(x,s)\|_H^2$ denotes $g^T(x,s)H(x,s)g(x,s)$. Then, we define a condition number for the LCP (M, q) as

$$\gamma(M, q, \epsilon) = \inf\{\|g(x,s)\|_H^2 : x^T s \geq \epsilon,\ \psi(x,s) \leq \psi^0 \text{ and } (x,s) \in \overset{\circ}{\mathcal{F}}\}. \quad (9.18)$$

We now derive Theorem 9.10.

Theorem 9.10 *The potential reduction algorithm with $\rho = \theta(n) > 0$ solves the LCP for which $\gamma(M, q, \epsilon) > 0$ in*

$$O\left(\frac{\psi^0 + \rho \log(1/\epsilon) - n\log n}{\xi(\gamma(M, q, \epsilon))}\right)$$

iterations and each iteration solves a system of linear equations in at most $O(n^3)$ operations, where

$$\xi(\gamma(M, q, \epsilon)) = \min\left\{\frac{\gamma(M, q, \epsilon)}{2(n+\rho+2)}, \frac{1}{2(n+\rho+2)}\right\}.$$

Proof. Since $\overset{\circ}{\mathcal{F}}$ is nonempty, by solving a linear program in polynomial time, we can find an approximate analytic center (x^0, s^0) of $\mathcal{F}$. Due to (9.15), (9.16), and (9.17) the potential function is reduced by $O(\xi(\gamma(M,q,\epsilon)))$ at each iteration. Hence, in total

$$O((\psi^0 + \rho \log(1/\epsilon) - n\log n)/\xi(\gamma(M,q,\epsilon)))$$

iterations we have $\psi(x^k, s^k) < \rho\log\epsilon + n\log n$ and $(x^k)^T s^k < \epsilon$.

□

Corollary 9.11 *An instance (M,q) of the LCP is solvable in polynomial time if $\gamma(M, q, \epsilon) > 0$ and if $1/\gamma(M, q, \epsilon)$ is bounded above by a polynomial in $\log(1/\epsilon)$ and n.*

The condition number $\gamma(M, q, \epsilon)$ represents the degree of difficulty for the potential reduction algorithm in solving the LCP (M, q). The larger the condition number, the easier the LCP problem. We know that some LCPs are very hard, and some are easy. Here, the condition number builds a connection from easy LCPs to hard LCPs. In other words, the degree of difficulty continuously shifts from easy LCPs to hard LCPs.

9.2. LINEAR COMPLEMENTARITY PROBLEM

9.2.2 A class of LCPs

Instances of LCP can be separated into various classes that can be characterized by their condition number. Beginning with a class of problems that are simple to solve, we describe a sequence of propositions concerning the condition number of progressively more difficult problems.

Proposition 9.12 *Let $\rho \geq n$. Then, for M being a diagonal and PSD matrix, and any $q \in \mathcal{R}^n$,*
$$\gamma(M, q, \epsilon) \geq n.$$

Proof. If M is diagonal and PSD, then the matrix
$$I - (XM^T - S)(S^2 + MX^2M^T)^{-1}(MX - S)$$
is diagonal. It is also PSD since the jth diagonal component is
$$1 - \frac{(M_{jj}x_j - s_j)^2}{s_j^2 + M_{jj}^2 x_j^2} = \frac{2M_{jj}x_j s_j}{s_j^2 + M_{jj}^2 x_j^2} \geq 0.$$

Therefore, for all $(x, s) \in \mathcal{\mathring{F}}$ and $\rho \geq n$,
$$\gamma(M, q, \epsilon) \geq \|g(x,s)\|^2 \geq \frac{\rho^2}{n} \geq n.$$

□

Proposition 9.13 *Let $\rho \geq n + \sqrt{2n}$. Then, for any PSD matrix M and any $q \in \mathcal{R}^n$,*
$$\gamma(M, q, \epsilon) \geq 1.$$

We leave its proof to the reader.

Instances with PSD matrices are examples of a larger class of problems whose condition number is greater than 1. To define this class, we start with the following lemma.

Lemma 9.14 $\|p^k\| < 1$ *implies*
$$s^k + M^T \pi^k > 0; \qquad x^k - \pi^k > 0$$
and
$$\frac{2n - \sqrt{2n}}{n + \rho} \Delta^k < \bar{\Delta} < \frac{2n + \sqrt{2n}}{n + \rho} \Delta^k,$$
where $\bar{\Delta} = (x^k)^T(s^k + M^T \pi^k) + (s^k)^T(x^k - \pi^k)$.

Proof. The proof is by contradiction. Let $\bar{s} = s^k + M^T \pi^k$ and $\bar{x} = x^k - \pi^k$. It is obvious that if $\bar{s} \not> 0$ or $\bar{x} \not> 0$, then

$$\|p^k\|^2 \geq 1.$$

On the other hand, we have

$$\begin{aligned}\|p^k\|^2 &= \left(\frac{n+\rho}{\Delta^k}\right)^2 \left\| \begin{pmatrix} X^k \bar{s} \\ s^k \bar{x} \end{pmatrix} - \frac{\bar{\Delta}}{2n} e \right\|^2 + \left\| \frac{(n+\rho)\bar{\Delta}}{2n\Delta^k} e - e \right\|^2 \\ &\geq \left(\frac{(n+\rho)\bar{\Delta}}{2n\Delta^k} - 1\right)^2 2n.\end{aligned}$$

Hence, the following must be true

$$\left(\frac{(n+\rho)\bar{\Delta}}{2n\Delta^k} - 1\right)^2 2n < 1,$$

that is,

$$\frac{2n - \sqrt{2n}}{n+\rho} \Delta^k < \bar{\Delta} < \frac{2n + \sqrt{2n}}{n+\rho} \Delta^k.$$

$\square$

$\bar{\Delta}$ can be further expressed as

$$\bar{\Delta} = 2\Delta^k - q^T \pi^k.$$

Now let

$\Sigma^+(M, q)$
$= \{\pi : x^T s - q^T \pi < 0, \ x - \pi > 0 \text{ and } s + M^T \pi > 0 \text{ for some } (x, s) \in \overset{\circ}{\mathcal{F}}\}.$

Then, using Lemma 9.14 we have the following propositions:

Proposition 9.15 *Let $\Sigma^+(M,q)$ be empty for an LCP (M,q). Then, for $\rho \geq n + \sqrt{2n}$,*

$$\gamma(M, q, \epsilon) \geq 1.$$

Proposition 9.16 *Let*

$$\{\pi : x^T s - q^T \pi > 0, \ x - \pi > 0 \text{ and } s + M^T \pi > 0 \text{ for some } (x,s) \in \overset{\circ}{\mathcal{F}}\}$$

be empty for an LCP (M,q). Then, for $0 < \rho \leq n - \sqrt{2n}$

$$\gamma(M, q, \epsilon) \geq 1.$$

9.2. LINEAR COMPLEMENTARITY PROBLEM

Proof. The proof again results from Lemma 9.14.

□

Now, let
$$\mathcal{G} = \{(M,q) : \overset{\circ}{\mathcal{F}} \text{ is nonempty and } \Sigma^+(M,q) \text{ is empty}\}.$$

It may not be possible in polynomial time to tell if an LCP problem (M,q) is an element of $\mathcal{G}$ (this is also true for some other LCP classes published so far). However, the co-problem, to tell whether an LCP problem (M,q) is not in $\mathcal{G}$, can be solved in polynomial time. We can simply run the potential reduction algorithm for the LCP problem. In polynomial time the algorithm either gives the solution or concludes that (M,q) is not in $\mathcal{G}$.

We see that the new class $\mathcal{G}$ has the same bound on the condition number as the PSD class, that is, $\gamma(M,q,\epsilon) \geq 1$. Here, we list several existing types of LCPs that belong to $\mathcal{G}$.

1. M is positive semi-definite and q is arbitrary.

   We have if Σ^+ is not empty, then
   $$0 < (x - \pi)^T(s + M^T\pi) = x^Ts - q^T\pi - \pi^T M^T \pi$$
   which implies
   $$x^Ts - q^T\pi > \pi^T M^T \pi \geq 0,$$
   a contradiction.

2. M is copositive and $q \geq 0$.

   We have
   $$x^Ts - q^T\pi = x^T M x + q^T(x - \pi).$$
   Thus, $x > 0$ and $x - \pi > 0$ imply $x^Ts - q^T\pi \geq 0$, that is, Σ^+ is empty.

3. M^{-1} is copositive and $M^{-1}q \leq 0$.

   We have
   $$x^Ts - q^T\pi = s^T M^{-T} s - (M^{-1}q)^T(s + M^T\pi).$$
   Thus, $s > 0$ and $s + M^T\pi > 0$ implies $x^Ts - q^T\pi \geq 0$, that is, Σ^+ is empty.

Although a trivial solution may exist for the last two classes (e.g., $x = 0$ and $s = q$ for the second class), our computational experience indicates that the potential reduction algorithm usually converges to a nontrivial solution if multiple solutions exist.

Example 9.2

$$M = \begin{pmatrix} 0 & -1 \\ 1 & -1 \end{pmatrix} \quad and \quad q = \begin{pmatrix} 2 \\ 0 \end{pmatrix}.$$

For this example the potential reduction algorithm constantly generates the solution

$$x = (2;2) \quad and \quad s = (0;0)$$

from virtually any interior starting point, avoiding the trivial solution $x = 0$ and $s = q$.

Another nonconvex LCP also belongs to $\mathcal{G}$.

Example 9.3

$$M = \begin{pmatrix} 1 & -1 \\ 2 & 0 \end{pmatrix}, \quad and \quad q = \begin{pmatrix} -1 \\ -1 \end{pmatrix}.$$

$\overset{\circ}{\mathcal{F}}$ is nonempty since $x = (3;1)$ is an interior feasible point; Σ^+ is empty since $x_1 - x_2 > 1$, $x_1 - \pi_1 > 0$, $x_2 - \pi_2 > 0$, $x_1 - x_2 - 1 + \pi_1 + 2\pi_2 > 0$ and $2x_1 - 1 - \pi_1 > 0$ imply

$$\begin{aligned}
x^T s - q^T \pi & \\
&= x^T(Mx + q) - q^T \pi \\
&= x_1(x_1 - x_2) + 2x_1 x_2 - x_1 - x_2 + \pi_1 + \pi_2 \\
&= x_1^2 + x_1 x_2 - 2x_1 - x_2 + 1 + (x_1 - x_2 - 1 + \pi_1 + 2\pi_2) + (x_2 - \pi_2) \\
&> x_1^2 + x_1 x_2 - 2x_1 - x_2 + 1 \\
&= (x_1 - 1)^2 + x_2(x_1 - 1) > 0.
\end{aligned}$$

As a by-product, we have

$$\mathcal{G} \subset \{(M, q) : |\mathcal{S}(M, q)| \geq 1\},$$

where $\mathcal{S}(M,q)$ represents the solution set of the LCP and $|\mathcal{S}(M,q)|$ denotes the number of solutions. In fact, any LCP (M,q) with $\gamma(M,q,\epsilon) > 0$ belongs to $\{(M,q) : |\mathcal{S}(M,q)| \geq 1\}$. Furthermore, if $\gamma(M,q,\epsilon) > 0$ for all $q \in \mathcal{R}^n$, then $M \in \mathcal{Q}$, a matrix class where the LCP (M,q) has at least one solution for all $q \in \mathcal{R}^n$. How to calculate $\gamma(M,q,\epsilon)$ or a lower bound for $\gamma(M,q,\epsilon)$ in polynomial time is a further research topic.

9.2.3 P-matrix LCP

A class of problems, called P-matrix LCP, have a smaller lower bound on the condition number.

9.2. LINEAR COMPLEMENTARITY PROBLEM

Definition 9.1 *A matrix M is a P matrix if and only if its every principal submatrix has a positive determinant.*

Proposition 9.17 *Let $\rho \geq 2n + \sqrt{2n}$. Then, for M being a P-matrix and any $q \in \mathcal{R}^n$,*
$$\gamma(M, q, \epsilon) \geq \min\{n\theta(M)/|\lambda(M)|, 1\},$$
where $\lambda(M)$ is the least eigenvalue of $(M+M^T)/2$, and $\theta(M)$ is the positive P-matrix number of M^T, i.e.,
$$\theta(M) = \min_{x \neq 0}\left\{\max_j \frac{x_j(M^T x)_j}{\|x\|^2}\right\}.$$

Thus, the P-matrix LCP can be solved in
$$O\left(n^2 \max\left\{\frac{-\lambda}{\theta n}, 1\right\} \log(1/\epsilon)\right)$$
iterations and each iteration solves a system of linear equations in at most $O(n^3)$ arithmetic operations. This bound indicates that the algorithm is a polynomial-time algorithm if $|\lambda|/\theta$ is bounded above by a polynomial in $\log(1/\epsilon)$ and n.

It has been shown that $\|p^k\|^2 \geq 1$ if M is positive semi-definite (that is, if $\lambda \geq 0$). Thus, in this section we assume that $\lambda < 0$. We also fix
$$\rho = 2n + \sqrt{2n}. \tag{9.19}$$

We first prove the following lemma:

Lemma 9.18
$$(a+b)^2 + (a+c)^2 \geq a^2 - 2bc.$$

Proof.
$$\begin{aligned}
(a+b)^2 + (a+c)^2 &= 2a^2 + 2a(b+c) + b^2 + c^2 \\
&= a^2 - 2bc + a^2 + 2a(b+c) + (b+c)^2 \\
&= a^2 - 2bc + (a+b+c)^2 \geq a^2 - 2bc.
\end{aligned}$$

$\square$

Now, we have the following lemma:

Lemma 9.19 *Given any $(x^k, s^k) \in \overset{\circ}{\mathcal{F}}$, let p^k be the scaled gradient projection computed from (9.13) and (9.14). Then,*
$$\|p^k\|^2 \geq \min\left\{\frac{\theta n}{|\lambda|}, 1\right\}.$$

Proof. Let $\bar{s} = s^k + M^T\pi$, $\bar{x} = x^k - \pi$ and $\bar{\Delta} = (x^k)^T\bar{s} + (s^k)^T\bar{x}$. Then, it is obvious that if $\bar{s} \not> 0$ or $\bar{x} \not> 0$, then

$$\|p^k\|^2 \geq 1. \tag{9.20}$$

Therefore, we assume that $\bar{s} > 0$ and $\bar{x} > 0$ in the rest of the proof. Note that from Lemma 9.14

$$\|p^k\|^2 \geq \left(\frac{(n+\rho)\bar{\Delta}}{2n\Delta^k} - 1\right)^2 2n$$

and

$$\bar{\Delta} = \Delta^k + \bar{x}^T\bar{s} + \pi^T M^T\pi > \Delta^k + \pi^T M^T\pi.$$

Thus, if

$$\pi M^T\pi \geq -\frac{n\Delta^k}{n+\rho},$$

then from (9.19)

$$\|p^k\|^2 \geq \left(\frac{(n+\rho)\bar{\Delta}}{2n\Delta^k} - 1\right)^2 2n \geq 1. \tag{9.21}$$

Otherwise, we have

$$\lambda\|\pi\|^2 \leq \pi^T M^T\pi < -\frac{n\Delta^k}{n+\rho},$$

i.e.,

$$\|\pi\|^2 \geq \frac{n\Delta^k}{(n+\rho)|\lambda|}. \tag{9.22}$$

Since M is a P-matrix, there exists an index j such that $\pi_j(M^T\pi)_j \geq \theta\|\pi\|^2 > 0$. Using Lemma 9.18 and (9.22), we have

$$\begin{aligned}
\|p^k\|^2 &\geq \left(\frac{n+\rho}{\Delta^k}x_j^k s_j^k - 1 + \frac{n+\rho}{\Delta^k}x_j^k(M^T\pi)_j\right)^2 \\
&\quad + \left(\frac{n+\rho}{\Delta^k}x_j^k s_j^k - 1 - \frac{n+\rho}{\Delta^k}s_j^k\pi_j\right)^2 \\
&\geq \left(\frac{n+\rho}{\Delta^k}x_j^k s_j^k - 1\right)^2 + 2x_j^k s_j^k \pi_j(M^T\pi)_j \left(\frac{n+\rho}{\Delta^k}\right)^2 \\
&\geq \left(\frac{n+\rho}{\Delta^k}x_j^k s_j^k - 1\right)^2 + 2\frac{n+\rho}{\Delta^k}x_j^k s_j^k \theta\|\pi\|^2 \frac{n+\rho}{\Delta^k} \\
&\geq \left(\frac{n+\rho}{\Delta^k}x_j^k s_j^k - 1\right)^2 + 2\frac{n+\rho}{\Delta^k}x_j^k s_j^k \frac{\theta n}{|\lambda|}. \tag{9.23}
\end{aligned}$$

9.2. LINEAR COMPLEMENTARITY PROBLEM

If
$$\frac{\theta n}{|\lambda|} \geq 1,$$
then again
$$\|p^k\|^2 \geq (\frac{n+\rho}{\Delta^k}x_j^k s_j^k)^2 + 1 \geq 1; \qquad (9.24)$$
otherwise,
$$\|p^k\|^2 \geq 2\frac{\theta n}{|\lambda|} - \frac{(\theta n)^2}{\lambda^2} \geq \frac{\theta n}{|\lambda|} \qquad (9.25)$$
since the quadratic term of (9.23) yields the minimum at
$$\frac{n+\rho}{\Delta^k}x_j^k s_j^k = 1 - \frac{\theta n}{|\lambda|}.$$
From (9.20), (9.21), (9.24) and (9.25), we have the desired result.

□

The result leads to the following theorem:

Theorem 9.20 Let $\psi(x^0, s^0) \leq O(n \log n)$ and M be a P-matrix. Then, the potential reduction algorithm terminates at $(x^k)^T s^k \leq \epsilon$ in
$$O(n^2 \max\{|\lambda|/(\theta n), 1\} \log(1/\epsilon))$$
iterations and each iteration uses at most $O(n^3)$ arithmetic operations.

Finally, we consider a very broad class of LCPs that contain some difficult problems.

Definition 9.2 A matrix M is row-sufficient if and only if for every vector ξ, $diag(\xi)M^T\xi \leq 0$ implies $diag(\xi)M^T\xi = 0$. A matrix M is column-sufficient if and only if M^T is row-sufficient. A matrix M is a sufficient matrix if and only if it is both row- and column-sufficient.

Note that the class of row-sufficient matrices contains some popular matrices such as PSD and P matrices.

Proposition 9.21 Let $\rho > 0$ and be fixed. Then, for M being a row-sufficient matrix and $\{(x, s) \in \overset{\circ}{\mathcal{F}} \colon \psi(x, s) \leq \psi^0\}$ being bounded,
$$\gamma(M, q, \epsilon) > 0.$$

Proof. It is easy to show that for any $(x,s) \in \overset{\circ}{\mathcal{F}}$,
$$\|g(x,s)\|_H^2 > 0.$$
Moreover, for all $(x,s) \in \overset{\circ}{\mathcal{F}}$, $x^T s \geq \epsilon$ and $\psi(x,s) \leq \psi^0$,

$$\begin{aligned}
\psi^0 &\geq \psi(x,s) \\
&= (n+\rho)\log(x^T s) - \sum_{j=1}^n \log(x_j s_j) \\
&= (\rho+1)\log(x^T s) + (n-1)\log(x^T s) - \sum_{j \neq i}\log(x_j s_j) - \log(x_i y_i) \\
&\geq (\rho+1)\log(x^T s) + (n-1)\log(x^T s - x_i y_i) \\
&\quad - \sum_{j \neq i}\log(x_j s_j) - \log(x_i s_i) \\
&\geq (\rho+1)\log(x^T s) + (n-1)\log(n-1) - \log(x_i s_i) \\
&\geq -(\rho+1)\log(1/\epsilon) + (n-1)\log(n-1) - \log(x_i s_i),
\end{aligned}$$

where $i \in \{1,2,\ldots,n\}$. Thus,
$$\log(x_i s_i) \geq -(\rho+1)\log(1/\epsilon) + (n-1)\log(n-1) - \psi^0,$$
that is, $x_i s_i$ is bounded away from zero for every i. Since $\{(x,s) \in \overset{\circ}{\mathcal{F}}: \psi(x,s) \leq \psi^0\}$ is bounded, there must exist a positive number $\bar{\epsilon}$, independent of (x,s), such that
$$x_i \geq \bar{\epsilon} \quad \text{and} \quad s_i \geq \bar{\epsilon}, \quad i = 1,2,\ldots,n$$
for all (x,s) such that $x^T s \geq \epsilon$, $\psi(x,s) \leq \psi^0$ and $(x,s) \in \overset{\circ}{\mathcal{F}}$. Therefore,

$$\begin{aligned}
\gamma(M,q,\epsilon) &= \inf\{\|g(x,s)\|_H^2 : x^T s \geq \epsilon,\ \psi(x,s) \leq \psi^0 \text{ and } (x,s) \in \overset{\circ}{\mathcal{F}}\} \\
&\geq \inf\{\|g(x,s)\|_H^2 : x \geq \bar{\epsilon}e,\ s \geq \bar{\epsilon}e,\ \psi(x,s) \leq \psi^0 \text{ and } (x,s) \in \mathcal{F}\} \\
&> 0.
\end{aligned}$$

The last inequality holds since the infimum is taken in a compact set where $\|g(x,s)\|_H^2$ is always positive.
□

Since the condition number is bounded away from 0, the potential algorithm will solve row-sufficient matrix LCPs.

9.3 Generalized Linear Complementarity Problem

In this section we consider a generalized linear complementarity problem:

$$(GLCP) \quad \text{minimize} \quad x^T s$$
$$\text{s.t.} \quad Ax + Bs + Cz = q, \quad (x, s, z) \geq 0.$$

Let $\mathcal{F}$ denote the feasible set. It is evident that a solution, with $x^T s = 0$, to the GLCP may not exist even when the problem is feasible. However, a finite stationary or KKT point of the GLCP, which is defined as a point satisfying the first order optimality conditions of (GLCP), must exist, since the objective function is quadratic and bounded from below so that it has a finite minimum.

More precisely, a KKT point, $(\bar{x}, \bar{s}, \bar{z}) \in \mathcal{F}$, of the GLCP is represented by

$$\bar{s}^T \bar{x} + \bar{x}^T \bar{s} \leq \bar{s}^T x + \bar{x}^T s \quad \text{for all} \quad (x, s, z) \in \mathcal{F}.$$

In other words, $(\bar{x}, \bar{s}, \bar{z})$ is a minimal solution for the related linear program

$$\text{minimize} \quad \bar{s}^T x + \bar{x}^T s$$
$$\text{s.t.} \quad Ax + Bs + Cz = q,$$
$$(x, s, z) \geq 0,$$

where its dual is

$$\text{maximize} \quad q^T \pi$$
$$\text{s.t.} \quad \bar{s} - A^T \pi \geq 0,$$
$$\bar{x} - B^T \pi \geq 0,$$
$$-C^T \pi \geq 0.$$

Thus, $(\bar{x}, \bar{s}, \bar{z}) \in \mathcal{F}$ is a KKT point if and only if there exists $\bar{\pi} \in \mathcal{R}^m$ such that

$$\bar{s} - A^T \bar{\pi} \geq 0, \quad \bar{x} - B^T \bar{\pi} \geq 0 \quad \text{and} \quad -C^T \bar{\pi} \geq 0,$$

and

$$\bar{x}^T (\bar{s} - A^T \bar{\pi}) = 0, \quad \bar{s}^T (\bar{x} - B^T \bar{\pi}) = 0 \quad \text{and} \quad \bar{z}^T (-C^T \bar{\pi}) = 0.$$

We see that finding such a KKT point itself is a GLCP. We also note that a solution to the GLCP, (x^*, s^*, z^*), can be viewed as a special KKT point with $\bar{\pi} = 0$.

The concept of the fully polynomial-time approximation scheme (FP-TAS) was introduced in combinatorial optimization. Given an instance of an optimization problem and an $\epsilon > 0$, it returns an ϵ-approximate solution within a time period bounded by a polynomial both in the length of the instance and $1/\epsilon$. For some combinatorial optimization problems, the theory of NP-completeness can be applied to prove not only that they cannot be solved exactly by polynomial-time algorithms (unless $P = NP$), but also that they do not have ϵ-approximate algorithms, for various ranges of ϵ, again unless $P = NP$. Furthermore, approximation algorithms are widely used and accepted in practice.

In this paper, we develop a fully polynomial-time approximation scheme for generating an ϵ-KKT point of the GLCP—a point $(\hat{x}, \hat{s}, \hat{z}) \in \mathcal{F}$ and $\hat{\pi} \in \mathcal{R}^m$ with

$$\hat{s} - A^T\hat{\pi} \geq 0, \quad \hat{x} - B^T\hat{\pi} \geq 0 \quad \text{and} \quad -C^T\hat{\pi} \geq 0, \qquad (9.26)$$

and

$$\frac{\hat{x}^T(\hat{s} - A^T\pi) + \hat{s}^T(\hat{x} - B^T\pi) + \hat{z}^T(-C^T\pi)}{\hat{x}^T\hat{s}} \leq \epsilon. \qquad (9.27)$$

In other words, $(\hat{x}, \hat{s}, \hat{z}, \hat{\pi})$ is feasible and the sum of the complementary slackness vectors (or the primal-dual objective gap) relative to the (primal) objective value is less than ϵ. Thus, the algorithm is actually a polynomial approximation algorithm for solving a class of GLCPs in which every KKT point is a solution. This class includes the LCP with the row-sufficient matrix.

We assume at this moment that

Assumption 9.5 *The interior of $\mathcal{F}$, $\overset{\circ}{\mathcal{F}} = \{(x, s, z) \in \mathcal{F} : x > 0, s > 0 \text{ and } z > 0\}$ is nonempty, and an interior point, (x^0, s^0, z^0) of $\mathcal{F}$, is available;*

and

Assumption 9.6 *Each component of z is bounded by R in the feasible set $\mathcal{F}$.*

Both assumptions will be *removed* later, so that the results should hold for general cases.

9.3.1 Potential reduction algorithm

Let $\rho \geq n + d$ and define the potential function

$$\psi(x, s, z) := (n + \rho)\log(x^T s) - \sum_{j=1}^{n}\log(x_j) - \sum_{j=1}^{n}\log(s_j) - \sum_{j=1}^{d}\log(z_j)$$

9.3. GENERALIZED LINEAR COMPLEMENTARITY PROBLEM

to associate with a feasible point $(x, s, z) \in \overset{\circ}{\mathcal{F}}$. Using the arithmetic-geometric mean inequality, we have

$$\rho \log(x^T s) - \sum_{j=1}^{d} \log(z_j) \leq \psi(x, s, z) - n \log(n) \leq \psi(x, s, z).$$

On the other hand, $\sum_{j=1}^{d} \log(z_j) \leq d \log R$ from the boundedness assumption. Thus,

$$\psi(x, s, z) < n \log n - d \log R + \rho \log \epsilon \Longrightarrow x^T s < \epsilon. \qquad (9.28)$$

The basic idea of the potential reduction algorithm works as follows. Given an interior point $(x^0, s^0, z^0) \in \overset{\circ}{\mathcal{F}}$, we generate a sequence of interior feasible solutions $\{x^k, s^k, z^k\} \in \overset{\circ}{\mathcal{F}}$ and $\{\pi^k\} \in \mathcal{R}^m$ with the following property: Unless

$$s^k - A^T \pi^k > 0; \quad x^k - B^T \pi^k > 0; \quad -C^T \pi^k > 0$$

and

$$\frac{(x^k)^T (s^k - A^T \pi^k) + (s^k)^T (x^k - B^T \pi^k) + (z^k)^T (-C^T \pi^k)}{(x^k)^T s^k}$$
$$< (2n + d + \sqrt{2n + d})/(n + \rho),$$

we always have

$$\psi(x^{k+1}, s^{k+1}, z^{k+1}) \leq \psi(x^k, s^k, z^k) - O(1/(n + \rho)).$$

Thus, if we choose $n + \rho = (2n + d + \sqrt{2n + d})/\epsilon$, the algorithm requires at most $(n + \rho)^2 (\log(1/\epsilon) + \log R)$ iterations, which is a polynomial in $1/\epsilon$, to generate either a solution or an ϵ-KKT point of the GLCP. Note that each iteration involves $O(n^3)$ arithmetic operations.

There are many ways to achieve a potential reduction. We again use the scaled gradient projection method. The gradient vectors $\nabla \psi_x$ and $\nabla \psi_s$ are identical to those in the preceding section, and the one with respect to z is

$$\nabla \psi_z = -Z^{-1} e,$$

where $\Delta = x^T s$. Now, we solve the following linear program subject to an ellipsoidal constraint at the kth iteration:

$$\begin{aligned}
\text{maximize} \quad & \nabla^T \psi_{x^k} d_x + \nabla^T \psi_{s^k} d_s + \nabla^T \psi_{z^k} d_z \\
\text{s.t.} \quad & A d_x + B d_s + C d_z = 0, \\
& \|(X^k)^{-1} d_x\|^2 + \|(S^k)^{-1} d_s\|^2 + \|(Z^k)^{-1} d_z\|^2 \leq \alpha^2 < 1,
\end{aligned}$$

and denote by d_x, d_s and d_z its minimal solutions. Then, we have

$$\begin{pmatrix} (X^k)^{-1}d_x \\ (S^k)^{-1}d_s \\ (Z^k)^{-1}d_z \end{pmatrix} = \frac{-\alpha p^k}{\|p^k\|},$$

where p^k is the projection of the scaled gradient vector

$$(\nabla^T \psi_{x^k} X^k, \nabla^T \psi_{s^k} S^k, \nabla^T \psi_{z^k} Z^k)$$

onto the null space of the scaled constraint matrix (AX^k, BS^k, CZ^k), i.e.,

$$p^k = \begin{pmatrix} p_x^k \\ p_s^k \\ p_z^k \end{pmatrix} = \begin{pmatrix} \frac{n+\rho}{\Delta^k} X^k(s^k - A^T \pi^k) - e \\ \frac{n+\rho}{\Delta^k} S^k(x^k - B^T \pi^k) - e \\ \frac{n+\rho}{\Delta^k} Z^k(-C^T \pi^k) - e \end{pmatrix},$$

$$\pi^k = \frac{\Delta^k}{n+\rho} (\bar{A}\bar{A}^T)^{-1} \bar{A} (\nabla^T \psi_{x^k} X^k, \nabla^T \psi_{s^k} S^k, \nabla^T \psi_{z^k} Z^k)^T,$$

$$\bar{A} = (AX^k, BS^k, CZ^k), \quad \text{and} \quad \Delta^k = (x^k)^T s^k.$$

Let $x^{k+1} = x^k + d_x$, $s^{k+1} = s^k + d_s$ and $z^{k+1} = z^k + d_z$. Then, similar to inequality (9.15) (also see Exercise 9.4), we have

$$\psi(x^{k+1}, s^{k+1}, z^{k+1}) - \psi(x^k, s^k, z^k) \le -\alpha \|p^k\| + \frac{\alpha^2}{2} \left(n + \rho + \frac{1}{1-\alpha} \right).$$

Therefore, choosing α as in (9.16), we have

$$\psi(x^{k+1}, s^{k+1}, z^{k+1}) - \psi(x^k, s^k, z^k) \le -\min \left\{ \frac{\|p^k\|^2}{2(n+\rho+2)}, \frac{1}{2(n+\rho+2)} \right\}. \quad (9.29)$$

In practice, the step-size α can be determined by a line search along the direction p^k to minimize the potential function.

9.3.2 Complexity analysis

We further study $\|p^k\|$ by introducing the following lemma.

Lemma 9.22 *The scaled gradient projection $\|p^k\| < 1$ implies*

$$s^k - A^T \pi^k > 0; \quad x^k - B^T \pi^k > 0; \quad -C^T \pi^k > 0,$$

and

$$\frac{2n + d - \sqrt{2n+d}}{n+\rho} \Delta^k < \bar{\Delta} < \frac{2n + d + \sqrt{2n+d}}{n+\rho} \Delta^k,$$

where $\bar{\Delta} = (x^k)^T(s^k - A^T\pi^k) + (s^k)^T(x^k - B^T\pi^k) + (z^k)^T(-C^T\pi^k)$.

9.3. GENERALIZED LINEAR COMPLEMENTARITY PROBLEM

Proof. The proof is by contradiction. It is obvious that if $s^k - A^T\pi^k \not> 0$ or $x^k - B^T\pi^k \not> 0$ or $-C^T\pi^k \not> 0$, then

$$\|p^k\| \geq \|p^k\|_\infty \geq 1.$$

On the other hand, we have

$$
\begin{aligned}
\|p^k\|^2 &= (\frac{n+\rho}{\Delta^k})^2 \| \begin{pmatrix} X^k(s^k - A^T\pi^k) \\ S^k(x^k - B^T\pi^k) \\ Z^k(-C^T\pi^k) \end{pmatrix} - \frac{\bar{\Delta}}{2n+d}e\|^2 + \|\frac{(n+\rho)\bar{\Delta}}{(2n+d)\Delta^k}e - e\|^2 \\
&\geq \|\frac{(n+\rho)\bar{\Delta}}{(2n+d)\Delta^k}e - e\|^2 \\
&= (\frac{(n+\rho)\bar{\Delta}}{(2n+d)\Delta^k} - 1)^2(2n+d).
\end{aligned} \tag{9.30}
$$

(Note that the dimension of e is $2n+d$ here.) Hence, the following must be true:

$$\left(\frac{(n+\rho)\bar{\Delta}}{(2n+d)\Delta^k} - 1\right)^2 (2n+d) < 1,$$

that is,

$$\frac{2n+d-\sqrt{2n+d}}{n+\rho}\Delta^k < \bar{\Delta} < \frac{2n+d+\sqrt{2n+d}}{n+\rho}\Delta^k.$$

□

Now we can prove the following theorem. For simplicity, we assume that $\psi(x^0, s^0, z^0) \leq O(n\log n)$.

Theorem 9.23 *For any given $0 < \epsilon \leq 1$, let $n+\rho = (2n+d+\sqrt{2n+d})/\epsilon$. Then, under Assumptions 9.5 and 9.6, the potential reduction algorithm terminates in at most $O((n+\rho)^2 \log(1/\epsilon) + (n+\rho)d\log R)$ iterations. The algorithm generates an ϵ-KKT point $(x^k, s^k, z^k) \in \mathcal{F}$ and $\pi^k \in \mathcal{R}^m$ of the GLCP, either*

$$(x^k)^T s^k \leq \epsilon$$

or

$$s^k - A^T\pi^k > 0; \quad x^k - B^T\pi^k > 0; \quad -C^T\pi^k > 0$$

and

$$\frac{(x^k)^T(s^k - A^T\pi^k) + (s^k)^T(x^k - B^T\pi^k) + (z^k)^T(-C^T\pi^k)}{(x^k)^T s^k} < \epsilon.$$

Proof. The proof directly follows Lemma 9.22. If $\|p^k\| \geq 1$ for all k, then from (9.29)

$$\psi(x^{k+1}, s^{k+1}, z^{k+1}) - \psi(x^k, s^k, z^k) \leq -O(1/(n+\rho)).$$

Therefore, in at most $O((n+\rho)^2 \log(1/\epsilon) + (n+\rho)d\log R)$ iterations

$$\psi(x^k, s^k, z^k) \leq \rho \log \epsilon - d \log R + n \log n,$$

and, from (9.28)

$$(x^k)^T s^k \leq \epsilon.$$

As we mentioned before, in this case (x^k, s^k, z^k) is a special stationary point with $\bar{\pi} = 0$. Otherwise, we have $\|p^k\| < 1$ for some $k \leq O((n+\rho)^2 \log(1/\epsilon) + (n+\rho)d \log R)$. This implies the relations (9.26) and (9.27) from Lemma 9.22. □

Theorem 9.23 indicates that the potential reduction algorithm is a fully polynomial-time approximation scheme for computing an ϵ-approximate KKT point of the GLCP. In the following, we present a sufficient condition to show that a solution to the GLCP always exists, and the potential reduction algorithm solves it in polynomial time under the assumptions. Moreover, we have

Theorem 9.24 *Let BA^T be negative semi-definite. Furthermore, let $\rho = n + d + \sqrt{2n+d}$. Then, the potential reduction algorithm generates a solution to the GLCP in $O((2n+d)^2 \log(1/\epsilon) + (2n+d)d\log R)$ iterations.*

Proof. Basically, we show that $\|p^k\| \geq 1$ for all k if BA^T is negative semi-definite and $\rho \geq n + d + \sqrt{2n+d}$. We prove it by contradiction. Suppose $\|p^k\| < 1$, then

$$s^k - A^T \pi^k > 0 \quad x^k - B^T \pi^k > 0 \quad \text{and} \quad -C^T \pi^k > 0.$$

Thus,

$$(x^k - B^T \pi^k)^T (s^k - A^T \pi^k) > 0,$$

that is

$$(x^k)^T s^k - (x^k)^T A^T \pi^k - (s^k)^T B^T \pi^k + (\pi^k)^T B A^T \pi^k > 0.$$

Also note

$$-(z^k)^T C^T \pi^k > 0.$$

9.3. GENERALIZED LINEAR COMPLEMENTARITY PROBLEM

Combining the above two inequalities, we have

$$(x^k)^T s^k - (x^k)^T A^T \pi^k - (s^k)^T B^T \pi^k - (z^k)^T C^T \pi^k > -(\pi^k)^T B A^T \pi^k \geq 0$$

or

$$\begin{aligned}
\bar{\Delta} &= (x^k)^T(s^k - A^T\pi^k) + (s^k)^T(x^k - B^T\pi^k) + (z^k)^T(-C^T\pi^k) \\
&= (x^k)^T s^k + (x^k)^T s^k - (x^k)^T A^T \pi^k - (s^k)^T B^T \pi^k - (z^k)^T C^T \pi^k \\
&> (x^k)^T s^k = \Delta^k.
\end{aligned}$$

From (9.30), we have

$$\|p^k\|^2 \geq \left(\frac{(n+\rho)\bar{\Delta}}{(2n+d)\Delta^k} - 1\right)^2 (2n+d) \geq 1,$$

which is a contradiction.

□

Clearly, the result for the LCP with positive semi-definite matrix is a special application of Theorem 9.24.

9.3.3 Further discussions

We now remove Assumptions 9.5 and 9.6 that were used in the above theorems and show that our main results remain valid. Note first that the z-boundedness assumption is automatically unnecessary for LCP.

We now remove the assumption of availability of the initial point. We apply the linear programming Phase I procedure to find a feasible solution for the system

$$(A, B, C)u = q \quad \text{and} \quad u \geq 0.$$

In polynomial time, an interior-point algorithm either declares that the system is infeasible, or generates a max-feasible solution $\bar{u}$. Thus, we have detected those variables that must be zero in every feasible solutions of (GLCP) (in this case, the feasible region has empty interior). Then, we eliminate those variables from the system. For example, if x_1 is zero in every feasible solution of (GLCP), we can eliminate x_1 and then move s_1 into z; if both x_1 and s_1 are zero in every feasible solution, we can eliminate both of them. Thus, we will have a reduced system where the feasible region has a nonempty interior, and a feasible solution is at hand.

Hence, Theorems 9.23 and 9.24 hold without the assumption. It has been shown that every KKT point of the LCP with a row-sufficient matrix is a solution of the LCP. Therefore, we have Corollary 9.25.

Corollary 9.25 *The potential reduction algorithm is a fully polynomial-time approximation scheme for generating an ϵ-approximate stationary point of the LCP with row-sufficient matrices, where every KKT point is a solution of the LCP.*

We give three examples of the GLCP and try to illustrate the convergence behavior of the algorithm. These experiments are very preliminary.

Example 9.4

$$A = \begin{pmatrix} 0 & 1 & 10 \\ 0 & 1 & 1 \\ 0 & 0 & 2 \end{pmatrix}, \quad B = -I, \quad C = \emptyset, \quad \text{and} \quad q = e.$$

The starting point is $x^0 = (2; 2; 2)$. The algorithm consistently generates the solution to the LCP, $x^* = (0; 0.5; 0.5)$ and $s^* = (4.5; 0; 0)$. In this example, A is a so called P_0 matrix, and it is indefinite.

Example 9.5

$$A = \begin{pmatrix} 0 & 1 & 10 \\ 0 & 0 & 1 \\ 0 & 0 & 2 \end{pmatrix}, \quad B = -I, \quad C = \emptyset, \quad \text{and} \quad q = e.$$

The starting point is again $x^0 = (2; 2; 2)$. The algorithm consistently generates a KKT point of the LCP, $\bar{x} = (0; \alpha; 1)$, $\bar{s} = (9 + \alpha; 0; 1)$ and $\bar{\pi} = (0; -3; 1)$ for some $\alpha > 0$. Note that there is no solution to the GLCP in this example.

Example 9.6

$$A = \begin{pmatrix} 0 & 1 & 10 \\ 0 & 0 & 1 \\ 0 & 0 & 0 \end{pmatrix}, \quad B = -I, \quad C = \emptyset, \quad \text{and} \quad q = \begin{pmatrix} 1 \\ 1 \\ -1 \end{pmatrix}.$$

The starting point is again $x^0 = (2; 2; 2)$. The algorithm consistently generates a KKT point of the LCP, $\bar{x} = (0; \alpha; 1)$, $\bar{s} = (9 + \alpha; 0; 1)$ and $\bar{\pi} = (0; -1; 1)$ for some $\alpha > 0$. Again, there is no solution to the GLCP in this example.

9.4 Indefinite Quadratic Programming

Consider the quadratic programming (QP) problem and its dual (QD) in Section 1.3.6. Denote by $\mathcal{F}_d$ the (dual) feasible set $\{(x, y, s) : A^T y + s - Qx = c, \ x, s \geq 0\}$. For simplicity we assume that A has full row-rank.

9.4. INDEFINITE QUADRATIC PROGRAMMING

If Q is positive semi-definite in the null space of A, meaning that $H := N^T Q N$ is positive semi-definite where $N \in R^{n \times (n-m)}$ is an orthonormal basis spanning the null space of A, then (QP) is a convex optimization problem and it can be solved as a monotone linear complementarity problem. The algorithm presented in this section handles general QP problems: convex or nonconvex. For the simplicity of our analysis, throughout this section we let (QP) be a non-convex problem, e.g., Q have at least one negative eigenvalue in the null space of A. Then, (QP) becomes a hard problem, an NP-complete problem.

No time complexity bounds have been developed for various QP methods. (Of course, an enumerative search approach will solve (QP) but it possesses an exponential time bound.) These algorithms generally generate a sequence of points that converges to a stationary or KKT point associated with (QP), which satisfies

$$x^T s = 0, \quad x \in \mathcal{F}_p, \quad \text{and} \quad (x,y,s) \in \mathcal{F}_d.$$

For any $x \in \mathcal{F}_p$ and $(x,y,s) \in \mathcal{F}_d$, the quantity $x^T s = q(x) - d(x,y)$ is the complementarity gap.

Here we assume that the feasible set, $\mathcal{F}_p$, of (QP) has a strictly positive feasible solution. For any given (A,b), to see if $\mathcal{F}_p$ possesses a strictly positive feasible solution can be solved as a single linear program in polynomial time, so that the above assumption is without of loss of any generality. We make an additional assumption that the feasible region is bounded. With this assumption (QP) has a minimizer and a maximizer. Let $\underline{z}$ and $\bar{z}$ be their minimal and maximal objective values, respectively.

An ϵ-minimal solution or ϵ-minimizer, $\epsilon \in (0,1)$, for (QP) is defined as an $x \in \mathcal{F}_p$ such that: [1]

$$\frac{q(x) - \underline{z}}{\bar{z} - \underline{z}} \leq \epsilon.$$

Similarly, we define an ϵ-KKT solution for (QP) as an (x,y,s) such that $x \in \mathcal{F}_p$, $(x,y,s) \in \mathcal{F}_d$, and

$$\frac{x^T s}{\bar{z} - \underline{z}} = \frac{q(x) - d(x,y)}{\bar{z} - \underline{z}} \leq \epsilon.$$

Note that the minimizer of (QP) is a special KKT point such that $q(x) = d(x,y) = \underline{z}$.

In this section we extend the potential reduction techniques described earlier to compute an ϵ-KKT point in $O((\frac{n^6}{\epsilon} \log \frac{1}{\epsilon} + n^4 \log n)(\log \frac{1}{\epsilon} + \log n))$

---

[1] Vavasis [451] discussed the importance to have the term $(\bar{z} - \underline{z})$ in the criterion for continuous optimization.

arithmetic operations. We also show that Q is positive semi-definite in the null space of the active set at the limit of this point, indicating that the limit satisfies the second-order necessary condition to be a local minimal solution. The result is the first approximation algorithm whose running time is almost linear in $\frac{1}{\epsilon}$, which was an open question in the area of nonlinear optimization complexity.

9.4.1 Potential reduction algorithm

The potential function used to solve (QP) is

$$\mathcal{P}(x) = \mathcal{P}_{n+\rho}(x, z) := (n+\rho)\log(q(x) - z) - \sum_{j=1}^{n} \log(x_j),$$

where $0 < x \in \mathcal{F}_p$, parameter $\rho > 0$, and $z \leq \underline{z}$. Unlike solving LP, here z is unchanged during the iterative process.

Starting from a point $0 < x^0 \in \mathcal{F}_p$, the potential reduction algorithm will generate a sequence of $\{x^k\} \in \mathcal{F}_p$ such that $\mathcal{P}(x^{k+1}) < \mathcal{P}(x^k)$. For simplicity and convenience, we assume $x^0 = e$, and x^0 is the analytic center of $\mathcal{F}_p$. Our results hold even if x^0 is replaced by an approximate center. Therefore, this assumption is also without loss of generality.

To determine how much the potential function must be reduced to generate an ϵ-KKT point, recall from Chapter 2 that x^0 is the analytic center of $\mathcal{F}_p$,

$$\begin{aligned} \mathcal{F}_p &\supset \mathcal{E}_{in} := \{x \in \mathcal{F}_p : \|(X^0)^{-1}(x - x^0)\| \leq 1\} \\ \mathcal{F}_p &\subset \mathcal{E}_{out} := \{x \in \mathcal{F}_p : \|(X^0)^{-1}(x - x^0)\| \leq n\}. \end{aligned} \quad (9.31)$$

In other words, $\mathcal{F}_p$ is inscribed by ellipsoid $\mathcal{E}_{in}$ and outscribed by ellipsoid $\mathcal{E}_{out}$. The two ellipsoids are concentric, where the ratio of their radii is n.

Thus, for any $x \in \mathcal{F}_p$

$$\sum_{j=1}^{n} \log(x_j^0/x_j) \geq -n\log(n+1). \quad (9.32)$$

Note that if

$$(n+\rho)\log(q(x^k) - z) - (n+\rho)\log(q(x^0) - z) \leq (n+\rho)\log \epsilon$$

or

$$\mathcal{P}(x^k) - \mathcal{P}(x^0) \leq (n+\rho)\log \epsilon + \sum_{j=1}^{n} \log(x_j^0/x_j^k) \quad (9.33)$$

9.4. INDEFINITE QUADRATIC PROGRAMMING

we must have

$$\frac{q(x^k) - z}{\bar{z} - z} \le \frac{q(x^k) - z}{q(x^0) - z} \le \frac{q(x^k) - z}{q(x^0) - z} \le \epsilon,$$

which implies that x^k is an ϵ-minimizer and, thereby, an ϵ-KKT point. From relations (9.32) and (9.33), x^k becomes an ϵ-minimizer as soon as

$$\mathcal{P}(x^k) - \mathcal{P}(x^0) \le (n + \rho) \log \epsilon - n \log(n+1). \tag{9.34}$$

Given $0 < x \in \mathcal{F}_p$, let $\Delta = q(x) - z$ and let $d_x \in \mathcal{N}(A)$, be a vector such that $x^+ := x + d_x > 0$. Then

$$(n+\rho)\log(q(x^+) - z) - (n+\rho)\log(q(x) - z)$$
$$= (n+\rho)\log\left(\Delta + \frac{1}{2}d_x^T Q d_x + (Qx+c)^T d_x\right) - (n+\rho)\log \Delta$$
$$= (n+\rho)\log\left(1 + \left(\frac{1}{2}d_x^T Q d_x + (Qx+c)^T d_x\right)/\Delta\right)$$
$$\le \frac{n+\rho}{\Delta}\left(\frac{1}{2}d_x^T Q d_x + (Qx+c)^T d_x\right).$$

On the other hand, if $\|X^{-1}d_x\| \le \alpha < 1$ then

$$-\sum_{j=1}^n \log(x_j^+) + \sum_{j=1}^n \log(x_j) \le -e^T X^{-1} d_x + \frac{\alpha^2}{2(1-\alpha)}.$$

Thus, if $\|X^{-1}d_x\| \le \alpha < 1$ then $x^+ = x + d_x > 0$ and

$$\mathcal{P}(x^+) - \mathcal{P}(x)$$
$$\le \frac{n+\rho}{\Delta}\left(\frac{1}{2}d_x^T Q d_x + \left(Qx + c - \frac{\Delta}{n+\rho}X^{-1}e\right)^T d_x\right)$$
$$+ \frac{\alpha^2}{2(1-\alpha)}. \tag{9.35}$$

To achieve a potential reduction, we minimize a quadratic function subject to an ellipsoid constraint. We solve the following problem at the kth iteration:

$$\text{minimize} \quad \tfrac{1}{2}d_x^T Q d_x + (Qx^k + c - \tfrac{\Delta^k}{n+\rho}(X^k)^{-1}e)^T d_x$$

$$\text{s.t.} \quad A d_x = 0,$$
$$\|(X^k)^{-1} d_x\|^2 \le \alpha^2.$$

Let
$$Q^k = X^k Q X^k, \ c^k = X^k Q x^k + X^k c - \frac{\Delta^k}{n+\rho} e, \ A^k = A X^k, \text{ and } d'_x = (X^k)^{-1} d_x.$$

Then the above problem becomes

$$(BQP) \quad \text{minimize} \quad q'(d'_x) := \tfrac{1}{2}(d'_x)^T Q^k d'_x + (c^k)^T d'_x$$

$$\text{s.t.} \quad A^k d'_x = 0,$$

$$\|d'_x\|^2 \le \alpha^2.$$

Let $N^k \in R^{n \times (n-m)}$ be an orthonormal basis spanning the null space of A^k, where $(N^k)^T N^k = I$, and let $H^k = (N^k)^T Q^k N^k \in R^{(n-m) \times (n-m)}$ and $g^k = (N^k)^T c^k \in R^{n-m}$. Then $d'_x = N^k v$ for some $v \in R^{n-m}$ and problem (BQP) can be rewritten as

$$(BHP) \quad \text{minimize} \quad \tfrac{1}{2} v^T H^k v + (g^k)^T v$$

$$\text{s.t.} \quad \|v\| \le \alpha^2.$$

This is the so-called ball-constrained quadratic problem in Section 1.5.5. We assume that, for now, this problem can be solved efficiently (we will establish this fact later).

The solution d'_x of problem (BQP) satisfies the following necessary and sufficient conditions:

$$\begin{array}{rcl}
(Q^k + \mu^k I) d'_x - (A^k)^T y(\mu^k) &=& -c^k \text{ for some } y(\mu^k), \\
A^k d'_x &=& 0, \\
\mu^k &\ge& \max\{0, -\lambda^k\}, \\
\text{and} \quad \|d'_x\| &=& \alpha,
\end{array} \quad (9.36)$$

or, equivalently, the solution v of (BHP) satisfies the following necessary and sufficient conditions:

$$\begin{array}{rcl}
(H^k + \mu^k I) v &=& -g^k, \\
\mu^k &\ge& \max\{0, -\lambda^k\}, \\
\text{and} \quad \|v\| &=& \alpha,
\end{array} \quad (9.37)$$

where $\lambda^k = \lambda(H^k)$, and $\lambda(H)$ denotes the least eigenvalue of matrix H. Since Q is not positive semi-definite in the null space of A and x^k is strictly positive, we must have $\lambda^k < 0$ (Exercise 9.8).

9.4. INDEFINITE QUADRATIC PROGRAMMING

Let

$$s(\mu^k) = Q(x^k + d_x) + c - A^T y(\mu^k) \text{ and}$$
$$p^k = Q^k d'_x + c^k - (A^k)^T y(\mu^k)) = X^k s(\mu^k) - \frac{\Delta^k}{n+\rho} e. \qquad (9.38)$$

Then,

$$\mu^k = \|p^k\|/\alpha, \qquad d'_x = -\frac{\alpha p^k}{\|p^k\|} \qquad (9.39)$$

and

$$\begin{aligned}
q'(d'_x) &= \frac{1}{2}(d'_x)^T Q^k d'_x + (c^k)^T d'_x \\
&= (d'_x)^T (Q^k d'_x + c^k) - \frac{1}{2}(d'_x)^T Q^k d'_x \\
&= (d'_x)^T (Q^k d'_x + c^k - (A^k)^T y(\mu^k)) - \frac{1}{2}(d'_x)^T Q^k d'_x \\
&= -\alpha^2 \mu^k - \frac{1}{2}(d'_x)^T Q^k d'_x \\
&= -\alpha^2 \mu^k - \frac{1}{2} v^T (N^k)^T Q^k N^k (N^k)^T v \\
&\leq -\alpha^2 \mu^k - \frac{1}{2} \lambda((N^k)^T Q^k N^k) \|v\|^2 \\
&= -\alpha^2 \mu^k + \frac{\alpha^2}{2} |\lambda^k| \\
&\leq -\frac{\alpha^2 \mu^k}{2} = -\frac{\alpha \|p^k\|}{2}. \qquad (9.40)
\end{aligned}$$

This implies that

$$\frac{n+\rho}{\Delta} \left(\frac{1}{2} d_x^T Q d_x + \left(Qx^k + c - \frac{\Delta^k}{n+\rho} X^{-1} e \right)^T d_x \right) = \frac{\Delta^k}{n+\rho} q'(d'_x)$$
$$\leq -\frac{\alpha}{2} \frac{n+\rho}{\Delta^k} \|p^k\|.$$

Thus, not only we have $x^{k+1} := x^k + d_x > 0$ but also, from (9.35),

$$\mathcal{P}(x^{k+1}) - \mathcal{P}(x^k) \leq -\frac{\alpha}{2} \frac{n+\rho}{\Delta^k} \|p^k\| + \frac{\alpha^2}{2(1-\alpha)}. \qquad (9.41)$$

Here we see that if

$$\frac{n+\rho}{\Delta^k} \alpha \mu^k = \frac{n+\rho}{\Delta^k} \|p^k\| \geq \frac{3}{4}$$

and if α is chosen around $1/4$, then

$$\mathcal{P}(x^{k+1}) - \mathcal{P}(x^k) \leq -\frac{5}{96}.$$

Therefore, according to the implication of (9.34) we have

Theorem 9.26 *Let μ^k and $\alpha = 1/4$ be in condition (9.36). Then, if $\frac{n+\rho}{\Delta^k}\alpha\mu^k = \frac{n+\rho}{\Delta^k}\|p^k\| \geq \frac{3}{4}$ for all k, the algorithm returns an ϵ-minimizer of (QP) in $O((n+\rho)\log\frac{1}{\epsilon} + n\log n)$ iterations.*

The question is what can we say if $\frac{n+\rho}{\Delta^k}\|p^k\| < \frac{3}{4}$ at some k. The next section will show how to calculate a suitable z such that using $\rho = \frac{2n^2(n+\sqrt{n})}{\epsilon} - n$ guarantees that x^{k+1} must be an ϵ-KKT point whenever $\frac{n+\rho}{\Delta^k}\|p^k\| < 1$. Thus, the number of total iterations to generate an ϵ-minimal or ϵ-KKT solution is bounded by $O(\frac{n^3}{\epsilon}\log\frac{1}{\epsilon} + n\log n)$. The algorithm can be simply stated as follows:

Algorithm 9.3 *Let $n + \rho = 2n^2(n+\sqrt{n})/\epsilon$, $\alpha = 1/4$, $z \leq \underline{z}$, and x^0 be an approximate analytic center of $\mathcal{F}_p$. Set $k := 0$.*
  While $\frac{n+\rho}{\Delta^k}\alpha\mu^k = \frac{n+\rho}{\Delta^k}\|p^k\| < \frac{3}{4}$ in (9.36) or $\frac{q(x^k)-z}{q(x^0)-z} < \epsilon$ do

1. *Solve (BQP).*

2. *Let $x^{k+1} = x^k + X^k d'_x$.*

3. *Let $k := k + 1$ and return to Step 1.*

9.4.2 Generating an ϵ-KKT point

We first illustrate how to compute a suitable lower bound, z, for the (QP) minimal value $\underline{z}$. A z can be generated by solving, again, a ball-constrained problem:

$$\text{minimize} \quad q(x) - q(x^0) = \tfrac{1}{2}(x-x^0)^T Q(x-x^0) + (Qx^0 + c)^T(x-x^0)$$

$$\text{s.t.} \quad A(x - x^0) = 0,$$

$$x \in \mathcal{E}_{in}.$$

Let $\hat{x}$ be the minimizer. Then, according to Exercise 9.10 and relation (9.31), i.e., the ratio of the radii of the inscribing and circumscribing ellipsoids to $\mathcal{F}_p$ is $1/n$, we have

$$q(x^0) - q(\hat{x}) \geq \frac{1}{n^2}(q(x^0) - \underline{z}).$$

9.4. INDEFINITE QUADRATIC PROGRAMMING

Thus, we can assign

$$z := q(x^0) - n^2(q(x^0) - q(\hat{x})). \tag{9.42}$$

Note that an approximate x, say $q(x^0) - q(x) \geq (q(x^0) - q(\hat{x}))/1.1$, would establish a lower bound $z := q(x^0) - 1.1n^2(q(x^0) - q(x))$. This bound is perfectly acceptable for establishing our complexity result as well.

We now back to the case that $\frac{n+\rho}{\Delta^k}\|p^k\| < \frac{3}{4}$. Actually we shall address a weaker case that $\frac{n+\rho}{\Delta^k}\|p^k\| < 1$, that is,

$$\|\frac{n+\rho}{\Delta^k} X^k s(\mu^k) - e\| < 1.$$

First, we must have

$$s(\mu^k) = Q(x^k + d_x) + c - A^T y(\mu^k) = Qx^{k+1} + c - A^T y(\mu^k) > 0.$$

Furthermore,

$$\|\frac{n+\rho}{\Delta^k} X^k s(\mu^k) - e\|^2$$
$$= \left(\frac{n+\rho}{\Delta^k}\right)^2 \|X^k s(\mu^k) - \frac{(x^k)^T s(\mu^k)}{n} e\|^2 + \|\frac{(n+\rho)(x^k)^T s(\mu^k)}{n\Delta^k} e - e\|^2$$
$$\geq \|\frac{(n+\rho)(x^k)^T s(\mu^k)}{n\Delta^k} e - e\|^2$$
$$\geq \left(\frac{(n+\rho)(x^k)^T s(\mu^k)}{n\Delta^k} - 1\right)^2 n.$$

Hence, $\frac{n+\rho}{\Delta^k}\|p^k\| < 1$ implies

$$\frac{n - \sqrt{n}}{n+\rho} \leq \frac{(x^k)^T s(\mu^k)}{\Delta^k} \leq \frac{n + \sqrt{n}}{n+\rho}.$$

Moreover,

$$(x^{k+1})^T s(\mu^k) = (x^k)^T (X^k)^{-1} X^{k+1} s(\mu^k)$$
$$\leq \|(X^k)^{-1} X^{k+1}\| (x^k)^T s(\mu^k)$$
$$\leq (1+\alpha)(x^k)^T s(\mu^k) \leq 2(x^k)^T s(\mu^k).$$

Therefore, we have

$$\frac{(x^{k+1})^T s(\mu^k)}{q(x^k) - z} = \frac{(x^{k+1})^T s(\mu^k)}{\Delta^k} \leq \frac{2(n+\sqrt{n})}{n+\rho} \leq \frac{\epsilon}{n^2}.$$

or
$$\frac{(x^{k+1})^T s(\mu^k)}{q(x^k) - q(x^0) + n^2(q(x^0) - q(\hat{x}))} \leq \frac{\epsilon}{n^2}.$$

Consequently, if $q(x^k) \geq q(x^0)$ then
$$\frac{(x^{k+1})^T s(\mu^k)}{n^2(q(x^k) - q(\hat{x}))} \leq \frac{\epsilon}{n^2};$$

otherwise
$$\frac{(x^{k+1})^T s(\mu^k)}{n^2(q(x^0) - q(\hat{x}))} \leq \frac{\epsilon}{n^2}.$$

Both of them imply that
$$\frac{(x^{k+1})^T s(\mu^k)}{\bar{z} - q(\hat{x})} \leq \epsilon$$

since $q(x^0) \leq \bar{z}$ and $q(x^k) \leq \bar{z}$, which further implies that
$$\frac{(x^{k+1})^T s(\mu^k)}{\bar{z} - \underline{z}} \leq \epsilon,$$

since $q(\hat{x}) \geq \underline{z}$. That is, x^{k+1} is an ϵ-KKT point for (QP). To summarize, we have

Theorem 9.27 *Let z and ρ be chosen as above and let $\alpha < 1$ in condition (9.36). Then, $\mu^k \|d'_x\| = \|p^k\| < \frac{\Delta^k}{n+\rho}$ implies that $x^{k+1} := x^k + X^k d'_x$ is an ϵ-KKT point for (QP).*

9.4.3 Solving the ball-constrained QP problem

We now analyze solving (BQP), or (BHP) equivalently, in each iteration of the algorithm. It may be impossible to solve them exactly, but we can efficiently solve them approximately to guarantee both Theorems 9.26 and 9.27 hold. We give a complexity bound on obtaining such a sufficient solution.

Consider the necessary and sufficient conditions (9.36) or (9.37). It has been shown that μ^k is unique in these conditions and (see Exercise 9.9)
$$\frac{\alpha^2}{2} \mu^k \leq -\frac{1}{2}(d'_x)^T Q^k d'_x - (c^k)^T d'_x.$$

9.4. INDEFINITE QUADRATIC PROGRAMMING

Note that

$$\begin{aligned}
-\frac{1}{2}(d'_x)^T Q^k d'_x - (c^k)^T d'_x &= q(x^k) - q(x^{k+1}) + \frac{\Delta^k}{\rho} e^T d'_x \\
&\leq q(x^k) - q(x^{k+1}) + \frac{\Delta^k}{\rho} \|e\| \|d'_x\| \\
&= q(x^k) - q(x^{k+1}) + \frac{q(x^k) - z}{\rho} \sqrt{n} \alpha \\
&\leq (q(x^k) - z)\left(1 + \frac{\alpha \epsilon}{2n\sqrt{n}(n + \sqrt{n})}\right) \\
&= \Delta^k \left(1 + \frac{\alpha \epsilon}{2n\sqrt{n}(n + \sqrt{n})}\right)
\end{aligned}$$

Let

$$R := \frac{2}{\alpha^2} + \frac{\epsilon}{\alpha n \sqrt{n}(n + \sqrt{n})} \ . \tag{9.43}$$

Then

$$0 \leq \mu^k \leq R^k := R\Delta^k,$$

where R^k is a computable upper bound at each iteration. Note that for $\alpha = 1/4$, $\epsilon < 1$ and $n \geq 1$, $R \leq 34$.

For any given μ, denote solutions of the top linear equations by $d'_x(\mu)$ in conditions (9.36) and $v(\mu)$ in conditions (9.37). It can be also shown that $\|d'_x(\mu)\| = \|v(\mu)\|$ is a decreasing function for any $\mu \geq |\lambda^k|$. Besides, for any given μ we can check to see if $\mu \geq |\lambda^k|$ by checking the definiteness of matrix $H^k + \mu I$, which can be solved as a LDL^T decomposition. These facts lead to a bisection method to search for the root of $\|d'_x(\mu)\| = \|v(\mu)\| = \alpha$ while $\mu \in [|\lambda^k|, R^k] \subset [0, R^k]$. Obviously, for a given $\epsilon' \in (0, 1)$, a μ, such that $0 \leq \mu - \mu^k \leq \epsilon'$, can be obtained in $\log(R^k/\epsilon') = \log(R\Delta^k/\epsilon')$ bisection steps, and the cost of each step is $O(n^3)$ arithmetic operations.

The remaining question is what ϵ' would be sufficient to carry out our main algorithm: we generate an x^{k+1} such that either the potential function is decreased by a constant or x^{k+1} is an ϵ-KKT point. We discuss the condition of ϵ' in two cases: Case (A) of $\|p^k\| < \frac{3\Delta^k}{4\rho}$, i.e., the KKT point or termination case; and Case (B) of otherwise.

(A) The KKT point case.

Let us consider the termination iteration where

$$\|p^k\| < \frac{3\Delta^k}{4(n + \rho)} \ .$$

From relation (9.39) we see

$$0 < \mu^k < \frac{3\Delta^k}{4(n+\rho)\alpha}.$$

Let the right endpoint of the interval generated by the bisection search be denoted as μ. If we require μ such that

$$0 \le \mu - \mu^k \le \frac{\Delta^k}{4(n+\rho)\alpha},$$

then

$$\mu^k \le \mu < \frac{\Delta^k}{(n+\rho)\alpha}.$$

Also note $\|d'_x(\mu)\| \le \alpha$ since $\mu \ge \mu^k$. This leads to that

$$\|\mu d'_x(\mu)\| \le \mu\alpha < \frac{\Delta^k}{n+\rho},$$

and, from Theorem 9.27, $x^+ := x^k + X^k d'_x(\mu) > 0$ must be an ϵ-KKT point. Thus, noticing the choice of ρ and $\alpha = 1/4$, in this case we can choose

$$\epsilon' \le \frac{\Delta^k}{2n^2(n+\sqrt{n})} \qquad (9.44)$$

to meet the requirement.

(B) The non-KKT point case.

Now let us consider the non-KKT point iteration, i.e.,

$$\|p^k\| \ge \frac{3\Delta^k}{4(n+\rho)}.$$

In this iteration from (9.39) and (9.40) we have

$$\mu^k \ge \frac{3\Delta^k}{4(n+\rho)\alpha} \quad \text{and} \quad q'(d'_x(\mu^k)) \le -\frac{3\Delta^k \alpha}{8(n+\rho)}.$$

Thus, if we could generate a d'_x such that

$$q'(d'_x) - q'(d'_x(\mu^k)) \le \frac{3\Delta^k \alpha}{24(n+\rho)} \le \frac{\alpha\|p^k\|}{6} = \frac{\alpha^2 \mu^k}{6} \qquad (9.45)$$

and

$$\|d'_x\| \le \alpha,$$

9.4. INDEFINITE QUADRATIC PROGRAMMING

then
$$q'(d'_x) \le -\frac{\alpha\|p^k\|}{2} + \frac{\alpha\|p^k\|}{6} = -\frac{\alpha\|p^k\|}{3},$$
which implies that
$$\frac{n+\rho}{\Delta}\left(\frac{1}{2}d_x^T Q d_x + \left(Qx^k + c - \frac{\Delta^k}{n+\rho}X^{-1}e\right)^T d_x\right) \le \frac{-\alpha}{3}\frac{n+\rho}{\Delta^k}\|p^k\|,$$

and we would still have $x^k + X^k d'_x > 0$ and, from (9.35) and $\alpha = 1/4$,
$$\mathcal{P}(x^k + X^k d'_x) - \mathcal{P}(x^k) \le -\frac{2}{96}.$$

In other words, d'_x is an acceptable approximation to $d'_x(\mu^k)$ to reduce the potential function by a constant.

We analyze the complexity bounds to compute such an approximation. Again, let the right endpoint of the interval generated by the bisection search be denoted as μ. Then, $\mu \ge \mu^k$. If $\mu = \mu^k$, then we get an exact solution. Thus, we assume $\mu > \mu^k \ge |\lambda^k|$. We consider two sub-cases, the one (B.1) of $\mu^k \ge |\lambda^k| + 5\epsilon'$ and the opposite case (B.2) of $\mu^k < |\lambda^k| + 5\epsilon'$.

(B.1). In this case, we have
$$\mu^k \ge |\lambda^k| + 5\epsilon'.$$

Note that
$$\begin{aligned}
&\|d'_x(\mu^k)\|^2 - \|d'_x(\mu)\|^2 \\
&= \|v(\mu^k)\|^2 - \|v(\mu)\|^2 \\
&= v^T(\mu^k)(I - (H^k + \mu^k I)(H^k + \mu I)^{-2}(H^k + \mu^k I))v(\mu^k) \\
&= v^T(\mu^k)\left(2(\mu - \mu^k)(H^k + \mu I)^{-1} - (\mu - \mu^k)^2 (H^k + \mu I)^{-2}\right)v(\mu^k) \\
&\le \|v(\mu^k)\|^2 \left(\frac{2(\mu - \mu^k)}{(\mu - |\lambda^k|)} - \frac{(\mu - \mu^k)^2}{((\mu - |\lambda^k|))^2}\right) \\
&= \|v(\mu^k)\|^2 \left(\frac{2(\mu - \mu^k)}{(\mu - \mu^k) + (\mu^k - |\lambda^k|)} - \frac{(\mu - \mu^k)^2}{((\mu - \mu^k) + (\mu^k - |\lambda^k|))^2}\right) \\
&= \alpha^2 \frac{(\mu - \mu^k)^2 + 2(\mu - \mu^k)(\mu^k - |\lambda^k|)}{((\mu - \mu^k) + (\mu^k - |\lambda^k|))^2} \\
&= \alpha^2 \left(1 - \frac{(\mu^k - |\lambda^k|)^2}{((\mu - \mu^k) + (\mu^k - |\lambda^k|))^2}\right) \\
&\le \alpha^2 \left(1 - \frac{(5\epsilon')^2}{((\mu - \mu^k) + 5\epsilon')^2}\right).
\end{aligned}$$

In $O(\log(R^k/\epsilon'))$ bisection steps, we have $\mu - \mu^k \leq \epsilon'$. Then,

$$\|d'_x(\mu^k)\|^2 - \|d'_x(\mu)\|^2 \leq \frac{11\alpha^2}{36}.$$

On the other hand,

$$\begin{aligned}
&q'(d'_x(\mu)) - q'(d'_x(\mu^k)) \\
&= \frac{1}{2}v(\mu)^T H^k v(\mu) + (g^k)^T v(\mu) - \frac{1}{2}v(\mu^k)^T H^k v(\mu^k) - (g^k)^T v(\mu^k) \\
&= \frac{1}{2}(H^k v(\mu) + g^k)^T (v(\mu) - v(\mu^k)) + \frac{1}{2}(H^k v(\mu^k) + g^k)^T (v(\mu) - v(\mu^k)) \\
&= -\frac{1}{2}\mu v(\mu)^T (v(\mu) - v(\mu^k)) - \frac{1}{2}\mu^k v(\mu^k)^T (v(\mu) - v(\mu^k)) \\
&= -\frac{1}{2}(\mu - \mu^k)v(\mu)^T(v(\mu) - v(\mu^k)) - \frac{1}{2}\mu^k(\|v(\mu)\|^2 - \|v(\mu^k)\|^2) \\
&\leq \epsilon'\alpha^2 + \frac{11\alpha^2 \mu^k}{72}.
\end{aligned}$$

Thus, if we impose the condition

$$\epsilon'\alpha^2 \leq \frac{1}{72}\frac{3\Delta^k \alpha}{4(n+\rho)} \leq \frac{\alpha^2 \mu^k}{72},$$

then (9.45) holds, i.e.,

$$q'(d'_x(\mu)) - q'(d'_x(\mu^k)) \leq \frac{\alpha^2 \mu^k}{72} + \frac{11\alpha^2 \mu^k}{72} = \frac{\alpha^2 \mu^k}{6}$$

and $d'_x(\mu)$ will be an acceptable approximation to $d'_x(\mu^k)$. Note that choosing

$$\epsilon' \leq \frac{\epsilon \Delta^k}{48n^2(n+\sqrt{n})} \tag{9.46}$$

will meet the condition, due to the selection of ρ and α.

(B.2). In this case, we have

$$\mu^k < |\lambda^k| + 5\epsilon'.$$

Thus, in $O(\log(R^k/\epsilon'))$ bisection steps, we have $\mu - \mu^k < \epsilon'$ so that $\mu - |\lambda^k| < 6\epsilon'$. However, unlike Case (B.1) we find $d'_x(\mu)$ (or $v(\mu)$) is not a sufficient approximation to $d'_x(\mu^k)$ (or $v(\mu^k)$). When we observe this fact, we do the following computation.

Let q^k, $\|q^k\| = 1$, be an eigenvector corresponding the λ^k, the least eigenvalue of matrix H^k. Then, one of the unit vector e_j, $j = 1, ..., n-m$,

9.4. INDEFINITE QUADRATIC PROGRAMMING

must have $|e_j^T q^k| \geq 1/\sqrt{n-m}$. (In fact, we can use any unit vector q to replace e_j as long as $q^T q^k \geq 1/\sqrt{n-m}$.) Now solve for y from

$$(H^k + \mu I)y = e_j$$

and let

$$v = v(\mu) + \delta y,$$

where δ is chosen such that $\|v\| = \alpha$. Note we have

$$(H^k + \mu I)v = -g^k + \delta e_j,$$

and in the computation of $v(\mu)$ and y, matrix $H^k + \mu I$ only needs to be factorized once.

We have that

$$\|y\| \geq \frac{1}{\sqrt{n-m}(\mu - |\lambda^k|)}$$

and

$$|\delta| \leq 2\alpha(\mu - |\lambda^k|)\sqrt{n-m} \leq 2\alpha(6\epsilon')\sqrt{n-m}.$$

Let $d'_x = N^k v$. Then, we have

$$\begin{aligned}
&q'(d'_x) - q'(d'_x(\mu^k)) \\
&= \frac{1}{2}v^T H^k v + (g^k)^T v - \frac{1}{2}v(\mu^k)^T H^k v(\mu^k) - (g^k)^T v(\mu^k) \\
&= \frac{1}{2}(H^k v + g^k)^T (v - v(\mu^k)) + \frac{1}{2}(H^k v(\mu^k) + g^k)^T (v - v(\mu^k)) \\
&= \frac{1}{2}(H^k v + g^k - \delta e_j)^T (v - v(\mu^k)) + \frac{1}{2}\delta e_j^T (v - v(\mu^k)) \\
&\quad - \frac{1}{2}\mu^k v(\mu^k)^T (v - v(\mu^k)) \\
&= -\frac{1}{2}\mu v^T (v - v(\mu^k)) + \frac{1}{2}\delta e_j^T (v - v(\mu^k)) - \frac{1}{2}\mu^k v(\mu^k)^T (v - v(\mu^k)) \\
&= -\frac{1}{2}(\mu v + \mu^k v(\mu^k))^T (v - v(\mu^k)) + \frac{1}{2}\delta e_j^T (v - v(\mu^k)) \\
&= -\frac{1}{2}(\mu - \mu^k)v^T (v - v(\mu^k)) + \frac{1}{2}\delta e_j^T (v - v(\mu^k)) \\
&\leq \epsilon' \alpha^2 + 2(6\epsilon')\alpha^2 \sqrt{n-m}.
\end{aligned}$$

Thus, if we impose the condition

$$\epsilon'\alpha^2 + 12\epsilon'\alpha^2 \sqrt{n-m} \leq \frac{3\Delta^k \alpha}{24(n+\rho)},$$

then (9.45) holds, i.e.,

$$q'(d'_x) - q'(d'_x(\mu^k)) \le \frac{3\Delta^k \alpha}{24(n+\rho)} \le \frac{\alpha \|p^k\|}{6},$$

and d'_x must be an acceptable approximate to $d'_x(\mu^k)$. Note that choosing

$$\epsilon' \le \frac{\epsilon \Delta^k}{52n^2(n+\sqrt{n})\sqrt{n-m}}, \qquad (9.47)$$

will meet the condition, due to the selection of ρ and α.

Comparing (9.44), (9.46), and (9.47), we choose

$$\epsilon' = \frac{\epsilon \Delta^k}{52n^2(n+\sqrt{n})\sqrt{n-m}},$$

will meet the conditions for all three cases. Hence, our bisection method will terminate in at most $O(\log(R/\epsilon) + \log n)$ steps, where constant $R \le 34$ is defined by (9.43), and it either finds an ϵ-KKT point in Case A of the termination iteration, or calculates a sufficient approximation $d'_x(\mu)$ in Case (B.1) of the non-termination iteration, or spends additional $O((n-m)^3)$ arithmetic operations to generate a sufficient approximation d'_x in Case (B.2) of the non-termination iteration. Thus, the running time of the bisection method in each iteration is bounded by $O(n^3(\log(1/\epsilon) + \log n))$ arithmetic operations. To summarize:

Theorem 9.28 *The total running time of the potential reduction algorithm is bounded by*

$$O\left(\left(\frac{n^6}{\epsilon}\log\frac{1}{\epsilon} + n^4 \log n\right)\left(\log\frac{1}{\epsilon} + \log n\right)\right)$$

arithmetic operations.

One comment is about z, the lower bound for $\underline{z}$, used in our algorithm. If we somehow know $\underline{z}$ (we probably do in solving many combinatorial optimization problems), then we can choose $n + \rho = 2(n + \sqrt{n})/\epsilon$ and reduce the overall complexity bound by a factor n^2.

Finally, when $\epsilon \to 0$, we must have $\mu^k \to 0$, so that $|\lambda^k| \to 0$. At the limit, λ^k represent the least eigenvalue of Q in the null space of all active constraints: $Ax = b$ plus $x_j = 0$ for every j that $x_j^k \to 0$. This implies that Q is positive semi-definite in the null space of all active constraints of the limit. This is the second order necessary condition for the limit being a local minimal solution.

9.5 Approximating Quadratic Programming

In this section we consider approximating the *global minimizer* of the following QP problem with certain quadratic constraints:

$$\underline{q}(Q) := \begin{array}{ll} \text{minimize} & q(x) := x^T Q x \\ \text{s.t.} & \sum_{j=1}^{n} a_{ij} x_j^2 = b_i, \ i = 1, \ldots, m, \\ & -e \leq x \leq e, \end{array}$$

where symmetric matrix $Q \in \mathcal{M}^n$, $A = \{a_{ij}\} \in \mathcal{R}^{m \times n}$ and $b \in \mathcal{R}^m$ are given and $e \in \mathcal{R}^n$ is, again, the vector of all ones. We assume that the problem is feasible and denote by $\underline{x}(Q)$ its global minimizer.

Normally, there is a linear term in the objective function:

$$q(x) = x^T Q x + c^T x.$$

However, the problem can be homogenized as

$$\begin{array}{ll} \text{minimize} & x^T Q x + t \cdot c^T x \\ \text{s.t.} & \sum_{j=1}^{n} a_{ij} x_j^2 = b_i, \ i = 1, \ldots, m, \\ & -e \leq x \leq e, \ -1 \leq t \leq 1 \end{array}$$

by adding a scalar variable t. There always is an optimal solution $(\underline{x}, \underline{t})$ for this problem in which $\underline{t} = 1$ or $\underline{t} = -1$. If $\underline{t} = 1$, then $\underline{x}$ is also optimal for the original problem; if $\underline{t} = -1$, then $-\underline{x}$ is optimal for the original problem. Thus, without loss of generality, we can assume $q(x) = x^T Q x$ throughout this section.

The function $q(x)$ has a *global maximizer* over the bounded feasible set as well. Let $\bar{q} := -\underline{q}(-Q)$ and $\underline{q} := \underline{q}(Q)$ denote their maximal and minimal objective values, respectively. We now present a "fast" algorithm to compute a 4/7-approximate minimizer, that is, to compute a feasible $\hat{x}$, such that

$$\frac{q(\hat{x}) - \underline{q}}{\bar{q} - \underline{q}} \leq 4/7.$$

9.5.1 Positive semi-definite relaxation

The algorithm for approximating the QP minimizer is to solve a positive semi-definite programming relaxation problem:

$$\underline{p}(Q) := \begin{array}{ll} \text{minimize} & Q \bullet X \\ \text{s.t.} & A_i \bullet X = b_i, \ i = 1, \ldots, m, \\ & d(X) \leq e, \ X \succeq 0. \end{array} \quad (9.48)$$

Here, $A_i = \text{diag}(a_i)$, $a_i = (a_{i1}, \ldots, a_{in})$, and unknown $X \in \mathcal{M}^n$ is a symmetric matrix. Furthermore, $d(X)$ is a vector containing the diagonal components of X. Note that $d(X) \le e$ can be written as $I_j \bullet X \le 1, j = 1, \ldots, n$, where I_j is the all-zero matrix except the jth diagonal component equal to 1.

The dual of the relaxation is

$$\begin{array}{rl} \underline{p}(Q) = & \text{maximize} \quad e^T z + b^T y \\ & \text{s.t.} \quad Q \succeq D(z) + \sum_{i=1}^m y_i A_i, \; z \le 0, \end{array} \qquad (9.49)$$

where $D(z)$ is the diagonal matrix such that $d(D(z)) = z \in \mathcal{R}^n$. Note that the relaxation is feasible and its dual has an interior feasible point so that there is no duality gap between the primal and dual. Denote by $\underline{X}(Q)$ and $(\underline{y}(Q), \underline{z}(Q))$ an optimal solution pair for the primal (9.48) and dual (9.49). For simplicity, in what follows we let $\underline{x} = \underline{x}(Q)$ and $\underline{X} = \underline{X}(Q)$.

We have the following relations between the QP problem and its relaxation:

Proposition 9.29 *Let $\underline{q} := \underline{q}(Q)$, $\bar{q} := -\underline{q}(-Q)$, $\underline{p} := \underline{p}(Q)$, $\bar{p} := -\underline{p}(-Q)$, $(\bar{y}, \bar{z}) = (-\underline{y}(-Q), -\underline{z}(-Q))$. Then,*

i) *$\bar{q}$ is the maximal objective value of $x^T Q x$ in the feasible set of the QP problem;*

ii) *$\bar{p} = e^T \bar{z} + b^T \bar{y}$ and it is the maximal objective value of $Q \bullet X$ in the feasible set of the relaxation, and $D(\bar{z}) + \sum_{i=1}^m \bar{y}_i A_i - Q \succeq 0$;*

iii)
$$\underline{p} \le \underline{q} \le \bar{q} \le \bar{p}.$$

Proof. The first and second statements are straightforward to verify. Let $X = \underline{x}\underline{x}^T \in \mathcal{M}^n$. Then $X \succeq 0$, $d(X) \le e$,

$$A_i \bullet X = \underline{x}^T A_i \underline{x} = \sum_{j=1}^n a_{ij}(\underline{x}_j)^2 = b_i, \; i = 1, \ldots, m,$$

and $Q \bullet X = \underline{x}^T Q \underline{x} = q(\underline{x}) = \underline{q}(Q)$. Thus, we have $\underline{q}(Q) \ge \underline{p}(Q)$, or $\underline{p} \le \underline{q}$. Similarly, we can prove $\underline{q}(-Q) \ge \underline{p}(-Q)$, or $\bar{p} \ge \bar{q}$.

□

Since $\underline{X}$ is positive semi-definite, there is a matrix factorization $\underline{V} = (\underline{v}_1, \ldots, \underline{v}_n) \in \mathcal{R}^{n \times n}$, i.e., $\underline{v}_j$ is the jth column of $\underline{V}$, such that $\underline{X} = \underline{V}^T \underline{V}$.

9.5. APPROXIMATING QUADRATIC PROGRAMMING

Then, after obtaining $\underline{X}$ and $\underline{V}$, we generate a random vector u uniformly distributed on the unit sphere in $\mathcal{R}^n$ and assign

$$\hat{x} = \underline{D}\sigma(\underline{V}^T u), \tag{9.50}$$

where

$$\underline{D} = \text{diag}(\|\underline{v}_1\|, \ldots, \|\underline{v}_n\|) = \text{diag}(\sqrt{\underline{x}_{11}}, \ldots, \sqrt{\underline{x}_{nn}}),$$

and, for any $x \in \mathcal{R}^n$, $\sigma(x) \in \mathcal{R}^n$ is the vector whose jth component is $\text{sign}(x_j)$:

$$\text{sign}(x_j) = \begin{cases} 1 & \text{if } x_j \geq 0 \\ -1 & \text{otherwise}. \end{cases}$$

It is easily see that $\hat{x}$ is a feasible point for the QP problem and we will show later that its expected objective value, $E_u q(\hat{x})$, satisfies

$$\frac{E_u q(\hat{x}) - \underline{q}}{\bar{q} - \underline{q}} \leq \frac{\pi}{2} - 1 \leq \frac{4}{7}.$$

That is, $\hat{x}$ is a 4/7-approximate minimizer for the QP problem expectantly. One can generate u repeatedly and choose the best $\hat{x}$ in the process. Thus, we will almost surely generate a $\hat{x}$ that is a 4/7-approximate minimizer.

9.5.2 Approximation analysis

First, we present a lemma which will be only used in our analysis.

Lemma 9.30 *Let u be uniformly distributed on the unit sphere in $\mathcal{R}^n$. Then,*

$$\begin{aligned} \underline{q}(Q) = \quad \text{minimize} \quad & E_u(\sigma(V^T u)^T DQD\sigma(V^T u)) \\ \text{s.t.} \quad & A_i \bullet (V^T V) = b_i, \ i = 1, \ldots, m, \\ & \|v_j\| \leq 1, \ j = 1, \ldots, n, \end{aligned}$$

where

$$D = \text{diag}(\|v_1\|, \ldots, \|v_n\|).$$

Proof. Since, for any feasible V, $D\sigma(V^T u)$ is a feasible point for the QP problem, we have

$$\underline{q}(Q) \leq E_u(\sigma(V^T u)^T DQD\sigma(V^T u)).$$

On the other hand, for any fixed u with $\|u\| = 1$, we have

$$E_u(\sigma(V^T u)^T DQD\sigma(V^T u)) = \sum_{i=1}^n \sum_{j=1}^n q_{ij} \|v_i\| \|v_j\| E_u(\sigma(v_i^T u)\sigma(v_j^T u)). \tag{9.51}$$

Let us choose $v_i = \frac{x_i}{\|\underline{x}\|}\underline{x}$, $i = 1, \ldots, n$. Note that V is feasible for the above problem. Then

$$E_u(\sigma(v_i^T u)\sigma(v_j^T u)) = \begin{cases} 1 & \text{if } \sigma(\underline{x}_i) = \sigma(\underline{x}_j) \\ -1 & \text{otherwise.} \end{cases}$$

Thus,

$$\|v_i\|\|v_j\|E_u(\sigma(v_i^T u)\sigma(v_j^T u)) = \underline{x}_i\underline{x}_j$$

which implies that for this particular feasible V

$$\underline{q}(Q) = q(\underline{x}) = E_u(\sigma(V^T u)^T DQD\sigma(V^T u)).$$

These two relations give the desired result.

□

For any function of one variable $f(t)$ and $X \in \mathcal{R}^{n \times n}$, let $f[X] \in \mathcal{R}^{n \times n}$ be the matrix with the components $f(x_{ij})$. We have the next technical lemma whose proof is an exercise.

Lemma 9.31 *Let $X \succeq 0$ and $d(X) \le 1$. Then $\arcsin[X] \succeq X$.*

Now we are ready to prove the following theorem, where we use "infimum" to replace "minimum," since for simplicity we require X to be positive definite in our subsequent analysis.

Theorem 9.32

$$\begin{array}{rl} \underline{q}(Q) = & \text{infimum} \quad \frac{2}{\pi}Q \bullet (D\arcsin[D^{-1}XD^{-1}]D) \\ & \text{s.t.} \quad A_i \bullet X = b_i, \ i = 1, \ldots, m, \\ & \quad\quad d(X) \le e, \ X \succ 0, \end{array}$$

where

$$D = \text{diag}(\sqrt{x_{11}}, \ldots, \sqrt{x_{nn}}).$$

Proof. For any $X = V^T V \succ 0$, $d(X) \le e$, we have

$$\begin{aligned} E_u(\sigma(v_i^T u)\sigma(v_j^T u)) &= 1 - 2\Pr\{\sigma(v_i^T u) \neq \sigma(v_j^T u)\} \\ &= 1 - 2\Pr\left\{\sigma\left(\frac{v_i^T u}{\|v_i\|}\right) \neq \sigma\left(\frac{v_j^T u}{\|v_j\|}\right)\right\}. \end{aligned}$$

We have the following relation: [2]

$$\Pr\left\{\sigma\left(\frac{v_i^T u}{\|v_i\|}\right) \neq \sigma\left(\frac{v_j^T u}{\|v_j\|}\right)\right\} = \frac{1}{\pi}\arccos\left(\frac{v_i^T v_j}{\|v_i\|\|v_j\|}\right),$$

---

[2] Lemma 1.2 of Goemans and Williamson [141].

9.5. APPROXIMATING QUADRATIC PROGRAMMING

as illustrated in Figure 9.2. Then, noting $\arcsin(t) + \arccos(t) = \frac{\pi}{2}$ we have

$$E_u(\sigma(v_i^T u)\sigma(v_j^T u)) = \frac{2}{\pi}\arcsin\left(\frac{v_i^T v_j}{\|v_i\|\|v_j\|}\right),$$

and using equality (9.51) we further have

$$\begin{aligned}E_u(\sigma(V^T u)^T DQD\sigma(V^T u)) &= \sum_{i=1}^{n}\sum_{j=1}^{n}q_{ij}\|v_i\|\|v_j\|\frac{2}{\pi}\arcsin\left(\frac{v_i^T v_j}{\|v_i\|\|v_j\|}\right)\\&= \frac{2}{\pi}Q\bullet(D\arcsin[D^{-1}XD^{-1}]D).\end{aligned}$$

Finally, Lemma 9.30 gives us the desired result.

□

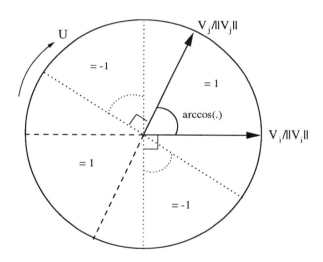

Figure 9.2. Illustration of the product $\sigma(\frac{v_i^T u}{\|v_i\|})\cdot\sigma(\frac{v_j^T u}{\|v_j\|})$ on the 2-dimensional unit circle. As the unit vector u is uniformly generated along the circle, the product is either 1 or -1.

Theorem 9.32 and Lemma 9.31 lead to our main result:

Theorem 9.33 *We have*
i)

$$\bar{p} - \underline{q} \geq \frac{2}{\pi}(\bar{p} - \underline{p}).$$

ii)

$$\bar{q} - \underline{p} \geq \frac{2}{\pi}(\bar{p} - \underline{p}).$$

iii)

$$\bar{p} - \underline{p} \geq \bar{q} - \underline{q} \geq \frac{4-\pi}{\pi}(\bar{p} - \underline{p}).$$

Proof. We prove (i). Recall $\bar{z} = -\underline{z}(-Q) \geq 0$, $\bar{y} = -\underline{y}(-Q)$, $\bar{p} = -\underline{p}(-Q) = e^T \bar{z} + b^T \bar{y}$, and $D(\bar{z}) + \sum_{i=1}^{m} \bar{y}_i A_i - Q \succeq 0$. Thus, for any $X \succ 0$, $d(X) \leq e$ and $D = \text{diag}(\sqrt{x_{11}}, \ldots, \sqrt{x_{nn}})$, from Theorem 9.32

$$\frac{\pi}{2}\underline{q} = \frac{\pi}{2}\underline{q}(Q)$$
$$\leq Q \bullet (D \arcsin[D^{-1}XD^{-1}]D)$$
$$= \left(Q - D(\bar{z}) - \sum_{i=1}^{m} \bar{y}_i A_i + D(\bar{z}) + \sum_{i=1}^{m} \bar{y}_i A_i\right) \bullet (D \arcsin[D^{-1}XD^{-1}]D)$$
$$= \left(Q - D(\bar{z}) - \sum_{i=1}^{m} \bar{y}_i A_i\right) \bullet (D \arcsin[D^{-1}XD^{-1}]D)$$
$$+ \left(D(\bar{z}) + \sum_{i=1}^{m} \bar{y}_i A_i\right) \bullet (D \arcsin[D^{-1}XD^{-1}]D)$$
$$\leq \left(Q - D(\bar{z}) - \sum_{i=1}^{m} \bar{y}_i A_i\right) \bullet (DD^{-1}XD^{-1}D)$$
$$+ \left(D(\bar{z}) + \sum_{i=1}^{m} \bar{y}_i A_i\right) \bullet (D \arcsin[D^{-1}XD^{-1}]D)$$

(since $Q - D(\bar{z}) - \sum_{i=1}^{m} \bar{y}_i A_i \preceq 0$ and $\arcsin[D^{-1}XD^{-1}] \succeq D^{-1}XD^{-1}$.)

$$= \left(Q - D(\bar{z}) - \sum_{i=1}^{m} \bar{y}_i A_i\right) \bullet X$$
$$+ \left(D(\bar{z}) + \sum_{i=1}^{m} \bar{y}_i A_i\right) \bullet (D \arcsin[D^{-1}XD^{-1}]D)$$
$$= Q \bullet X - \left(D(\bar{z}) + \sum_{i=1}^{m} \bar{y}_i A_i\right) \bullet X$$

9.5. APPROXIMATING QUADRATIC PROGRAMMING

$$+ \left(D(\bar{z}) + \sum_{i=1}^{m} \bar{y}_i A_i\right) \bullet (D \arcsin[D^{-1}XD^{-1}]D)$$

$$= Q \bullet X - \bar{z}^T d(X) - \sum_{i=1}^{m} \bar{y}_i a_i^T d(X)$$

$$+ \bar{z}^T d(D \arcsin[D^{-1}XD^{-1}]D) + \sum_{i=1}^{m} \bar{y}_i a_i^T d(D \arcsin[D^{-1}XD^{-1}]D)$$

$$= Q \bullet X - \bar{z}^T d(X) - \bar{y}^T b + \bar{z}^T (\frac{\pi}{2} d(X)) + \bar{y}^T (\frac{\pi}{2} b)$$

(since $d(D \arcsin[D^{-1}XD^{-1}]D) = \frac{\pi}{2} d(X)$ and $a_i^T d(X) = b_i$)

$$= Q \bullet X + (\frac{\pi}{2} - 1) \bar{z}^T d(X) + (\frac{\pi}{2} - 1) \bar{y}^T b$$

$$\leq Q \bullet X + (\frac{\pi}{2} - 1)(\bar{z}^T e + \bar{y}^T b)$$

(since $0 \leq d(X) \leq e$ and $\bar{z} \geq 0$)

$$= Q \bullet X + (\frac{\pi}{2} - 1) \bar{p}.$$

Let $X \succ 0$ converge to $\underline{X}$, then $Q \bullet X \to \underline{p}$ and we prove (i).
  Replacing Q with $-Q$ proves (ii) in the theorem.
  Adding the first two inequalities gives (iii) of the theorem.

□

The result indicates that the positive semi-definite relaxation value $\bar{p} - \underline{p}$ is a constant approximation of $\bar{q} - \underline{q}$. Similarly, the following corollary can be devised:

Corollary 9.34 Let $X = V^T V \succ 0$, $d(X) \leq e$, $A_i \bullet X = b_i$ ($i = 1, \ldots, m$), $D = \text{diag}(\sqrt{x_{11}}, \ldots, \sqrt{x_{nn}})$, and $\hat{x} = D\sigma(V^T u)$ where u with $\|u\| = 1$ is a random vector uniformly distributed on the unit sphere. Moreover, let $X \succ 0 \to \underline{X}$. Then,

$$\lim_{X \to \underline{X}} \mathrm{E}_u(q(\hat{x})) = \lim_{X \to \underline{X}} \frac{2}{\pi} Q \bullet (D \arcsin[D^{-1}XD^{-1}]D) \leq \frac{2}{\pi}\underline{p} + (1 - \frac{2}{\pi})\bar{p}.$$

Finally, we have the following theorem:

Theorem 9.35 Let $\hat{x}$ be randomly generated from $\underline{X}$. Then

$$\frac{\mathrm{E}_u q(\hat{x}) - \underline{q}}{\bar{q} - \underline{q}} \leq \frac{\pi}{2} - 1 < 4/7.$$

Proof. Since

$$\bar{p} \geq \bar{q} \geq \frac{2}{\pi}\bar{p} + (1 - \frac{2}{\pi})\underline{p} \geq (1 - \frac{2}{\pi})\bar{p} + \frac{2}{\pi}\underline{p} \geq \underline{q} \geq \underline{p},$$

we have, from Corollary 9.34,

$$\begin{aligned}
\frac{\mathrm{E}_u q(\hat{x}) - \underline{q}}{\bar{q} - \underline{q}} &\leq \frac{\frac{2}{\pi}\underline{p} + (1 - \frac{2}{\pi})\bar{p} - \underline{q}}{\bar{q} - \underline{q}} \\
&\leq \frac{\frac{2}{\pi}\underline{p} + (1 - \frac{2}{\pi})\bar{p} - \underline{q}}{\frac{2}{\pi}\bar{p} + (1 - \frac{2}{\pi})\underline{p} - \underline{q}} \\
&\leq \frac{\frac{2}{\pi}\underline{p} + (1 - \frac{2}{\pi})\bar{p} - \underline{p}}{\frac{2}{\pi}\bar{p} + (1 - \frac{2}{\pi})\underline{p} - \underline{p}} \\
&= \frac{(1 - \frac{2}{\pi})(\bar{p} - \underline{p})}{\frac{2}{\pi}(\bar{p} - \underline{p})} \\
&= \frac{1 - \frac{2}{\pi}}{\frac{2}{\pi}} = \frac{\pi}{2} - 1.
\end{aligned}$$

□

9.6 Notes

In this chapter we have extended a potential reduction algorithm to solving the fractional programming problem, general linear complementarity problem, and quadratic programming problem.

For the NEG problem, see Gale [130], Kemeny, Morgenstern and Thompson [220], Jagannathan and Schaible[197], and Robinson [363]. A recent description of the problem can be found in Schrijver [373], and Tan and Freund [399]. It is easy to show that Assumption 9.3 implies Assumption 9.4; see Theorem 2 of Robinson [363].

The method described here is based on the paper of Ye [474]. Similar methods for the fractional programming over polyhedral and nonpolyhedral cones were developed by Boyd and Ghaoui [71], Freund and Jarre [121, 120], Nesterov and Nemirovskii [328], and Nemirovskii [320].

The LCP potential algorithm described in this chapter is due to Kojima, Megiddo and Ye [228]. Other algorithms can be found in Kojima, Mizuno and Noma [229] and Noma [333]. The description of $\mathcal{G}$ is technically similar to Eaves' class and Garcia's class. These two classes and some others have been extensively studied; see Cottle, Pang and Stone [91]. Since every KKT point of the LCP with a row-sufficient matrix is a solution (Cottle et al.

9.6. NOTES

[91]), the algorithm is actually a polynomial approximation algorithm for solving the class of LCPs with row-sufficient matrices.

We have shown that the algorithm is a fully polynomial-time approximation scheme for computing an ϵ-approximate stationary or KKT point, which itself is a (nonconvex) linear complementarity problem. (The concept of the fully polynomial-time approximation scheme (FPTAS) was introduced in combinatorial optimization; for example, see Papadimitriou and Steiglitz [337].) The result is the first approximation algorithm, whose running time is almost linear in $\frac{1}{\epsilon}$, which was an open question in the area of nonlinear optimization complexity; see Vavasis [450]. We would also like to mention that algorithms, similar to the one described in this paper, have actually been used in practice, and they seem to work very well (e.g., Kamarth, Karmarkar, Ramakrishnan, and Resende [213, 214]).

If (QP) is a convex optimization problem, then it can be solved in polynomial time, e.g., see Vavasis [450] and references therein. If Q have at least one negative eigenvalue in the null space of A, then (QP) becomes a hard problem–an NP-complete problem ([131], [338], [368], and [450]). Some proposed algorithms for solving general QP problems include the principal pivoting method of Lemke-Cottle-Dantzig (e.g., [90]), the active-set method (e.g., [138]), the interior-point algorithm (e.g., [213, 214] and [469]), and other special-case methods (e.g., [309] and [338]). Other interior-point methods for nonconvex optimization can be found in Bonnans and Bouhtou [68] and Pardalos and Ye [469, 478].

Even finding a local minimum and checking the existence of a KKT point are NP-complete problems (see, e.g., Murty and Kabadi [311], Horst, Pardalos and Thoai [187], Johnson, Papadimitriou and Yannakakis [209], and Pardalos and Jha [339]). Finding even an ϵ-minimal or ϵ-KKT point are hard problems. Bellare and Rogaway [49] showed that there exists a constant, say, $\frac{1}{4}$, such that no polynomial-time algorithm exists to compute an $\frac{1}{4}$-minimal solution for (QP), unless $P = NP$. Vavasis [451] and Ye [469] developed a polynomial-time algorithm to compute an $(1 - \frac{1}{n^2})$-minimal solution. Using a steepest-descent-type method, Vavasis [450] also proved an arithmetic operation upper bound, $O(n^3(\frac{R}{\epsilon})^2)$, for computing an ϵ-KKT point of a box-constrained QP problem, where R is a fixed number depending on the problem data. Other results can be found in Fu, Luo and Ye [127] and Pardalos and Rosen [338].

Now consider (BQP), or (BHP) equivalently. First, a brief history of this problem. There is a class of nonlinear programming algorithms called model trust region methods. In these algorithms, a quadratic function is used as an approximate model of the true objective function around the current iterate. The next step is to minimize the model function. In general, however, the model is expected to be accurate or trusted only in a neighborhood of the current iterate. Accordingly, the quadratic model is minimized in a 2-norm neighborhood, which is a ball, around the current iterate.

The model-trust region problem, (BQP), is due to Levenberg [243] and Marquardt [263]. These authors considered only the case when Q^k is positive definite. Moré [308] proposed an algorithm with a convergence proof for this case. Gay [132] and Sorenson [387] proposed algorithms for the general case, also see Dennis and Schnable [96]. These algorithms work very well in practice, but no complexity result was established for this problem then.

A simple polynomial bisection method was proposed by Ye [469] and Vavasis and Zippel [453]. Recently, Rendl and Wolkowicz [357] showed that (BQP) can be reformulated as a positive-semidefinite problem, which is a *convex* nonlinear problem. There are polynomial interior-point algorithms (see Nesterov and Nemirovskii [327]) to compute an d'_x such that $q'(d'_x) - q(d'_x(\mu^k)) \leq \epsilon'$ in $O(n^3 \log(R^k/\epsilon'))$ arithmetic operations. This will also establish an

$$O\left(\left(\frac{n^6}{\epsilon}\log\frac{1}{\epsilon} + n^4 \log n\right)\left(\log\frac{1}{\epsilon} + \log n\right)\right)$$

arithmetic operation bound for our algorithm.

In addition, Ye [473] developed a Newton-type method for solving (BQP) and established an arithmetic operation bound $O(n^3 \log(\log(R^k/\epsilon')))$ to yield a μ such that $0 \leq \mu - \mu^k \leq \epsilon'$. The method can be adapted into our each iteration. We first find an approximate $\underline{\mu}$ to the absolute value of the least eigenvalue $|\lambda^k|$ and an approximate eigenvector q to the true q^k, such that $0 \leq \underline{\mu} - |\lambda^k| \leq \epsilon'$ and $q^T q^k \geq 1/\sqrt{n-m}$. This approximation can be done in $O(\log(\log(R^k/\epsilon')))$ arithmetic operations. Then, we will use q to replace e_j in Case (B.2) of the non-termination iteration (i.e., $\|v(\underline{\mu})\| < \alpha$) to enhance $v(\underline{\mu})$ and generate a desired approximation. Otherwise, we know $\mu^k > \underline{\mu}$ and, using the method in Ye [473], we will generate a $\mu \in (\underline{\mu}, R^k)$ such that $|\mu - \mu^k| \leq \epsilon'$ in $O(n^3 \log(\log(R^k/\epsilon')))$ arithmetic operations. This shall establish an

$$O\left(\left(\frac{n^6}{\epsilon}\log\frac{1}{\epsilon} + n^4 \log n\right)\log\left(\log\frac{1}{\epsilon} + \log n\right)\right)$$

arithmetic operation bound for our algorithm.

Most recently, there have been several remarkable results on approximating specific quadratic problems using positive semi-definite programming. Goemans and Williamson [141] proved an approximation result for the Maxcut problem where $\epsilon \le 1 - 0.878$. Nesterov [323] extended their result to approximating a boolean QP problem

$$\text{maximize} \quad q(x) = x^T Q x$$

$$\text{Subject to} \quad |x_j| = 1, \ j = 1, \ldots, n,$$

where $\epsilon \le 4/7$.

The positive semi-definite relaxation was first proposed by Lovász and Shrijver [247]; also see recent papers by Fujie and Kojima [128] and Polijak, Rendl and Wolkowicz [342]. The material on approximating quadratic programming in Section 9.5 is a further generalization of these results. The approximated problem has many applications in combinatorial optimization; see, e.g., Gibbons, Hearn and Pardalos [137].

9.7 Exercises

9.1 *Show that Assumption 9.3 implies Assumption 9.4.*

9.2 *Prove Lemma 9.6.*

9.3 *Prove Proposition 9.5.*

9.4 *Prove in Sections 9.2.1 and 9.3.1*
1.
$$\frac{d_x^T d_s}{x^T s} \le \|(X^k)^{-1} d_x\| \|(S^k)^{-1} d_s\| \le \frac{\|(X^k)^{-1} d_x\|^2 + \|(S^k)^{-1} d_s\|^2}{2}.$$

2.
$$(n+\rho)\log((x^{k+1})^T s^{k+1}) - (n+\rho)\log((x^k)^T s^k)$$
$$\le \frac{n+\rho}{\Delta^k}((s^k)^T d_x + (x^k)^T d_s) + \frac{(n+\rho)\alpha^2}{2}.$$

3. Inequalities (9.17) and (9.29).

9.5 *For any $x, s \in \overset{\circ}{\mathcal{R}}_+^n$, prove that*

$$H(x, s) = 2I - (XM^T - S)(S^2 + MX^2M^T)^{-1}(MX - S)$$

is positive semi-definite.

9.6 *Prove Proposition 9.13.*

9.7 *In Proposition 9.21 show that for any* $(x,s) \in \overset{\circ}{\mathcal{F}}$
$$\|g(x,s)\|_H^2 > 0.$$

9.8 *If Q is not positive semi-definite in the null space of A, prove that DQD is not positive semi-definite in the null space of AD for any positive diagonal matrix D.*

9.9 *Prove that μ^k is unique in (9.37) and*
$$\frac{\alpha^2}{2}\mu^k \leq -\frac{1}{2}v^T H^k v - (g^k)^T v,$$
which implies
$$\frac{\alpha^2}{2}\mu^k \leq -\frac{1}{2}(d'_x)^T Q^k d'_x - (c^k)^T d'_x.$$

9.10 *(Theorem 4 of [469]) Given $r > 0$ and let $d(r)$ be the minimizer for*
$$\text{minimize} \quad q(d) := \tfrac{1}{2}d^T Q d + c^T d$$
$$\text{s.t.} \quad Ad = 0,$$
$$\|d\|^2 \leq r^2.$$
Then, for $0 < r \leq R$
$$q(0) - q(d(r)) \geq \frac{r^2}{R^2}(q(0) - q(d(R))).$$

9.11 *Given $\underline{z}$, the exact global minimal objective value of a QP problem, develop an approximation algorithm for finding an KKT point of the problem and analyze its complexity.*

9.12 *Let $X \succeq 0$ and $d(X) \leq 1$. Then $\arcsin[X] \succeq X$.*

Chapter 10

Implementation Issues

It is common to have a gap between a theoretical algorithm and its practical implementation: the theoretical algorithm makes sure that it works for all instances and never fails, while the practical implementation emphasizes average performance and uses many ad-hoc "tricks." In this chapter we discuss several effective implementation techniques frequently used in interior-point linear programming software, such as the presolver process, the sparse linear system solver, the high-order predictor-corrector method, the homogeneous and self-dual method, and the optimal basis finder. Our goal is to provide additional theoretical justification for using these techniques and to explain their practical effectiveness.

10.1 Presolver

One way to eliminate data error and to improve solution efficiency in solving linear programs is called the "presolver"—a preliminary process to check the data set (A, b, c) in order to detect inconsistency and to remove redundancy. This process could reduce the size of the problem as well, because users of many LP problems likely introduce superfluous variables and redundant constraints for simplicity and convenience in model formulation.

In general, detecting all inconsistency and redundancy in an LP problem is computationally intractable. Therefore all presolvers use an arsenal of simple inspection techniques. These techniques are applied repeatedly until the problem cannot be reduced any further. Below, we briefly present the most common inspection procedures.

- Remove empty or all-zero rows and columns.

- Eliminate a fixed variable, the variable has a fixed value, from the problem by substitution.

- Remove duplicate constraints. Two constraints are said to be duplicate if they are identical up to a scalar multiplier. One of the duplicate constraints is removed from the problem.

  Remove duplicate columns. Two columns are said to be duplicate if they are identical up to a scalar multiplier. (They make duplicate constraints in the dual.)

- Remove linearly dependent constraints. The presence of linearly dependent rows in A may lead to serious numerical problems in an interior-point methods, since it implies a rank deficiency in the Newton equation system.

- Remove a singleton row (only one non-zero coefficient in the row) by construction of a simple variable bound. For example, if the ith constraint is in the form $a_{i1}x_1 \leq b_i$, we can convert it to $x_1 \leq b_i/a_{i1}$ or $x_1 \geq b_i/a_{i1}$, depending on the sign of a_{i1},

- Remove a free and singleton column (only one non-zero coefficient in the column and the associated variable is free). For example, let free variable x_1 appears only in the ith constraint. Then, x_1 and the ith constraint can be eliminated, while the optimal value of x_1 can be recovered from the ith constraint by substitution of the optimal solution of the remaining LP problem.

  A nonnegative but unbounded variable, say, $0 \leq x_1 < +\infty$ in singleton column 1, can be used to generate a bound on dual variables y_i. Namely,
  $$a_{i1}y_i \leq c_1.$$
  This inequality can be used, depending on the sign of a_{i1}, to produce a lower or upper bound on y_i.

- Determine lower and upper limits for every constraint and detect infeasibility. For example, consider the ith (inequality) constraint
  $$\sum_j a_{ij}x_j \leq b_i,$$

10.1. PRESOLVER

and let each variable x_j lie on $[0, u_j]$. Then compute

$$\underline{b}_i = \sum_{\{j:\ a_{ij}<0\}} a_{ij}u_j \leq 0 \quad \text{and} \quad \overline{b}_i = \sum_{\{j:\ a_{ij}>0\}} a_{ij}u_j \geq 0. \quad (10.1)$$

Thus, we must have

$$\underline{b}_i \leq \sum_j a_{ij}x_j \leq \overline{b}_i. \quad (10.2)$$

If $\overline{b}_i \leq b_i$, then the ith constraint is *redundant* and can be removed. If $\underline{b}_i > b_I$, then the problem is infeasible. If $\underline{b}_i = b_i$, the i constraint becomes equality and will force all involved variables take values at their appropriate bounds.

The same technique can be applied to each of the dual constraints.

- Add implicit bound to a free primal variable. For example, suppose the ith constraint is

$$\sum_j a_{ij}x_j = b_i,$$

where $a_{i1} > 0$, x_1 is a free variable, and all other variables x_j lie on $[0, u_j]$. Then

$$x_1 \leq \left(b_i - \sum_{\{j \neq 1:\ a_{ij}<0\}} a_{ij}u_j \right) / a_{i1}$$

and

$$x_1 \geq \left(b_i - \sum_{\{j \neq 1:\ a_{ij}>0\}} a_{ij}u_j \right) / a_{i1}.$$

The same technique can be applied to a dual free variable.

- Improve the sparsity of A, i.e., reduce the non-zero elements in A. We could look for a nonsingular matrix $M \in \mathcal{R}^{m \times m}$ such that the matrix MA is as sparse as possible. Primal constraints can in such case be replaced with equivalent

$$MAx = Mb, \quad (10.3)$$

which may be more suitable for an interior-point solver. Exact solution of this *sparsity problem* is an NP-complete problem but efficient heuristics usually produce satisfactory non-zero reduction in A.

The application of the above presolver techniques often results in impressive size-reduction of an initial LP formulation. Thus, it is our hope that the reduced problem obtained after the presolver can be solved faster. Once a solution is found, it could be used to recover a complete primal and dual solution to the original LP problem. This phase is called the *postsolver*.

10.2 Linear System Solver

The major work in a single iteration of all interior-point algorithms is to solve a set of linear equations, such as (4.17). It can be reduced to the so-called KKT system:

$$\begin{pmatrix} D^{-2} & A^T \\ A & 0 \end{pmatrix} \begin{pmatrix} d_x \\ -d_y \end{pmatrix} = \begin{pmatrix} \bar{c} \\ \bar{b} \end{pmatrix}, \qquad (10.4)$$

The diagonal matrix D varies in different interior-point methods. Most general purpose codes use *direct* methods to solve the KKT system. Two competitive direct methods are: the *normal equation* approach and the *augmented system* approach. The former works with a smaller positive definite matrix, and the latter requires factorization of a symmetric indefinite matrix. They all use variants of the symmetric triangular $L\Lambda L^T$ decomposition, where L is a lower triangular matrix and Λ is a block diagonal matrix with blocks of dimension 1 or 2. (The QR decomposition of A uses an orthogonal transformation and guarantees high accuracy, but it cannot be used in practice due to its costly operations.)

10.2.1 Solving normal equation

The normal equation approach further reduces (10.4) to the normal equation:

$$(AD^2 A^T)d_y = \bar{b} - AD^2 \bar{c}. \qquad (10.5)$$

An advantage of this approach is that it works with a positive definite matrix $AD^2 A^T$ if A has full row rank. Thus the Choleski decomposition of this matrix exists for any D and numerical stability is assured in the pivoting process. Moreover, the sparsity pattern in the decomposition is independent of the value of D and hence it is invariant in all iterations. Consequently, once a good sparsity preserving pivoting order is chosen, it can be used throughout the entire iterative process. This argument has been used to justify the application of the normal equations approach in very first interior-point method implementation.

10.2. LINEAR SYSTEM SOLVER

The success of the Choleski factorization depends on a pivoting order for preserving sparsity in the Choleski factor L. Its goal is to find a permutation matrix P such that the factor of $PAD^2A^TP^T$ is as sparse as possible. In practice, heuristics are used to find such a permutation or ordering. (Finding an optimal permutation is an NP-complete problem.) After an ordering is found, the data structure and indirect index addressing of L are setup. This is referred to as the symbolic phase because no numerical computation is involved.

Two heuristic orderings, *minimum degree* and the *minimum local fill-in*, are particularly useful in implementing interior-point algorithms. They are both "local" or myopic, i.e. they select a pivot only from a set of currently best candidates.

Minimum degree ordering

Assume that, in the kth step of the Gaussian elimination, the ith column of the Schur complement contains d_i non-zero entries and its diagonal element becomes a pivot. The kth step of the elimination requires thus

$$l_i = (1/2)(d_i - 1)^2, \qquad (10.6)$$

floating-point operations or *flops* to be executed.

In what follows, keep in your mind the fact that the decomposed matrix AD^2A^T is positive definite so the pivot choice can be limited to the diagonal elements. In fact, only this choice preserves symmetry.

Note that if the ith diagonal element becomes a pivot, l_i evaluates flops and gives an overestimate of the fill-ins, the new non-zeros created in the Schur complement, which can result from the current elimination step. Thus, the "best" pivot at step k, in terms of the number of flops required to complete the kth elimination step, is the one that minimizes d_i among all diagonal elements in the Schur complement. Interpreting the elimination process as the corresponding incidence graph elimination, one can see that this strategy chooses a node (diagonal element) in the graph which has the minimum degree (d_i). This is how the strategy is named. This ordering procedure can be implemented efficiently both in terms of time speed and storage requirement.

There is also an *approximate* minimum degree ordering available. The method is faster while generates the same quality ordering.

Minimum local fill-in ordering

In general, l_i of (10.6) considerably overestimates the number of fill-ins in the kth step of the Gaussian elimination, because it does not take into account the fact that in many positions of the predicted fill-ins, non-zero entries already exist. It is possible that another pivot candidate, although may not minimize d_i, would produce least fill-ins in the remaining Schur complement. The minimum local fill-in ordering chooses such a pivot. Generally, the minimum local fill-in procedure produces an ordering resulting in a sparser factorization but at a higher cost, because it chooses the pivot that produces the minimum number of fill-ins among all remaining pivot candidates.

Pros and cons

Solving the normal equation is proved to be a reliable approach to solutions of most practical linear programs. However, it suffers two drawbacks. First, the normal equation behaves badly whenever a primal linear program contains free variables. In order to transform such a problem to the standard form, a free variable is usually replaced with the difference of two nonnegative variables: $x = x^+ - x^-$. Interior-point algorithms typically generate iterates in which both x^+ and x^- converge to ∞, although their difference is kept relatively close to the optimal value of x. This results in a serious ill-condition in the normal matrix and a loss of accuracy in solving (10.5). A remedy used in many implementations is to prevent excessive growth of x^+ and x^- by enforcing bounds on x^+ and x^-.

Second, a more serious drawback of the normal equation approach is that it looses sparsity from the presence of dense columns in A. The reason is that a single dense column in A with p non-zero elements creates a complete dense submatrix of size $p \times p$ in AD^2A^T after a symmetric row and column permutation. Special care has to be taken in this case.

Assume that
$$A = (A_S, \ A_D), \tag{10.7}$$

where $A_S \in \mathcal{R}^{m \times n-k}$ is sparse and $A_D \in \mathcal{R}^{m \times k}$ is dense. Then, we need to treat A_D separately. The most popular way in solving the normal equation employs the *Schur complement* mechanism. It is based on separating the normal matrix
$$AD^2A^T = A_S D_S^2 A_S^T \ + \ A_D D_D^2 A_D^T, \tag{10.8}$$

into the presumably sparse part $A_S D_S^2 A_S^T$ and the significantly denser symmetric rank-k matrix $A_D D_D^2 A_D^T$. A Choleski decomposition is computed for the sparse part and the dense rank-k matrix is then updated by the Sherman-Morrison-Woodbury formula (see Exercise 1.1).

This method is not guaranteed to work correctly because the sparse part may be rank deficient, since A_S may not have full row rank. Whenever this happens, the Choleski decomposition of $A_S D_S^2 A_S^T$ does not exist

10.2. LINEAR SYSTEM SOLVER

and the Sherman-Morrison-Woodbury update is not well defined. Therefore in a practical implementation diagonal elements are selectively added to $A_S D_S^2 A_S^T$ to make the decomposition exist. We observe that the rank deficiency of $A_S D_S^2 A_S^T$ cannot exceed k, the number of dense columns. This method usually works in satisfaction for a small number of dense columns.

This is how we do it. If unacceptably small pivots are encountered during the Choleski decomposition of $A_S D_S^2 A_S^T$, we add a "regularizing" diagonal term to each of them. Consequently, instead of computing the decomposition of $A_S D_S^2 A_S^T$, we compute the decomposition of another matrix $A_S D_S^2 A_S^T + \sigma E E^T$, where positive number σ is a regularizing term and E is a matrix built from unit columns where each non-zero appears in the row corresponding to regularized pivots, that is,

$$L \Lambda L^T = A_S D_S^2 A_S^T + \sigma E E^T. \tag{10.9}$$

L is used as a stable "working basis" in the Sherman-Morrison-Woodbury update of the Schur complement to compute

$$(AD^2 A^T)^{-1} = (L \Lambda L^T + (A_D D_D^2 A_D^T - \sigma E E^T))^{-1}.$$

In many cases, choosing $\sigma = 1$ seems sufficient.

Summing up, it is possible to overcome the dense column difficulty arisen in the normal equation approach. But there remains a question to decide which columns should be treated as dense ones. A naive selection rule, which is based on counting the number of non-zero elements in a column, does not necessarily identify all the "troubling" columns—the columns make the decomposition dense. This motivated researchers to directly solve the augmented system of the Newton equation (10.4), which allows more freedom in selecting pivots.

10.2.2 Solving augmented system

The augmented system approach is a well understood technique to solve a least-squares problem. It applies a factorization to a symmetric indefinite matrix

$$L \Lambda L^T = \begin{pmatrix} D^{-2} & A^T \\ A & 0 \end{pmatrix}, \tag{10.10}$$

where Λ is an indefinite block diagonal matrix where each block is either 1×1 or 2×2.

In contrast to solving the normal equation in which the sparsity ordering and the numerical factorization are separated, the factorization of (10.10) is computed dynamically. In other words, the choice of a pivot is concerned with both sparsity and stability of the triangular factor L. Thus, the factorization of the augmented system is at least as stable as that of the normal equation. Moreover, due to greater freedom in the choice of a pivot order, the augmented system factorization may produce a significantly sparser factor than that of the normal equation. Indeed the latter is actually a special case of (10.10) in which the first n pivots are chosen solely from D^2, regardless their stability and sparsity outcome.

The stable property of solving the augmented system has motivated many researchers to incorporate this approach into their implementation. There are other advantages for this approach, such as easy handling of free variables and dense columns, and its effortless extension to solving convex quadratic programming problems.

However, efficiency of the augmented system approach depends highly on keeping a consistent pivot ordering. One should avoid reordering pivots on every iteration and try to use the current pivot order in subsequent iterations as much as possible. The order is only updated occasionally when the KKT system has changed considerably.

One specific pivoting rule is again detecting "dense" columns in A and pivoting early those diagonal elements of D^{-2} which are not associated with the dense columns. One can set a density threshold to partition A into the sparse and dense parts as in (10.7).

A fixed threshold value approach works well only in a case when dense columns are easily identifiable, i.e., when the number of non-zero in each of them exceeds significantly the average number of entries in sparse columns. Whenever more complicated sparsity structure appears in A, a more sophisticated heuristic is needed.

Instead of the simple column partition (10.7), one may consider more complicated sparsity structure and the following partition of A:

$$A = \begin{pmatrix} A_{11} & A_{12} \\ A_{21} & A_{22} \end{pmatrix}. \qquad (10.11)$$

Here A_{11} is supposed sparse and is assumed to create a sparse normal matrix $A_{11}A_{11}^T$, A_{12} is a small set of "troubling" columns (either dense columns or columns associated with free variables), and $(A_{21} \; A_{22})$ represents a set of "troubling" rows.

Once the partition (10.11) is determined, (10.4) becomes

$$\begin{pmatrix} D_1^{-2} & & A_{11}^T & A_{21}^T \\ & D_2^{-2} & A_{12}^T & A_{22}^T \\ A_{11} & A_{12} & & \\ A_{21} & A_{22} & & \end{pmatrix} \begin{pmatrix} d_{x1} \\ d_{x2} \\ d_{y1} \\ d_{y2} \end{pmatrix} = \begin{pmatrix} \bar{c}_1 \\ \bar{c}_2 \\ \bar{b}_1 \\ \bar{b}_2 \end{pmatrix}.$$

The structure of this system shows immediately which block, such as D_1^{-2},

10.2. LINEAR SYSTEM SOLVER

can be inexpensively pivoted out, and which block, such as D_2^{-2}, should be pivoted lately.

The elimination of D_1^{-2} causes very limited fill-ins and reduces the KKT system to

$$\begin{pmatrix} D_2^{-2} & A_{12}^T & A_{22}^T \\ A_{12} & -A_{11}D_1^2 A_{11}^T & -A_{11}D_1^2 A_{21}^T \\ A_{22} & -A_{21}D_1^2 A_{11}^T & -A_{21}D_1^2 A_{21}^T \end{pmatrix}. \quad (10.12)$$

The elimination of D_2^{-2} should be delayed after all attractive pivot candidates from $A_{11}D_1^2 A_{11}^T$ and $A_{21}D_1^2 A_{21}^T$ blocks are exploited.

10.2.3 Numerical phase

So far we have extensively discussed the symbolic phase—the pivoting rule and pivoting order. Now we turn our attention to the numerical phase of a sparse symmetric system solver. This is a well developed area both in theory and in computational practice. Here we demonstrate several implementation techniques of the numerical factorization phase in the normal equation approach. These methods could be applied to the general symmetric decomposition of the augmented system as well.

Let $M = AD^2 A^T$ and consider its Choleski factorization $L\Lambda L^T = M$, where L is a lower triangular matrix and Λ is a diagonal matrix. The basic formulae for computing the column j of L, denoted by $L_{.j}$, and the pivot Λ_{jj} are:

$$\begin{aligned} \Lambda_{11} &= M_{11}, \\ L_{.1} &= \tfrac{1}{\Lambda_{jj}} M_{.1}, \\ \Lambda_{jj} &= M_{jj} - \sum_{k=1}^{j-1} L_{jk}^2 \quad j \geq 2, \\ L_{.j} &= \tfrac{1}{\Lambda_{jj}} \left(M_j - \sum_{k=1}^{j-1} (\Lambda_{kk} L_{jk}) L_k \right) \quad j \geq 2. \end{aligned} \quad (10.13)$$

Several methods have been developed to compute the factorization. They all exploit sparsity of the matrix but use different storage techniques in computations. These calculations can be organized either by rows or by columns. During the *row*-Choleski factorization the rows of the Choleski factor L are computed one by one.

The commonly used factorization is the *column*-Choleski factorization in which the columns of L are computed one by one as in (10.13). Its efficient implementations can be found, for example, in the Yale Sparse Matrix Package and Waterloo SPARSPAK. This method is also called *left-looking* factorization, because the computation of column $L_{.j}$ follows the left-to-right order. Its implementation uses dynamic linked lists to look at all "left" columns when computing the current pivot and column, and a double precision work array to accumulate the column modifications and to resolve the non-zero matching between different columns.

Another commonly used approach is the *submatrix*-Choleski factorization, also referred to as the *right-looking* factorization. In this approach, once a column $L_{.j}$ has been computed, we immediately update its contributions to all subsequent columns, i.e. to all columns on its right side using (10.13). In this method the matching of non-zero during the process is not trivial, but several solutions have been found. Interest in this method has been increased in the past few years because of its ability to better exploit high performance architecture and memory hierarchy.

We now present several numerical "tricks" that work very well in interior-point methods. These techniques all based on using matrix-vector operations in a "dense" mode (assuming matrices and vectors are complete dense) to reduce the overhead computation and book-keeping map in a sparse mode using indirect index addressing and sophisticated memory referencing.

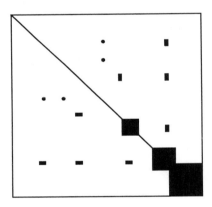

Figure 10.1. Illustration of dense sub-factors in a Choleski factorization.

Dense window

The most straightforward improvement of the factorization is exploitation of a *dense window*. In practice, some triangular sub-factors become completely dense near the end of the Choleski factorization; see Figure 10.1. Therefore, we can treat these blocks complete dense and use dense matrix factorization, even though there may still be some zeros in this block. This is called a dense window. In doing so we avoid the overhead of sparse computation, such as indirect index addressing and memory referencing.

10.2. LINEAR SYSTEM SOLVER

It might also be beneficial to treat some almost-dense columns complete dense and to include them in a dense window.

Supernode

It is often observed that several columns in L tend to have the same sparsity pattern below the diagonal. Such a block of columns is called a *supernode* and it can be treated as a dense submatrix. The supernode name comes from the elimination graph representation of the Choleski decomposition, because these nodes (columns) more-or-less share the same set of adjacent nodes and they can be grouped as a single "super" node (share the same addressing), as illustrated below:

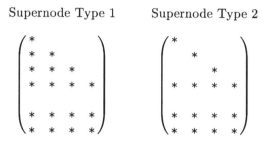

Both types of supernodes can be exploited in a similar manner within the numerical factorization. Similar to the dense window technique, the use of supernodes increases the portion of matrix-vector operations in the dense mode, and thereby saves on indirect addressing and memory referencing. Specifically, the following operations take advantage of supernodes:

- When column j is a member of a supernode, the operation of computing $L_{.j}$ and other columns of the supernode are done in the dense mode.

- When column j is not a member of a supernode but it has a summation term from a set of columns that belong to a supernode, a temporary work array is used to accumulate the sum from the whole supernode in the dense mode before the term is added to $L_{.j}$.

Sometime it is even beneficial to treat some zeros as non-zeros in L to create supernodes. The introduction of zeros does not necessarily lead to an increase in the memory allocation. This is due to the fact that only indexes of the last column in a supernode are booked, so there is a saving in index addressing.

Cache memory

Computers has a memory hierarchy consisting a slow and large main memory and a fast and small cache memory. Computation will be more efficient if memory references are stored in the cache memory so they can be fetched faster. Thus, it is advisable to set an upper bound on the number of nonzeros in each supernode to such that they can be all stored in the cache memory during the dense mode matrix-vector operations. One should leave 5 percent of the cache memory for overhead.

Of course, such partition of large supernodes leads to more overhead computation. An advise is to also impose a lower bound on the size of supernodes since the extra work in constructing the work array may not pay off if the size of the supernode is too small.

Block Choleski factorization

Another possibility is to partition L into smaller, presumably dense blocks. For example, try to divide L into block diagonal dense submatrices. This technique is very effective in some cases, because a typical Choleski factor contains many such blocks, the largest of which is usually the dense window located at the bottom of L. Consider the following matrix:

$$AD^2A^T = \begin{pmatrix} M_{11} & M_{21}^T \\ M_{21} & M_{22} \end{pmatrix},$$

with an additional simplifying assumption that the blocks L_{11} and L_{22} of L are dense matrices. The Choleski factorization of this matrix can be computed in the following steps:

1. Factorize $L_{11}\Lambda_{11}L_{11}^T = M_{11}$.
2. Compute $L_{21} = M_{21}(L_{11}^{-1})^T$.
3. Compute $\hat{M}_{22} = M_{22} - L_{21}\Lambda_{11}L_{21}^T$.
4. Factorize $L_{22}\Lambda_{22}L_{22}^T = \hat{M}_{22}$.

The advantage of this procedure is that steps 1 and 4 can be performed in the dense mode.

Loop unrolling

Dense mode computation can be further specialized to exploit a loop unrolling technique. Let a be the target column, b the source column, and α the multiplier kept in a single register. Then the steps performed by a

10.2. LINEAR SYSTEM SOLVER

computer to execute the transformation $a \leftarrow a + \alpha \cdot b$ can be written as follows:

1. Read $a(i)$ from the memory.
2. Read $b(i)$ from the memory.
3. Compute $a(i) + \alpha \cdot b(i)$.
4. Store the result in the memory.

Consequently, three memory references in steps 1, 2, and 4 are associated with only one arithmetic multiplication in step 3.

During a typical inner loop of the factorization, several multiple columns are added to a single column; see (10.13). This opens a possibility to unroll the loop over the multiple column transformation. Let a be the target column, b, c, d, e, f and g the source columns, and $\alpha(1), \ldots, \alpha(6)$ their multipliers kept in a single register. A loop rolling technique to compute

$$a \leftarrow a + \alpha(1)b + \alpha(2)c + \alpha(3)d + \alpha(4)e + \alpha(5)f + \alpha(6)g$$

is to execute the above procedure 6 times and uses total 18 memory references and 6 multiplications. However, a 6-step loop unrolling technique consists of first reading a, b, c, d, e, f, g, then performing 6 multiplications, and finally storing new a. This execution needs only eight memory references. Hence, 10 memory references have been saved compared with the loop rolling execution. The loop unrolling technique generally makes considerable time savings on many different computer architectures.

10.2.4 Iterative method

An alternative to solve the KKT system (10.4) or the normal equation (10.5) is the *iterative* method, e.g., the conjugate gradient method. This method automatically exploits the sparse structure of the system because it neither uses nor stores any inverse matrix. Its effectiveness highly depends on the selection of an appropriate and simple preconditioner. In solving general linear programs the iterative method seems not competitive with the direct method, but it becomes highly successful in solving special LP problems such as network-flow problems.

Consider the network-flow problem, where A matrix is the node-arc incidence matrix for a network with $m+1$ nodes and n arcs (For ease of notation, an arbitrary row in A is assumed to have been deleted so that A has full row rank m). Let $A = (A_B, A_N)$ where A_B is a basis of A. Then,

$$AD^2 A^T = \left(A_B D_B^2 A_B^T + A_N D_N^2 A_N^T\right).$$

If the diagonal components of D_B are all greater than or equal to the diagonal components of D_N, then we expect that $A_B D_B^2 A_B^T$ becomes a dominate block and it is a good estimation of AD^2A^T. The following theorem indicates "why", whose proof is derived from Exercise 10.1.

Theorem 10.1 *Choose π so that A_π contains a basis of A and $D_{ii} \geq D_{jj}$ for all $i \in \pi$ and $j \notin \pi$. Then,*

$$(2m^3+1)I \succeq (A_\pi D_\pi^2 A_\pi^T)^{-.5}(AD^2A^T)(A_\pi D_\pi^2 A_\pi^T)^{-.5} \succeq I.$$

Thus, $A_B D_B^2 A_B^T$ is a reasonable preconditioner for AD^2A^T, where B is a basis of π. Note that A_B can be reordered as a triangular matrix so that the Choleski factor of $A_B D_B^2 A_B^T$ is A_B itself after a permutation. Furthermore, B can be found by the maximum-spanning tree algorithm where D_{jj} is the weight of arc j. This algorithm is very cost-effective.

10.3 High-Order Method

If a direct approach is used to solve the KKT system, in each iteration a matrix factorization (10.4) or (10.5) is computed and followed by several backsolve steps. The factorization phase, $O(n^3)$ operations for a dense matrix, consumes the major amount of work, and the backsolve phase, at most $O(n^2)$ operations, is usually significantly easier in theory as well as in practice. An obvious idea, known from different applications of the Newton method, is to reuse the factorization in several subsequent iterations or, equivalently, to repeat several backsolves to generate a better next iterate. We call such an approach a *high-order method*. The goal is to reduce the total number of interior point iterations and therefore the total number of factorizations as well.

10.3.1 High-order predictor-corrector method

The second-order predictor-corrector strategy has two components: one is an adaptive choice of the barrier parameter γ and the other is the computation of a sort of second-order approximation to the central path. For simplicity we illustrate the algorithm with a feasible starting point.

The first step of the predictor-corrector strategy is to compute the predictor direction of the predictor-corrector algorithm in Section 4.5.1. Recall that the predictor direction solves the Newton equation system (4.17) for $\gamma = 0$ and is denoted with $d^p := d(x^k, s^k, 0)$. It is easy to show that if a step of size θ is taken along this direction, then the complementarity gap is reduced by the factor $(1-\theta)$. Therefore, the larger step can be made,

10.3. HIGH-ORDER METHOD

the more progress can be achieved. On the other hand, if the step-size in this direction is small, then the current point is probably too close to the boundary. In this case the barrier parameter should not be reduced too much in order to move a way from boundary like the corrector step.

Thus, it is reasonable to use this possible complementarity gap reduction in the predictor step to adjust the new barrier parameter γ. After the predictor direction is computed, the maximum step-sizes θ_p and θ_d along this direction in the primal and dual spaces are determined to preserve nonnegativity of $(x(\theta_p), s(\theta_d))$. The possible new complementarity gap

$$n\mu^+ := (x + \theta_p d_x^p)^T (s + \theta_d d_s^p).$$

Then, the barrier parameter is chosen using the heuristic

$$\gamma := \left(\frac{\mu^+}{\mu^k}\right)^2 \min\{\frac{\mu^+}{\mu^k}, \eta\} \tag{10.14}$$

for a constant $\eta \in (0, 1)$. We could come back to compute the actual direction $d(x^k, s^k, \gamma)$ from (4.17) where γ is given above. But we like to do more, which is the second component of the second-order predictor-corrector method.

Note that we ideally want to compute a direction such that the next iterate is perfectly centered for $\gamma\mu^k$, i.e.,

$$(X^k + D_x)(s^k + d_s) = \gamma\mu^k e.$$

The above system can be rewritten as

$$S^k d_x + X^k d_s = -X^k s^k + \gamma\mu^k e - D_x d_s. \tag{10.15}$$

Observe that in the "first-order" direction $d = d(x^k, s^k, \gamma)$ in equation (4.17), we have ignored the second order term $D_x d_s$ on the right-hand side and it becomes the residual error. This needs to be corrected: Instead of setting the second order term equal to zero, we would approximate $D_X d_s$ on the right-hand side using the available predictor direction $D_x^p d_s^p$. The actual direction d is then computed from system (10.15) with parameter γ chosen through (10.14). (Again, the matrix of the system is already factorized and it is "free" now.) We finally choose the next iterate

$$(x^{k+1}, s^{k+1}) = (x(\bar{\theta}_p), s(\bar{\theta}_d)) \in \mathcal{N}_\infty^-(\eta)$$

for η close to 1.

We should note here that the second-order predictor-corrector method basically tries to approximate the second-order Taylor expansion of the central path. A single iteration of the method needs two solves of the same large but sparse linear system for two different right hand sides. The benefit of the method is, we obtain a good estimate for the barrier parameter γ and a second-order approximation to the central path. Indeed computational practice shows that the additional solve cost of this method is more than offset by a reduction in the total number of iterations (factorizations).

Why not use even higher-order Taylor expansions? Indeed, in solving many large scale linear programs where the factorization is extremely expensive and the need to save on the number of factorizations becomes more important, a high-order predictor-corrector method is beneficial. We will explain the method in the next section.

10.3.2 Analysis of a high-order method

Now we would like to provide some theoretical justification for using the techniques involved in the high-order method.

One theoretical support of the method is already seen in Section 5.1, where we showed that $A(X^{k+1})^2 A^T$ only differs slightly from $A(X^k)^2 A^T$, and it is sufficient to inverse a matrix AD^2A^T to generate next iterate where D is still close to X^k. This justifies that the normal matrix could be used repeatedly.

Another support relates to the neighborhoods used in the high-order method. Among all existing path-following (infeasible or feasible) LP algorithms, the theoretical iteration complexity of small-neighborhood ($\mathcal{N}_2$) algorithms is $O(\sqrt{n}L)$, and the complexity of wide-neighborhood ($\mathcal{N}_\infty$ or $\mathcal{N}_\infty^-$) algorithms is at least $O(nL)$. In contrast, wide-neighborhood algorithms outperform small-neighborhood ones by a big margin in practice. It seems that smaller neighborhoods generally restrict all iterates moved by a short step and they might be too conservative for solving real LP problems.

To support using the wide-neighborhood and high-order Taylor expansions, we present a r-order Taylor expansion primal-dual path-following algorithm that is based on $\mathcal{N}_\infty^-(\beta)$ where β is any fixed constant in $(0,1)$. We show that its iteration complexity is $O(n^{\frac{r+1}{2r}}L)$ where $r \in [1, n]$. Again, each iteration uses $O(n^3)$ arithmetic operations. Note that if we let $r = O(n)$, then this iteration bound is asymptotical $O(\sqrt{n}L)$ as n increases.

Algorithm 10.1 *Given $(x^0, s^0) \in \mathcal{N}_\infty^-(\eta)$ with $\eta \in (0,1)$, and integer $r \geq 1$ and $\gamma \in (0,1)$. Set $k := 0$.*
   While $(x^k)^T s^k > \epsilon$ **do:**

1. *First-order step: Solve for the first order direction $d^{(1)} := d(x^k, s^k, \gamma)$ from (4.17).*

10.3. HIGH-ORDER METHOD

2. **High-order steps:** For $j = 2, 3, \ldots, r$, solve for the jth order direction from

$$\begin{aligned} A\, d_x^{(j)} &= 0, \\ -A^T\, d_y^{(j)} \quad -d_s^{(j)} &= 0, \end{aligned} \tag{10.16}$$

and

$$X^k d_s^{(j)} + S^k d_x^{(j)} = -\sum_{t=1}^{j-1} D_x^{(t)} d_s^{(j-t)}. \tag{10.17}$$

3. *Compute the largest θ^k so that*

$$\begin{aligned} x(\theta) &= x^k + \sum_{j=1}^{r} (\theta)^j d_x^{(j)}, \\ y(\theta) &= y^k + \sum_{j=1}^{r} (\theta)^j d_y^{(j)}, \\ s(\theta) &= s^k + \sum_{j=1}^{r} (\theta)^j d_s^{(j)}, \end{aligned}$$

lies in $\mathcal{N}_\infty^-(\eta)$ *for* $\theta \in [0, \theta^k]$. *Let*

$$(y^{k+1}, x^{k+1}, s^{k+1}) := (y(\theta^k), x(\theta^k), s(\theta^k)).$$

4. *Let $k := k+1$ and return to Step 1.*

Note that for $r = 1$, Algorithm 10.1 is identical to the wide-neighborhood algorithm in Section 4.5.2. For $r = 2$, it is close to the second-order predictor-corrector strategy described earlier; see Exercise 10.2.

In general, the step-size selection involves locating roots for each of $n+1$ polynomials with degree $2r$, which with specified error is in the complexity class NC and can be solved efficiently in theory. In practice, we need to locate only an approximate step-size. (Even for the case $r = 1$, one will never obtain the exact α^k since it is generally irrational.)

We will have a lower bound for θ^k:

$$\theta^k \geq \frac{const}{(n+1)^{\frac{r+1}{2r}}},$$

where

$$const = \frac{1-\eta}{4(1-\gamma)} \sqrt[r]{\eta\gamma}.$$

Thus, we need only to compute an approximate step-size, $\bar{\theta}$, such that

$$(x(\bar{\theta}), s(\bar{\theta})) \in \mathcal{N}_\infty^-(\eta),$$

and

$$\theta^k - \bar{\theta} \leq \frac{.001 const}{n+1} \leq 0.001\theta^k,$$

that is, $\bar{\theta}$ will be at least a fraction, .999, of the exact step-size θ^k, and it approaches above θ^k as $n \to \infty$.

We may compute such an approximate step-size using the bisection method. We know that the step-size must be in

$$\left[\frac{const}{(n+1)^{\frac{r+1}{2r}}}, \frac{1}{1-\gamma}\right].$$

Obviously, the total number of operations of this process is of order $nr(\log n)$. Even when $r = n$, the cost, $n^2(\log n)$, is well below n^3.

We now present the main complexity result.

Theorem 10.2 *Given any initial point in $\mathcal{N}_\infty^-(\eta)$ for any constant $\eta \in (0,1)$, Algorithm 10.1, with any constant $0 < \gamma < 1$, will terminate in $O(n^{\frac{r+1}{2r}} \log((x^0)^T s^0/\epsilon))$ iterations, where $r \in [1, n]$, and each iteration uses $O(n^3)$ arithmetic operations.*

As we can see that if $r = n$ and n increases, the iteration complexity of the algorithm tends to $O(\sqrt{n} \log((x^0)^T s^0/\epsilon))$ asymptotically. Furthermore, a popular choice for γ in practice is not a constant but $\gamma = O(1/n)$. Interestingly, the asymptotical iteration complexity of the algorithm for such a choice of γ is still $O(\sqrt{n} \log((x^0)^T s^0/\epsilon))$. More precisely, we have

A number of implications and points can be drawn from the main result:

- The high-order Taylor expansion method, where iterative points move along a high-order polynomial curve, has been used in practice and partially analyzed in theory. The main result indicates that the use of this method also significantly improves the worst-case iteration complexity. The result provides a further theoretical base for using this approach.

- The order r of Taylor expansion has a diminishing role in improving the worst-case complexity result. Thus, we probably expect only the first few order steps really make a difference in algorithm performance. This seems what is observed in practice.

- The result also provides a justification for using the wider neighborhood $\mathcal{N}_\infty^-$, coupled with a high-order method. The theoretical complexity based on wider neighborhoods is not much worse than that based on smaller neighborhoods. We hope this is a significant step to bridge the gap between theory and practice.

- The result also indicates how insensitive of the value of γ, the centering weight, is in high-order power-series methods. Virtually, γ can be set to any positive number if iterative points move along a high-order polynomial curve. This implies that the method has a sufficient self-centering function even γ is close to 0. Note that, when $\gamma = 0$, the algorithm becomes the pure Newton method for the LP optimality condition.

10.4 Homogeneous and Self-Dual Method

In Section 5.3 we described a homogeneous and self-dual method to solve (LP) and (LD) simultaneously. From the implementation point of view, each iteration of the method solves the linear system (5.14) and (5.15).

It can be shown that (Exercise 10.4)

$$d_\theta = \gamma - 1.$$

Then eliminating d_s and d_κ, we face the KKT system of linear equations:

$$\begin{pmatrix} X^k S^k & -X^k A^T & X^k c \tau^k \\ AX^k & 0 & -\tau^k b \\ -\tau^k c^T X^k & \tau^k b^T & \tau^k \kappa^k \end{pmatrix} \begin{pmatrix} (X^k)^{-1} d_x \\ d_y \\ (\tau^k)^{-1} d_\tau \end{pmatrix}$$

$$= \begin{pmatrix} \gamma \mu^k e - X^k s^k \\ 0 \\ \gamma \mu^k - \tau^k \kappa^k \\ 0 \end{pmatrix} + (1 - \gamma) \begin{pmatrix} -X^k \bar{c} \\ \bar{b} \\ \tau^k \bar{z} \end{pmatrix}.$$

Thus, the dimension of the system is increased only by 1 over the case when strictly feasible points for both (LP) and (LD) are known and used for starting primal-dual interior-point algorithms. (It seems that the benefit of knowing a starting interior point is not great.)

All implementation techniques discussed earlier for feasible-starting interior point algorithms can be used in the homogeneous and self-dual method. For example, If the second-order predictor-corrector scheme is used, it means that we have 3 solves instead of 2 for each factorization. Again, the additional solve cost is still more than offset by a reduction in the total number of iterations (factorizations), and all favorable features discussed in Section 5.3 of the method are retained.

It is also possible to take different step-sizes to update x and s. In doing so special attention should be paid to update τ since it couples both the primal and dual.

10.5 Optimal-Basis Identifier

Contrary to the simplex algorithm an interior-point algorithm never generates the exact optimal solution during its iterative process; instead it generates an infinite sequence converging towards an optimal solution. Thus, the algorithm discussed produces an approximate optimal basic solution only if the optimal solution is unique (which is very rare in practice). In fact, in the case that either multiple primal or dual solutions exist, the sequence converges to the analytic center of the optimal face as discussed before. Therefore, an important problem is to generate an optimal basic solution from of interior-point algorithms, which is desirable in solving many practical problems.

It can be shown that if a pair of exact primal and dual solutions is known, then an optimal basic solution can be produced in strongly polynomial time using a simplified a pivoting or simplex method. We now discuss a algorithm which combines the termination scheme in Section 5.2 and the pivoting method to produce an optimal basic solution.

Consider solving (LP). It is well-known that any optimal solution (x^*, y^*, z^*) must satisfy the complementarity slackness condition $x_j^* z_j^* = 0$ for each j. Moreover, it is known from Theorem 1.14 that there exists a strictly complementary solution that satisfies $x_j^* + z_j^* > 0$ for each j, and the complementarity partition (P^*, Z^*) is unique. The pair (P^*, Z^*), where $Z = \{1, \ldots, n\} \setminus P$ for any index set P, determines an optimal partition.

Recall that (B, N) denote a partition of the variables into basic and non-basic variables. (B, N) is an optimal basis, if B is non-singular and

$$x_B = A_B^{-1} b \geq 0; \quad x_N = 0$$

and

$$y = A_B^{-T} c_B; \quad s_B = c_B - A_B y = 0; \quad s_N = c_N - A_N^T y \geq 0.$$

10.5.1 A pivoting algorithm

Given a complementary solution pair, a pivoting algorithm can construct an optimal basis in less than n pivoting steps. Below we shall discuss the algorithm and its implementation. For convenience we assume that a set of artificial variables has been added to the problem (LP). Let $V = \{n+1, \ldots, n+m\}$ denote the set of artificial variables; naturally, we must have $x_V = 0$ in any optimal solution. Furthermore, we assume that a strictly complementary solution is known. Hence, we assume that:

- We know the complementarity partition (P^*, Z^*) and $V \subseteq Z^*$.

10.5. OPTIMAL-BASIS IDENTIFIER

- We know an optimal primal solution x^* such that $Ax^* = b$, $x^*_{Z^*} = 0$ and $x^*_{P^*} \geq 0$.

- We know an optimal dual solution (y^*, s^*) such that $A^T y^* + s^* = c$, $s^*_{Z^* \setminus V} \geq 0$ and $s^*_{P^*} = 0$.

The algorithm consists of a primal and a dual phase. We start with a description of the primal phase.

Let (B, N) be any partition of the variables of the problem (LP) into basic and non-basic parts. Let

$$x_B := A_B^{-1}(b - Nx_N^*) = x_B^* \geq 0.$$

Here solution x_B^* is called a *super-basic* solution since some of non-basic variables x_N^* may not be zero, and variables of x_N^* that are not zero are called super-non-basic variables. For each of super-non-basic variables, the primal phase is to either move it to zero or pivot it into basis B using the simplex (pivoting) step. The resulting basis will be primal optimal, because it is feasible and it is still complementary with respect to the dual optimal solution (y^*, s^*). Each moving or pivoting step reduces the number of super-non-basic variables at least by one. Since the number of super-non-basic variables cannot exceed $|P^*|$, the primal phase terminates after at most $|P^*|$ steps.

Now we will formally state the primal phase.

Algorithm 10.2

1. *Choose a basis B and let $x = x^*$.*
2. *While($\exists\, j \in P^* \setminus B : x_j \neq 0$)*
3. *Use a primal ratio test to move variable x_j to zero if we can keep $A_B^{-1}(b - Nx_N) \geq 0$, or pivot it into the basis.*
4. *Update x, or (B, N) and x.*
5. *end while*
6. *B is a primal optimal basis.*

It is always possible to choose an initial basis B in Step 1. One possible choice is $B = V$, the set of artificial variables. Algorithm 10.2 can be viewed as a simplified version of the primal simplex method, because there is no pricing step in selecting an incoming variable and those incoming candidates are predetermined from $x^*_{P^*}$.

The dual phase of the algorithm is similar to the primal phase because, in this case, a super-basic dual solution is known, which means that some of the reduced costs of s_B^* might not be zero. Similarly to the primal phase, those non-zero reduced costs in s_B^* can either be moved to zero or the corresponding primal variable has to be pivoted out of basis B. The dual phase can be stated as follows:

Algorithm 10.3

1. Choose a basis B and let $y = y^*$, $s = c - A^T y$.
2. $While(\exists\, j \in Z^* \cap B : s_j \neq 0)$
3. Use the dual ratio test to move variable s_j to zero if we can keep $c_N + N^T B^{-T}(s_B - c_B) \geq 0$, or pivot it out of the basis.
4. Update (y, s) or (B, N) and (y, s).
5. end while
6. B is a dual optimal basis.

If the initial basis B of the dual phase is primal optimal, i.e, $x_B^* := B^{-1} b \geq 0$ and $x_{Z^*}^* = 0$, then it remains primal optimal throughout all steps of Algorithm 10.3 because $x_N^* = 0$ and all pivots are primal degenerate. Once Algorithm 10.3 terminates, the final basis is both primal and dual feasible and hence optimal. Algorithm 10.3 can be viewed as a simplified version of the dual simplex method, because there is no pricing step in selecting an outgoing variable and those outgoing candidates are predetermined from $s_{Z^*}^*$. Furthermore, the number of moves or pivots in the dual phase cannot exceed $|Z^*|$.

In summary, Algorithms 10.2 and 10.3 generate an optimal basis after at most n moving or pivoting steps. In practice, the total number of steps is dependent on the level of primal and dual degeneracy of the problem.

10.5.2 Theoretical and computational issues

The algorithm presented in the previous subsection assumes that an exact optimal solution is known. This assumption is never met in practice, because the primal-dual algorithm only generates a sequence of solutions converging towards an optimal solution. Furthermore, due to the finite precision of computations, the solution returned by an interior-point algorithm is neither exactly feasible nor exactly complementary.

Let (x^k, y^k, z^k) be the iterate generated by an algorithm on iteration k and (P^k, Z^k) be a guess of the complementarity partition generated on iteration k. Now define the following perturbed problem:

$$\text{minimize} \quad (c^k)^T x \quad \text{s.t.} \quad Ax = b^k; \quad x \geq 0, \qquad (10.18)$$

where

$$b^k = P^k x_{P^k}^k; \quad c_{P^k}^k = (P^k)^T y^k \quad \text{and} \quad c_{Z^k}^k = (\bar{P}^k)^T y^k + z_{Z^k}^k.$$

10.6. NOTES

Assume that variables in (10.18) are reordered such that $x = (x_{P^k}, x_{Z^k})$ then the vector $(x, y, s) = ((x_{P^k}^k, 0), y^k, (0, z_{Z^k}^k))$ is a strictly complementary solution to (10.18). Moreover, if x^k converges towards an optimal primal solution and P^k converges towards P^*, then b^k converges towards b and, similarly, c^k converges towards c. Therefore the two problems (LP) and (10.18) will eventually become close and share some same optimal bases according to Exercises 10.3 and the following theorem.

Theorem 10.3 *Let B be an optimal basis for $LP(A, b^k, c^k)$. Then, there is $0 < \bar{t} < \infty$ such that B must be also an optimal basis for the original $LP(A, b, c)$ when $(x^k)^T s^k \leq 2^{-\bar{t}}$. Furthermore, if $LP(A, b, c)$ has rational data, then $\bar{t} \leq O(L)$.*

This advocates for an application of the above basis identification procedure to the perturbed problem (10.18), since an optimal complementary solution to problem (10.18) is known, and it will be an optimal basis for (LP) when problem (10.18 is near (LP).

An important practical issue is how to select P^k that equals to the complementarity partition P^*. A trivial one is

$$P^k = \{j : x_j^k \geq z_j^k\}. \tag{10.19}$$

A more practically effective choice is

$$P^k = \{j : |d_{x\,j}^d|/x_j^k \leq |d_{s\,j}^d|/s_j^k\}, \tag{10.20}$$

where (d_x^d, d_s^d) is the primal-dual predictor direction. These quantities are scaling invariant. It uses the relative variable change to indicate the optimal partition. This indicator is justified by the theory of Section 7.2, where they converges to 1 for $j \in P^*$ and to 0 otherwise.

Another question is the choice of the right time to start the pivoting procedure. According to Theorem 10.3 the generated basis can only be expected to be the correct optimal basis of (LP) if the interior point solution is almost optimal and P^k is a good guess for P^*. A reasonable and practical criterion is the moment when fast (quadratic) convergence of the primal-dual gap μ^k to zero occurs, which is also consistent to the theory of Section 7.2.

10.6 Notes

The use of a presolver is an old but effective idea, see for example, Brearley et al. [73]; its role was acknowledged in many simplex algorithm optimizers. The simplex method for LP works with sparse submatrices of A (bases)

[395] while any interior-point algorithm needs an inversion of a considerably denser matrix AA^T. Consequently, the potential savings resulting from an initial problem reduction may be larger in interior-point implementations. This is the reason why the presolve analysis has recently enjoyed great attention [5, 255, 15, 157, 14, 58, 258, 394]. An additional important motivation is that large-scale LP problems are solved routinely nowadays and the amount of redundancy is increasing with the size of the problem.

When discussing the disadvantages of the normal equations approach in Section 10.2.1, we have mentioned the negative consequences of splitting free variables. Sometimes it is possible to generate a finite explicit bound on a free variable [157] and avoid the need of splitting it. Subramanian et al. [393] and Andersen [14] report that in some cases the computational saving from removing the linearly dependent constraints are significant.

Exact solution of the *sparsity problem* of Section 10.1 is an NP–complete problem ([80]) but efficient heuristics [5, 80, 157] usually produce satisfactory non-zero reductions in A. The algorithm of [157], for example, looks for such a row of A that has a sparsity pattern being the subset of the sparsity pattern of other rows and uses it to pivot out non-zero elements from other rows. Also, the postsolver analysis has been discussed extensively in [15].

Most general purpose interior-point codes use the *direct* method [102] to solve the KKT system. Two competitive direct methods are: the *normal equation* approach [33, 34] and the *augmented system* approach. The former works with a smaller positive definite matrix, and the latter requires factorization of a symmetric indefinite matrix.

The normal equation approach was used among very first "professional" interior-point implementations [5, 218, 269]. The success of their application of the Choleski factorization relies on the quality of a pivoting order for preserving sparsity [102, 134]. To find an optimal order or permutation is an NP-complete problem [464]). Two effective heuristics described in this book, the *minimum degree* and the *minimum local fill-in* order rules, are due to Duff [102] and George and Liu [134, 135]. In the minimum-degree order l_i is actually the Markowitz merit function applied to a symmetric matrix [262]. For details, the reader is referred to an excellent summary in [135]. Another efficient technique to determine the pivot order has been proposed in Mészáros [283]. The remedy to the rank deficiency arising in the Schur complement mechanism is due to Andersen [20]. His approach employs an old technique due to Stewart [388].

The augmented system approach is an old and well understood technique to solve a least squares problem [33, 34, 61, 102]. It consists in the application of the Bunch-Parlett [74] factorization to the symmetric indefinite matrix. Mehrotra's augmented system implementation [119, 277], for

10.6. NOTES

example, is based on the Bunch-Parlett factorization [74] and on the use of the generalized Markowitz [262] count of type (10.6) for 2×2 pivots. Maros and Mészáros [264] give a detailed analysis of this issue as well. The stable condition of the augmented system approach motivated many researchers to incorporate it into their LP codes; see [103, 119, 264, 434, 447]. Other advantages include easy handling of free variables and dense columns, and a straightforward extension to solving convex quadratic programming problems [447, 277, 77].

In the numerical factorization, George and Liu [134] demonstrate how the Choleski factorization can be organized either by rows or by columns. Several enhancements can be found in [134, 245] and [102, 365]. The Yale Sparse Matrix Package is due to [107] and the Waterloo SPARSPAK Package is due to [134].

Lustig et al. [256] explored the supernode in their implementation. The effect of the supernodal method is highly hardware-dependent and several results can be found in the literature: the efficiency of the supernodal decomposition on the shared-memory multiprocessors is discussed by Esmond and Peyton [332], the exploitation of the cache memory on high-performance workstations is studied by Rothberg and Gupta [365] in the framework of the right-looking factorization, while the case of the left-looking factorization is investigated by Mészáros [282].

The iterative becomes highly successful in solving special LP problems, such as network-flow problems; see [212, 362, 346]. Theorem 10.1 is proved by Kaliski [212].

The first high-order method was incorporated into a dual affine-scaling method of AT&T's Korbx system [218]. An efficient high-order method was proposed by Mehrotra; his second-order predictor-corrector strategy [276] was incorporated in almost all primal-dual interior-point implementations. As shown in Mehrotra [278], the improvement from using orders higher than 2 seems very limited. Recently, Gondzio [156] has proposed a new way to exploit higher order information in a primal-dual algorithm and shown considerable improvement in solving large-scale problems. His approach applies multiple centrality corrections and combines their use with a choice of reasonable, well-centered targets that are supposed to be easier to reach than perfectly centered (but usually unreachable) analytic centers. The idea to use targets that are not analytic centers comes from Jansen, Roos, Terlaky and Vial [200]. They define a sequence of traceable weighted analytic centers, called *targets*, that go from an arbitrary interior point to a point close to the central path. The algorithm follows these targets and continuously (although very slowly) improves the centrality of subsequent iterates. The targets are defined in the space of the complementarity products.

Another high-order approach, due to Domich et al. [101] uses three independent directions and solves an auxiliary linear program in a three dimensional subspace to find a search direction. The method of Sonnevend et al. [386] uses subspaces spanned by directions generated by higher order derivatives of the feasible central path, or earlier computed points of it as a predictor step. This is later followed by one (or more) centering steps to take the next iterate sufficiently close to the central path. Hung and Ye [194] has studied theoretically higher order predictor-corrector techniques, incorporated them in the homogeneous self-dual algorithm, and proved Theorem 10.2.

The fact of Exercise 10.4 was proved by Xu et al. [461], who also first implemented the homogeneous and self-dual algorithm and presented favorable computational results in solving both feasible and infeasible LP problems. Extensive implementation results were recently given by Andersen and Andersen [16]. They even discussed how the solution resulted from the homogeneous and self-dual model can be used in diagnosing the cause of infeasibility.

Recovering an optimal basis from a near-optimal solution is necessary in solving integer programming problems. We would also like to note that there are LP applications in which an optimal interior-point solution is preferable; see, e.g., Christiansen and Kortanek [87] and Greenberg [167].

Bixby and Lustig solve the basis-recovering problem using a Big-M version of Megiddo's procedure [272]. Their procedure drives both complementarity and feasibility to zero. Andersen and Ye [17] propose an alternative solution to this problem, which is the perturbed problem construction described in this book. For a discussion of linear algebra issues related to implementing the simplex or pivoting algorithm we refer the reader to the papers [60, 17].

There are some open implementation issues. In many practical applications of linear programming, a sequence of closely related problems is solved, such as in branch and bound algorithms for integer programming an in column generation (cutting planes) methods. Obviously when two closely related problems are solved the previous optimal solution should be and could be used to solve the new problem faster. In the context of the simplex algorithm this aim is achieved by starting from the previous optimal basic solution, which is called the "warm-start." In the context of interior-point methods, an effective warm start procedure is difficult to find. Some hope comes from a particular application demonstrated in [146].

10.7 Exercises

10.1 Let A be the node-arc incidence matrix with $m+1$ nodes and n arcs (For ease of notation, an arbitrary row in A is assumed to have been deleted so that A has full row rank m), and let D be an $n \times n$ positive diagonal matrix. Choose π so that A_π contains a basis of A and $D_{ii} \geq \beta \geq D_{jj}$ for all $i \in \pi$ and $j \notin \pi$. Then,

$$\underline{\lambda}(A_\pi D_\pi^2 A_\pi^T) \geq \beta m^{-2}$$

and

$$\bar{\lambda}(AD^2 A^T - A_\pi D_\pi^2 A_\pi^T) \leq 2\beta m.$$

10.2 In Algorithm 10.1, let

$$\begin{aligned}
\Delta x &= x^{k+1} - x^k = \sum_{j=1}^{r} (\theta^k)^j d_x^{(j)}, \\
\Delta y &= y^{k+1} - y^k = \sum_{j=1}^{r} (\theta^k)^j d_y^{(j)}, \\
\Delta s &= s^{k+1} - s^k = \sum_{j=1}^{r} (\theta^k)^j d_s^{(j)}.
\end{aligned}$$

Show that $(\Delta x, \Delta s, \Delta y)$ satisfy

1.

$$X^k \Delta s + S^k \Delta x = \theta^k(\gamma \mu^k e - X^k s^k) - \sum_{j=2}^{r}(\theta^k)^j \left(\sum_{t=1}^{j-1} D_x^{(t)} d_s^{(j-t)}\right).$$

2.

$$(x^k)^T \Delta s + (s^k)^T \Delta x = \theta^k(\gamma - 1)(x^k)^T s^k,$$
$$\Delta x^T \Delta s = 0.$$

3.

$$\mu^{k+1} = [1 - \theta^k(1-\gamma)]\mu^k.$$

10.3 Let B be an optimal basis for $LP(A, b^k, c^k)$ of Section 10.5.2. There there is a positive number $\zeta(A, b, c) > 0$ such that when $\|b^k - b\| < \zeta(A, b, c)$ and $\|c^k - c\| < \zeta(A, b, c)$, B is also an optimal basis for $LP(A, b, c)$, i.e., both solutions $\bar{x}_B$ of $B\bar{x}_B = b$ and $\bar{s} = c - A^T \bar{y}$, where $B^T \bar{y} = c_B$, are nonnegative.

10.4 In solving system of equations (5.14) and (5.15) of Section 5.3, shown that

$$d_\theta = \gamma - 1.$$

Bibliography

[1] S. S. Abhyankar, T. L. Morin, and T. B. Trafalis. Efficient faces of polytopes : Interior point algorithms, parameterization of algebraic varieties, and multiple objective optimization. In J. C. Lagarias and M. J. Todd, editors, *Mathematical Developments Arising from Linear Programming*, volume 114 of *Contemporary Mathematics*, pages 319–341. American Mathematical Society, Providence, RI, 1990.

[2] I. Adler, R. M. Karp and R. Shamir. A simplex variant solving an $m \times d$ linear program in $O(\min(m^2, d^2))$ expected number of pivot steps. *J. Complexity*, 3:372–387, 1987.

[3] I. Adler and N. Megiddo. A simplex algorithm whose average number of steps is bounded between two quadratic functions of the smaller dimension. *Journal of the ACM*, 32:871–895, 1985.

[4] I. Adler, N. K. Karmarkar, M. G. C. Resende, and G. Veiga. An implementation of Karmarkar's algorithm for linear programming. *Math. Programming*, 44:297–335, 1989. Errata in *Math. Programming*, 50:415, 1991.

[5] I. Adler, N. Karmarkar, M. G. C. Resende, and G. Veiga. Data structures and programming techniques for the implementation of Karmarkar's algorithm. *ORSA J. on Comput.*, 1(2):84–106, 1989.

[6] I. Adler and R. D. C. Monteiro. Limiting behavior of the affine scaling continuous trajectories for linear programming problems. *Math. Programming*, 50:29–51, 1991.

[7] I. Adler and R. D. C. Monteiro. A geometric view of parametric linear programming. *Algorithmica*, 8:161–176, 1992.

[8] M. Akgül. On the exact solution of a system of linear homogeneous equations via a projective algorithm. *Arabian Journal for Science and Engineering*, 15(4):753–754, 1990.

[9] F. Alizadeh. *Combinatorial optimization with interior point methods and semi-definite matrices.* Ph.D. Thesis, University of Minnesota, Minneapolis, MN, 1991.

[10] F. Alizadeh. Optimization over the positive semi-definite cone: Interior-point methods and combinatorial applications. In P. M. Pardalos, editor, *Advances in Optimization and Parallel Computing*, pages 1–25. North Holland, Amsterdam, The Netherlands, 1992.

[11] F. Alizadeh. Interior point methods in semidefinite programming with applications to combinatorial optimization. *SIAM J. Optimization*, 5:13–51, 1995.

[12] F. Alizadeh, J.-P. A. Haeberly, and M. L. Overton. Complementarity and nondegeneracy in semidefinite programming. *Math. Programming*, 77:111–128, 1997.

[13] A. Altman and J. Gondzio. An efficient implementation of a higher order primal–dual interior point method for large sparse linear programs. *Archives of Control Sciences*, 1(2):23–40, 1992.

[14] E. D. Andersen. Finding all linearly dependent rows in large-scale linear programming. *Optimization Methods and Software*, 6:219–227, 1995.

[15] E. D. Andersen and K. D. Andersen. Presolving in Linear Programming. *Math. Programming*, 71:221–245, 1995.

[16] E. D. Andersen and K. D. Andersen. The APOS linear programming solver: an implementation of the homogeneous algorithm. Working Paper, Department of Management, Odense University, Denmark, 1997.

[17] E. D. Andersen and Y. Ye. Combining interior-point and pivoting algorithms for linear programming. *Management Science*, 42:1719–1731, 1996.

[18] E. D. Andersen and Y. Ye. On a homogeneous algorithm for the monotone complementarity problem. Working Paper, Department of Management Sciences, The University of Iowa, IA, 1995. To appear in *Math. Programming*.

[19] K. D. Andersen. An infeasible dual affine scaling method for linear programming. *COAL Newsletter*, 22:19–27, 1993.

[20] K. D. Andersen. A modified Schur complement method for handling dense columns in interior point methods for linear programming. *ACM Trans. Math. Software*, 22(3):348–356, 1996.

[21] K. D. Andersen and E. Christiansen. A newton barrier method for minimizing a sum of Euclidean norms subject to linear equality constraints. Preprint 95-07, Department of Mathematics and Computer Science, Odense University, Denmark, 1995.

[22] K. M. Anstreicher. A monotonic projective algorithm for fractional linear programming. *Algorithmica*, 1(4):483–498, 1986.

[23] K. M. Anstreicher. On long step path following and SUMT for linear and quadratic programming. *SIAM J. Optimization*, 6:33-46, 1996.

[24] K. M. Anstreicher. A standard form variant and safeguarded linesearch for the modified Karmarkar algorithm. *Math. Programming*, 47:337–351, 1990.

[25] K. M. Anstreicher. A combined phase I – phase II scaled potential algorithm for linear programming. *Math. Programming*, 52:429–439, 1991.

[26] K. M. Anstreicher. On Vaidya's volumetric cutting plane method for convex programming. *Mathematics of Operations Research*, 22(1):63–89, 1997.

[27] K. M. Anstreicher. On the performance of Karmarkar's algorithm over a sequence of iterations. *SIAM J. Optimization*, 1(1):22–29, 1991.

[28] K. M. Anstreicher and R. A. Bosch. Long steps in a $O(n^3 L)$ algorithm for linear programming. *Math. Programming*, 54:251–265, 1992.

[29] K. M. Anstreicher, D. den Hertog, C. Roos, and T. Terlaky. A long step barrier method for convex quadratic programming. *Algorithmica*, 10:365–382, 1993.

[30] K. M. Anstreicher, J. Ji, F. A. Potra, and Y. Ye. Average performance of a self–dual interior–point algorithm for linear programming. in P. Pardalos, editor, *Complexity in Numerical Optimization* page 1–15, World Scientific, New Jersey, 1993.

[31] K. M. Anstreicher, J. Ji, F. A. Potra, and Y. Ye. Probabilistic analysis of an infeasible primal–dual algorithm for linear programming. Reports on Computational Mathematics 27, Dept. of Mathematics, University of Iowa, Iowa City, IA, 1992.

[32] K. M. Anstreicher and J. P. Vial. On the convergence of an infeasible primal–dual interior–point method for convex programming. *Optimization Methods and Software*, 3:273-283, 1994.

[33] M. Arioli, J. W. Demmel, and I. S. Duff. Solving sparse linear systems with sparse backward error. *SIAM J. Mat. Anal. Appl.*, 10(2):165–190, 1989.

[34] M. Arioli, I. S. Duff, and P. P. M. de Rijk. On the augmented system approach to sparse least-squares problems. *Numer. Math.*, 55:667–684, 1989.

[35] M. D. Asic and V. V. Kovacevic-Vujcic. An interior semi–infinite programming method. *Journal of Optimization Theory and Applications*, 59:369–390, 1988.

[36] M. D. Asic, V. V. Kovacevic-Vujcic, and M. D. Radosavljevic-Nikolic. Asymptotic behavior of Karmarkar's method for linear programming. *Math. Programming*, 46:173–190, 1990.

[37] G. Astfalk, I. J. Lustig, R. E. Marsten, and D. Shanno. The interior-point method for linear programming. *IEEE Software*, 9(4):61–68, 1992.

[38] D. S. Atkinson and P. M. Vaidya. A cutting plane algorithm for convex programming that uses analytic centers. *Math. Programming*, 69:1–44, 1995.

[39] O. Bahn, J. L. Goffin, J. P. Vial, and O. Du Merle. Implementation and behavior of an interior point cutting plane algorithm for convex programming: An application to geometric programming. *Discrete Applied Mathematics*, 49:3–23, 1993.

[40] O. Bahn, O. Du Merle, J. L. Goffin, and J. P. Vial. A cutting plane method from analytic centers for stochastic programming. *Math. Programming*, 69:45–74, 1995.

[41] E.-W. Bai and Y. Ye. Bounded error parameter estimation: a sequential analytic center approach. Working Paper, Dept. of Electrical and Computer Engineering, University of Iowa, Iowa City, Iowa, 1997.

[42] I. Bárány and Füredi. Computing the volume is difficult. In *Proceedings of the 18th Annual ACM Symposium on Theory of Computing*, page 442-447, ACM, New York, 1986.

[43] E. R. Barnes. A variation on Karmarkar's algorithm for solving linear programming problems. *Math. Programming*, 36:174–182, 1986.

[44] E. R. Barnes, S. Chopra, and D. J. Jensen. The affine scaling method with centering. Technical Report, Dept. of Mathematical Sciences, IBM T. J. Watson Research Center, P. O. Box 218, Yorktown Heights, NY, 1988.

[45] D. A. Bayer and J. C. Lagarias. The nonlinear geometry of linear programming, Part I: Affine and projective scaling trajectories. *Transactions of the American Mathematical Society*, 314(2):499–526, 1989.

[46] D. A. Bayer and J. C. Lagarias. The nonlinear geometry of linear programming, Part II: Legendre transform coordinates. *Transactions of the American Mathematical Society*, 314(2):527–581, 1989.

[47] D. A. Bayer and J. C. Lagarias. Karmarkar's linear programming algorithm and Newton's method. *Math. Programming*, 50:291–330, 1991.

[48] M. S. Bazaraa, J. J. Jarvis, and H. F. Sherali. *Linear Programming and Network Flows*, chapter 8.4 : Karmarkar's projective algorithm, pages 380–394, chapter 8.5: Analysis of Karmarkar's algorithm, pages 394–418. John Wiley & Sons, New York, second edition, 1990.

[49] M. Bellare and P. Rogaway. The complexity of approximating a nonlinear program. *Math. Programming*, 69:429-442, 1995.

[50] M. Ben–Daya and C. M. Shetty. Polynomial barrier function algorithm for convex quadratic programming. *Arabian Journal for Science and Engineering*, 15(4):657–670, 1990.

[51] A. Ben–Tal and A. S. Nemirovskii. Interior point polynomial time method for truss topology design. *SIAM J. Optimization*, 4:596–612, 1994.

[52] D. P. Bertsekas and J. N. Tsitsiklis. *Introduction to Linear Optimization*. Athena Scientific, Belmont, MA, 1997.

[53] D. P. Bertsekas. *Nonlinear Programming.* Athena Scientific, Belmont, MA, 1995.

[54] D. Bertsimas and X. Luo. On the worst case complexity of potential reduction algorithms for linear programming. Working Paper 3558-93, Sloan School of Management, MIT, Cambridge, MA, 1993. To appear in *Math. Programming.*

[55] U. Betke and P. Gritzmann. Projection algorithms for linear programming. *European Journal of Operational Research,* 60:287–295, 1992.

[56] J. R. Birge and D. F. Holmes. Efficient solution of two stage stochastic linear programs using interior point methods. *Computational Optimization and Applications,* 1:245–276, 1992.

[57] J. R. Birge and L. Qi. Computing block–angular Karmarkar projections with applications to stochastic programming. *Management Science,* 34:1472–1479, 1988.

[58] R. E. Bixby. Progress in linear programming. *ORSA J. on Comput.,* 6(1):15–22, 1994.

[59] R. E. Bixby, J. W. Gregory, I. J. Lustig, R. E. Marsten, and D. F. Shanno. Very large–scale linear programming : A case study in combining interior point and simplex methods. *Operations Research,* 40:885–897, 1992.

[60] R. E. Bixby and M. J. Saltzman. Recovering an optimal LP basis from an interior point solution. *Operations Research Letters,* 15:169–178, 1994.

[61] A. Björk. Methods for sparse linear least squares problems. In J. R. Bunch and D. J. Rose, editors, *Sparse Matrix Computation,* pages 177–201. Academic Press INC., 1976.

[62] C. E. Blair. The iterative step in the linear programming algorithm of N. Karmarkar. *Algorithmica,* 1(4):537–539, 1986.

[63] R. G. Bland, D. Goldfarb and M. J. Todd. The ellipsoidal method: a survey. *Operations Research,* 29:1039–1091, 1981.

[64] L. Blum. Towards an asymptotic analysis of Karmarkar's algorithm. *Information Processing Letters,* 23:189–194, 1986.

[65] L. Blum. A new simple homotopy algorithm for linear programming I. *Journal of Complexity*, 4:124–136, 1988.

[66] L. Blum, M. Shub and S. Smale. On a theory of computations over the real numbers: NP-completeness, recursive functions and universal machines. *Proc. 29th Symp. Foundations of Computer Science*, 387–397, 1988.

[67] P. T. Boggs, P. D. Domich, J. R. Donaldson, and C. Witzgall. Algorithmic enhancements to the method of center for linear programming. *ORSA Journal on Computing*, 1:159–171, 1989.

[68] F. Bonnans and M. Bouhtou. The trust region affine interior point algorithm for convex and nonconvex quadratic programming. *RAIRO - Recherche Opérationnelle*, 29:195-217, 1995.

[69] F. Bonnans and C. Gonzaga. Convergence of interior-point algorithms for the monotone linear complementarity problem. *Mathematics of Operations Research*, 21:1–25, 1996.

[70] K. H. Borgwardt. *The Simplex Method: A Probabilistic Analysis.* Springer, Berlin, 1987.

[71] S. Boyd and L. E. Ghaoui. Methods of centers for minimizing generalized eigenvalues. *Linear Algebra and Its Applications, special issue on Linear Algebra in Systems and Control*, 188:63–111, 1993.

[72] S. Boyd, L. E. Ghaoui, E. Feron, and V. Balakrishnan. *Linear Matrix Inequalities in System and Control Science.* SIAM Publications. SIAM, Philadelphia, 1994.

[73] A. L. Brearley, G. Mitra, and H. P. Williams. Analysis of mathematical programming problems prior to applying the simplex algorithm. *Math. Programming*, 15:54–83, 1975.

[74] J. R. Bunch and B. N. Parlett. Direct methods for solving symmetric indefinite systems of linear equations. *SIAM J. Numer. Anal.*, 8:639–655, 1971.

[75] J. V. Burke and S. Xu. The global linear convergence of a noninterior path-following algorithm for linear complementarity problems. Preprint, Department of Mathematics, University of Washington, Seattle, December, 1996.

[76] R. Byrd, J. C. Gilbert and J. Nocedal. A trust region method based on interior point techniques for nonlinear programming. Report OTC 96/02 Optimization Technology Center, Northwestern University, Evanston IL 60208, 1996.

[77] T. J. Carpenter, I. J. Lustig, J. M. Mulvey, and D. F. Shanno. Separable quadratic programming via a primal-dual interior point method and its use in a sequential procedure. *ORSA J. on Comput.*, 5:182–191, 1993.

[78] T. J. Carpenter, I. J. Lustig, J. M. Mulvey, and D. F. Shanno. Higher order predictor–corrector interior point methods with application to quadratic objectives. *SIAM J. Optimization*, 3:696–725, 1993.

[79] T. M. Cavalier and A. L. Soyster. Some computational experience and a modification of the Karmarkar algorithm. Working Paper 85–105, Dept. of Industrial and Management Systems Engineering, Pennsylvania State University, University Park, PA, 1985.

[80] S. F. Chang and S. T. McCormick. A hierarchical algorithm for making sparse matrices sparse. *Math. Programming*, 56:1–30, 1992.

[81] B. Chen and P. T. Harker. A non-interior-point continuation method for linear complementarity problems. *SIAM J. Matrix Anal. Appl.*, 14:1168-1190, 1993.

[82] C. Chen and O. L. Mangasarian. A class of smoothing functions for nonlinear and mixed complementarity problems. *Comp. Optim. Appl.*, 5:97-138, 1996.

[83] X. Chen. Superlinear convergence of smoothing quasi-Newton methods for nonsmooth equations. Applied Mathematics Report, University of New South Wales, Sydney, 1996. To appear in *J. Comp. Appl. Math.*.

[84] I. C. Choi and D. Goldfarb. Exploiting special structure in a primal–dual path–following algorithm. *Math. Programming*, 58:33–52, 1993.

[85] I. C. Choi, C. L. Monma, and D. F. Shanno. Computational experience with a primal–dual interior point method for linear programming. Technical Report, RUTCOR Center of Operations Research, Rutgers University, New Brunswick, NJ, 1989.

[86] I. C. Choi, C. L. Monma, and D. F. Shanno. Further development of a primal–dual interior point method. *ORSA Journal on Computing*, 2:304–311, 1990.

[87] E. Christiansen and K. O. Kortanek. Computation of the collapse state in limit analysis using the LP primal affine scaling algorithm. *Journal of Computational and Applied Mathematics*, 34:47–63, 1991.

[88] S. A. Cook. The complexity of theorem-proving procedures. *Proc. 3rd ACM Symposium on the Theory of Computing*, 151–158, 1971.

[89] T. F. Coleman and Y. Li. A globally and quadratically convergent affine scaling method for linear L_1 problems. *Math. Programming*, 52:189–222, 1992.

[90] R. W. Cottle and G. B. Dantzig. Complementary pivot theory in mathematical programming. *Linear Algebra and its Applications*, 1:103-125, 1968.

[91] R. Cottle, J. S. Pang, and R. E. Stone. *The Linear Complementarity Problem*, chapter 5.9 : Interior–point methods, pages 461–475. Academic Press, Boston, 1992.

[92] J. Czyzyk, S. Mehrotra and S. Wright. PCx User Guide. Technical Report OTC 96/01, Optimization Technology Center, Argonne National Laboratory and Northwestern University, May, 1996.

[93] G. B. Dantzig. *Linear Programming and Extensions*. Princeton University Press, Princeton, New Jersey, 1963.

[94] L. Danzer, B. Grünbaum and V. Klee. Helly's theorem and its relatives. in V. Klee, editor, *Convexity*, page 101–180, AMS, Providence, RI, 1963.

[95] J. Dennis, A. M. Morshedi, and K. Turner. A variable metric variant of the Karmarkar algorithm for linear programming. *Math. Programming*, 39:1–20, 1987.

[96] J. E. Dennis and R. E. Schnabel. *Numerical Methods for Unconstrained Optimization and Nonlinear Equations*. Prentice–Hall, Englewood Cliffs, New Jersey, 1983.

[97] Z. Y. Diao. Karmarkar's algorithm and its modification. *Chinese Journal on Operations Research*, 7(1):73–75, 1988.

[98] I. I. Dikin. Iterative solution of problems of linear and quadratic programming. *Doklady Akademii Nauk SSSR*, 174:747–748, 1967. Translated in : *Soviet Mathematics Doklady*, 8:674–675, 1967.

[99] I. I. Dikin. On the convergence of an iterative process. *Upravlyaemye Sistemi*, 12:54–60, 1974. (In Russian).

[100] J. Ding and T. Y. Li. A polynomial–time predictor–corrector algorithm for a class of linear complementarity problems. *SIAM J. Computing*, 1(1):83–92, 1991.

[101] P. D. Domich, P. T. Boggs, J. E. Rogers, and C. Witzgall. Optimizing over 3–d subspaces in an interior point method. *Linear Algebra and Its Applications*, 152:315–342, 1991.

[102] I. S. Duff, A. M. Erisman, and J. K. Reid. *Direct methods for sparse matrices*. Oxford University Press, New York, 1989.

[103] I. S. Duff, N. I. M. Gould, J. K. Reid, J. A. Scott, and K. Turner. The factorization of sparse symmetric indefinite matrices. *IMA J. Numer. Anal.*, 11:181–204, 1991.

[104] M. E. Dyer and A. M. Frieze. The complexity of computing the volume of a polyhedron. *SIAM J. Comput.*, 17:967–974, 1988.

[105] M. E. Dyer, A. M. Frieze and R. Kannan. A random polynomial time algorithm for estimating volumes of convex bodies. *J. Assoc. Comp. Mach.*, 38:1–17, 1991.

[106] J. Edmonds. Systems of distinct representatives and linear algebra. *J. Res. Nat. Bur. Standards*, 71B, 1967, pages 241–245.

[107] S. C. Eisenstat, M. C. Gursky, M. H. Schultz, and A. H. Sherman. The Yale sparse matrix package, I. the symmetric code. *Internat. J. Numer. Methods Engrg.*, 18:1145–1151, 1982.

[108] A. S. El–Bakry, R. A. Tapia, and Y. Zhang. A study of indicators for identifying zero variables in interior–point methods. *SIAM Review*, 36:45–72, 1994.

[109] G. Elekes. A geometric inequality and the complexity of computing volume. *Discr. Comput. Geom.*, 1:289–292, 1986.

[110] J. Elzinga and T. G. Moore. A central cutting plane algorithm for the convex programming problem. *Math. Programming*, 8:134–145, 1975.

[111] S. C. Fang and S. Puthenpura. *Linear Optimization and Extensions*. Prentice–Hall, Englewood Cliffs, NJ, 1994.

[112] L. Faybusovich. Jordan algebras, Symmetric cones and Interior-point methods. Research Report, University of Notre Dame, Notre Dame, 1995.

[113] L. Faybusovich. Semidefinite programming: a path-following algorithm for a linear-quadratic functional. *SIAM J. Optimization*, 6:1007–1024, 1996.

[114] M. C. Ferris and A. B. Philpott. On affine scaling and semi–infinite programming. *Math. Programming*, 56:361–364, 1992.

[115] S. Filipowski. On the complexity of solving feasible systems of linear inequalities specified with approximate data. *Math. Programming*, 71:259–288, 1995.

[116] A. V. Fiacco and G. P. McCormick. *Nonlinear Programming : Sequential Unconstrained Minimization Techniques*. John Wiley & Sons, New York, 1968. Reprint : Volume 4 of *SIAM Classics in Applied Mathematics*, SIAM Publications, Philadelphia, PA, 1990.

[117] J. J. H. Forrest and D. Goldfarb. Steepest-edge simplex algorithms for linear programming. *Math. Programming*, 57:341–374, 1992.

[118] J. J. H. Forrest and J. A. Tomlin. Implementing the simplex method for optimization subroutine library. *IBM Systems J.*, 31(1):11–25, 1992.

[119] R. Fourer and S. Mehrotra. Solving symmetric indefinite systems in an interior–point method for linear programming. *Math. Programming*, 62:15–39, 1993.

[120] R. Freund and F. Jarre. An interior–point method for fractional programs with convex constraints. *Math. Programming*, 67:407–440, 1994.

[121] R. Freund and F. Jarre. An interior-point method for convex multi-fractional programming. Technical Report (in preparation), AT&T Bell Laboratories, Murray Hill, NJ, 1993.

[122] R. M. Freund. Dual gauge programs, with applications to quadratic programming and the minimum-norm problem. *Math. Programming*, 38:47–67, 1987.

[123] R. M. Freund. Polynomial–time algorithms for linear programming based only on primal scaling and projected gradients of a potential function. *Math. Programming*, 51:203–222, 1991.

[124] R. M. Freund. A potential-function reduction algorithm for solving a linear program directly from an infeasible "warm start." *Math. Programming*, 52:441–466, 1991.

[125] R. M. Freund and J. R. Vera. some characterizations and properties of the 'distance to ill-posedness' and the condition measure of a conic linear system. Technical Report, MIT, 1995. To appear in *Math. Programming*.

[126] K. R. Frisch. The logarithmic potential method for convex programming. Unpublished manuscript, Institute of Economics, University of Oslo, Oslo, Norway, 1955.

[127] M. Fu, Z.-Q. Luo and Y. Ye. Approximation algorithms for quadratic programming. manuscript, Department of Electrical and Computer Engineering, McMaster University, Hamilton, Ontario, CANADA L8S 4K1, 1996.

[128] T. Fujie and M. Kojima. Semidefinite programming relaxation for nonconvex quadratic programs. Research Report B-298, Dept. of Mathematical and Computing Sciences, Tokyo Institute of Technology, Meguro, Tokyo 152, May 1995. To appear in *Journal of Global Optimization*.

[129] S. A. Gabriel and J. J. Moré. Smoothing of mixed complementarity problems. Technical Report MCS-P541-0995, Mathematics and Computer Science Division, Argonne National Laboratory, Argonne, 1995.

[130] D. Gale. *The Theory of Linear Economic Models*. McGraw-Hill, New York, 1960.

[131] M. R. Garey and D. S. Johnson. *Computers and Intractability: A Guide to the Theory of NP-Completeness*. W. H. Freeman, San Francisco, 1979.

[132] D. M. Gay. Computing optimal locally constrained steps. *SIAM J. Sci. Statis. Comput.*, 2:186-197, 1981.

[133] D. M. Gay. A variant of Karmarkar's linear programming algorithm for problems in standard form. *Math. Programming*, 37:81–90, 1987. Errata in *Math. Programming*, 40:111, 1988.

[134] A. George and J. W. -H. Liu. *Computing Solution of Large Sparse Positive Definite Systems*. Prentice-Hall, Englewood Cliffs, NJ, 1981.

[135] A. George and J. W. -H. Liu. The evolution of the minimum degree ordering algorithm. *SIAM Rev.*, 31:1–19, 1989.

[136] G. de Ghellinck and J. P. Vial. A polynomial Newton method for linear programming. *Algorithmica*, 1(4):425–453, 1986.

[137] L. E. Gibbons, D. W. Hearn and P. M. Pardalos. A continuous based heuristic for the maximum clique problem. *DIMACS Series in Discrete Mathematics and Theoretical Computer Science*, 26:103-124, 1996.

[138] P. E. Gill, W. M. Murray, and M. H. Wright. *Practical Optimization.* Academic Press, London, 1981.

[139] P. E. Gill, W. Murray, M. A. Saunders, J. A. Tomlin, and M. H. Wright. On projected Newton barrier methods for linear programming and an equivalence to Karmarkar's projective method. *Math. Programming*, 36:183–209, 1986.

[140] V. L. Girko. On the distribution of solutions of systems of linear equations with random coefficients. *Theor. Probability and Math. Statist.*, 2:41–44, 1974.

[141] M. X. Goemans and D. P. Williamson. Improved approximation algorithms for maximum cut and satisfiability problems using semidefinite programming. *Journal of Assoc. Comput. Mach.*, 42:1115–1145, 1995.

[142] M. X. Goemans and D. P. Williamson. A general approximation technique for constrained forest problems. *SIAM J. Computing*, 24:296–317, 1995.

[143] J. L. Goffin, A. Haurie, and J. P. Vial. Decomposition and nondifferentiable optimization with the projective algorithm. *Management Science*, 38(2):284–302, 1992.

[144] J. L. Goffin, Z. Q. Luo, and Y. Ye. On the complexity of a column generation algorithm for convex or quasiconvex feasibility problems. In *Large Scale Optimization: State–of–the–Art*, W. W. Hager, D. W. Hearn and P. M. Pardalos (eds.), Kluwer Academic Publishers, Dordrecht, The Netherlands, 1993, pages 182–191.

[145] J. L. Goffin, Z. Q. Luo, and Y. Ye. Complexity analysis of an interior-point cutting plane method for convex feasibility problem. *SIAM J. Optimization*, 6(3):638–652, 1996.

[146] J. L. Goffin and J. P. Vial. Cutting planes and column generation techniques with the projective algorithm. *J. Optim. Theory Appl.*, 65:409–429, 1990.

[147] A. V. Goldberg, S. A. Plotkin, D. B. Shmoys, and E. Tardos. Interior–point methods in parallel computation. In *Proceedings of the 30th Annual Symposium on Foundations of Computer Science, Research Triangle Park, NC, USA, 1989*, pages 350–355. IEEE Computer Society Press, Los Alamitos, CA, USA, 1990.

[148] A. V. Goldberg, S. A Plotkin, D. B. Shmoys, and E. Tardos. Using interior–point methods for fast parallel algorithms for bipartite matching and related problems. *SIAM J. Computing*, 21(1):140–150, 1992.

[149] D. Goldfarb and S. Liu. An $O(n^3L)$ primal interior point algorithm for convex quadratic programming. *Math. Programming*, 49:325–340, 1990/91.

[150] D. Goldfarb and K. Scheinberg. Interior point trajectories in semidefinite programming. Technical Report, Department of IE/OR, Columbia University, New York, 1996.

[151] D. Goldfarb and M. J. Todd. Linear Programming. In G. L. Nemhauser, A. H. G. Rinnooy Kan, and M. J. Todd, editors, *Optimization*, volume 1 of *Handbooks in Operations Research and Management Science*, pages 141–170. North Holland, Amsterdam, The Netherlands, 1989.

[152] D. Goldfarb and D. Xiao. A primal projective interior point method for linear programming. *Math. Programming*, 51:17–43, 1991.

[153] A. J. Goldman and A. W. Tucker. Polyhedral convex cones. In H. W. Kuhn and A. W. Tucker, Editors, *Linear Inequalities and Related Systems*, page 19–40, Princeton University Press, Princeton, NJ, 1956.

[154] G. H. Golub and C. F. Van Loan. *Matrix Computations*, 2nd Edition. Johns Hopkins University Press, Baltimore, MD, 1989.

[155] J. Gondzio. Splitting dense columns of constraint matrix in interior point methods for large scale linear programming. *Optimization*, 24:285–297, 1992.

[156] J. Gondzio. Multiple centrality corrections in a primal-dual method for linear programming. *Computational Optimization and Applications*, 6:137–156, 1996.

[157] J. Gondzio. Presolve analysis of linear programs prior to applying the interior point method. *INFORMS Journal on Computing*, 9(1):73–91, 1997.

[158] C. C. Gonzaga. An algorithm for solving linear programming problems in $O(n^3L)$ operations. In N. Megiddo, editor, *Progress in Mathematical Programming : Interior Point and Related Methods*, Springer Verlag, New York, 1989, pages 1–28.

[159] C. C. Gonzaga. Conical projection algorithms for linear programming. *Math. Programming*, 43:151–173, 1989.

[160] C. C. Gonzaga. Polynomial affine algorithms for linear programming. *Math. Programming*, 49:7–21, 1990.

[161] C. C. Gonzaga. Large steps path–following methods for linear programming, Part I : Barrier function method. *SIAM J. Optimization*, 1:268–279, 1991.

[162] C. C. Gonzaga. Large steps path–following methods for linear programming, Part II : Potential reduction method. *SIAM J. Optimization*, 1:280–292, 1991.

[163] C. C. Gonzaga. Path following methods for linear programming. *SIAM Review*, 34(2):167–227, 1992.

[164] C. C. Gonzaga and R. A. Tapia. On the convergence of the Mizuno–Todd–Ye algorithm to the analytic center of the solution set. *SIAM J. Optimization*, 7:47–65, 1997.

[165] C. C. Gonzaga and R. A. Tapia. On the quadratic convergence of the simplified Mizuno–Todd–Ye algorithm for linear programming. *SIAM J. Optimization*, 7:46–85, 1997.

[166] C. C. Gonzaga and M. J. Todd. An $O(\sqrt{n}L)$–iteration large–step primal–dual affine algorithm for linear programming. *SIAM J. Optimization*, 2:349–359, 1992.

[167] H. J. Greenberg. The use of the optimal partition in a linear programming solution for postoptimal analysis. *Operations Research Letters*, 15(4):179–186, 1994.

[168] H. J. Greenberg, A. G. Holder, C. Roos and T. Terlaky. On the dimension of the set of rim perturbations for optimal partition invariance. Technical Report CCM No. 94, Center for Computational Mathematics, Mathematics Department, University of Colorado at Denver, Denver, 1996.

[169] P. Gritzmann and V. Klee. Mathematical programming and convex geometry. In P. Gruber and J. Wills, editors, *Hand Book of Convex Geometry*, North Holland, Amsterdam, 1993, page 627–674.

[170] M. Grötschel, L. Lovász and A. Schrijver. *Geometric Algorithms and Combinatorial Optimization*. Springer, Berlin, 1988.

[171] B. Grünbaum. *Convex Polytopes*. John Wiley & Sons, New York, NY, 1967.

[172] B. Grünbaum. Partitions of mass-distributions and of convex bodies by hyperplanes. *Pacific Journal of Mathematics*, 10:1257–1261, 1960.

[173] O. Güler. Limiting behavior of weighted central paths in linear programming. *Math. Programming*, 65:347–363, 1994.

[174] O. Güler. Existence of interior points and interior paths in nonlinear monotone complementarity problems. *Mathematics of Operations Research*, 18(1):128–147, 1993.

[175] O. Güler. Barrier functions in interior-point methods. *Mathematics of Operations Research*, 21(4):860–885, 1996.

[176] O. Güler and L. Tuncel. Characterizations of the barrier parameter of homogeneous convex cones. Research Report CORR 95-14, Department of Combinatorics and Optimization, University of Waterloo, Waterloo, Ontario, Canada, 1995. To appear in *Math. Programming*.

[177] O. Güler and Y. Ye. Convergence behavior of interior point algorithms. *Math. Programming*, 60:215–228, 1993.

[178] C. G. Han, P. M. Pardalos, and Y. Ye. Implementation of interior-point algorithms for some entropy optimization problems. *Optimization Methods and Software*, 1(1):71–80, 1992.

[179] P. T. Harker and B. Xiao. A polynomial–time algorithm for affine variational inequalities. *Applied Mathematics Letters*, 4(2):31–34, 1991.

[180] J. Hartmanis and R. E. Stearns, On the computational complexity of algorithms. *Trans. A.M.S*, 117:285–306, 1965.

[181] C. Helmberg, F. Rendl, R. J. Vanderbei and H. Wolkowicz. An interior point method for semidefinite programming. *SIAM J. Optimization*, 6:342–361, 1996.

[182] D. den Hertog. *Interior Point Approach to Linear, Quadratic and Convex Programming, Algorithms and Complexity.* Ph.D. Thesis, Faculty of Mathematics and Informatics, TU Delft, NL–2628 BL Delft, The Netherlands, 1992.

[183] D. den Hertog, F. Jarre, C. Roos, and T. Terlaky. A sufficient condition for self-concordance, with application to some classes of structured convex programming problems. *Math. Programming*, 69:75–88, 1995.

[184] D. den Hertog and C. Roos. A survey of search directions in interior point methods for linear programming. *Math. Programming*, 52:481–509, 1991.

[185] D. den Hertog, C. Roos, and T. Terlaky. A large–step analytic center method for a class of smooth convex programming problems. *SIAM J. Optimization*, 2:55–70, 1992.

[186] A. J. Hoffman. On approximate solutions of systems of linear inequalities. *Journal of Research of the National Bureau of Standards*, 49:263–265, 1952.

[187] R. Horst, P. M. Pardalos, and N. V. Thoai. *Introduction to Global Optimization.* Kluwer Academic Publishers, Boston, 1995.

[188] K. Hotta and A. Yoshise. Global convergence of a class of non-interior-point algorithms using Chen-Harker-Kanzow functions for nonlinear complementarity problems. Discussion Paper Series No. 708, Institute of Policy and Planning Sciences, University of Tsukuba, Tsukuba, Ibaraki 305, Japan, December, 1996.

[189] S. Huang and K. O. Kortanek. A note on a potential reduction algorithm for LP with simultaneous primal–dual updating. *Operations Research Letters*, 10:501–507, 1991.

[190] P. Huard. Resolution of mathematical programming with nonlinear constraints by the method of centers. In J. Abadie, editor, *Nonlinear Programming*, pages 207–219. North Holland, Amsterdam, The Netherlands, 1967.

[191] P. Huard and B. T. Liêu. La méthode des centres dans un espace topologique. *Numeriche Methematik*, 8:56-67, 1966.

[192] H. Fujita, H. Konno, and K. Tanabe. *Optimization Method*. Iwanami Applied Mathematics Book Series, Iwanami-Shoten, Tokyo, 1994.

[193] P-F. Hung. *An asymptotical $O(\sqrt{n}L)$-Iteration Path-Following Linear Programming Algorithm that Uses Wider Neighborhood and its Implementation*. Ph.D. Thesis, Applied Mathematical and Computational Sciences, University of Iowa, 1994.

[194] P. -F. Hung and Y. Ye. An asymptotical $O(\sqrt{n}L)$-iteration path-following linear programming algorithm that uses wide neighborhoods. *SIAM J. Optimization*, 6:570–586, 1996.

[195] M. Iri and H. Imai. A multiplicative barrier function method for linear programming. *Algorithmica*, 1(4):455–482, 1986.

[196] S. Ito, C. T. Kelley, and E. W. Sachs. Inexact primal-dual interior point iteration for linear programs in function spaces. *Computational Optimization and Applications*, 4:189–201, 1995.

[197] R. Jagannathan and S. Schaible. Duality in generalized fractional programming via Farkas' lemma. *Journal of Optimization Theory and Applications*, 41:417–424, 1983.

[198] B. Jansen. *Interior Point Techniques in Optimization*. Ph.D. Thesis, Faculty of Mathematics and Informatics, TU Delft, NL–2628 BL Delft, The Netherlands, 1995.

[199] B. Jansen, C. Roos, and T. Terlaky. A polynomial primal-dual Dikin-type algorithm for linear programming. *Mathematics of Operations Research*, 21:341–344, 1996.

[200] B. Jansen, C. Roos, T. Terlaky, and J. P. Vial. Primal–dual target following algorithms for linear programming. *Annals of Operations Research*, 62:197–232, 1996.

[201] B. Jansen, T. Terlaky, and C. Roos. The theory of linear programming: Skew symmetric self-dual problems and the central path. *Optimization*, 29:225–233, 1994.

[202] F. Jarre. On the convergence of the method of analytic centers when applied to convex quadratic programs. *Math. Programming*, 49:341–358, 1990/91.

[203] F. Jarre. An interior–point method for minimizing the largest eigenvalue of a linear combination of matrices. *SIAM J. Control and Optimization*, 31:1360–1377, 1993.

[204] F. Jarre. Interior–point methods for convex programming. *Applied Mathematics & Optimization*, 26:287–311, 1992.

[205] J. Ji, F. A. Potra, and S. Huang. Predictor-corrector method for linear complementarity problems with polynomial complexity and superlinear convergence. *Journal of Optimization Theory and Applications*, 84:187–199, 1995.

[206] J. Ji, F. A. Potra and R. Sheng On the local convergence of a predictor-corrector method for semidefinite programming. Reports on Computational Mathematics, No. 98/1997, Department of Mathematics, University of Iowa, Iowa City, IA, 1997.

[207] J. Ji and Y. Ye. A complexity analysis for interior–point algorithms based on Karmarkar's potential function. *SIAM J. Optimization*, 4:512–520, 1994.

[208] F. John. Extremum problems with inequalities as subsidiary conditions. In *Sduties and Essays, presented to R. Courant on his 60th Birthday, January 8, 1948*, Interscience, New York, 1948, page 187-204.

[209] D. S. Johnson, C. Papadimitriou and M. Yannakakis. How easy is local search. *J. Comp. Syst. Sci.*, 37:79-100, 1988.

[210] M. J Kaiser, T. L. Morin and T. B. Trafalis. Centers and invariant points of convex bodies. In P. Gritzmann and B. Sturmfels, editors, *Applied Geometry and Discrete Mathematics: The Victor Klee Festschrift*, Amer. Math. Soc. and Ass. Comput. Mach., Providence, RI, 1991.

[211] B. Kalantari. Karmarkar's algorithm with improved steps. *Math. Programming*, 46:73–78, 1990.

[212] J. A. Kaliski and Y. Ye. A short-cut potential reduction algorithm for linear programming. *Management Science*, 39:757–776, 1993.

[213] A. P. Kamath and N. K. Karmarkar. A continuous method for computing bounds in integer quadratic optimization problems. *Journal of Global Optimization*, 2(3):229–241, 1992.

[214] A. P. Kamath, N. K. Karmarkar, K. G. Ramakrishnan, and M. G. C. Resende. A continuous approach to inductive inference. *Math. Programming*, 57:215–238, 1992.

[215] C. Kanzow. Some noninterior continuation methods for linear complementarity problems. To appear in *SIAM J. Matrix Anal. Appl.*, 1997.

[216] S. Kapoor and P. M. Vaidya. Fast algorithms for convex quadratic programming and multicommodity flows. *Proceedings of the 18th Annual ACM Symposium on Theory of Computing*, 1986, pages 147–159.

[217] N. K. Karmarkar. A new polynomial–time algorithm for linear programming. *Combinatorica*, 4:373–395, 1984.

[218] N. K. Karmarkar, J. C. Lagarias, L. Slutsman, and P. Wang. Power series variants of Karmarkar–type algorithms. *AT&T Tech. J.*, 68:20–36, 1989.

[219] R. M. Karp. Reducibility among combinatorial problems. R. E. Miller and J. W. Thatcher, eds., *Complexity of Computer Computations*, Plenum Press, New York, 1972, pages 85–103.

[220] J. Kemeny, O. Morgenstern, and G. Thompson. A generalization of the von Neumann model of an expanding economy. *Econometrica*, 24:115–135, 1956.

[221] L. G. Khachiyan. A polynomial algorithm for linear programming. *Doklady Akad. Nauk USSR*, 244:1093–1096, 1979. Translated in *Soviet Math. Doklady*, 20:191–194, 1979.

[222] L. G. Khachiyan and M. J. Todd. On the complexity of approximating the maximal inscribed ellipsoid for a polytope. *Math. Programming*, 61:137–160, 1993.

[223] V. Klee and G. J. Minty. How good is the simplex method. In O. Shisha, editor, *Inequalities III*, Academic Press, New York, NY, 1972.

[224] E. de Klerk, C. Roos and T. Terlaky. Initialization in semidefinite programming via a self–dual skew–symmetric embedding. Report 96–10, Faculty of Technical Mathematics and Computer Science, Delft University of Technology, Delft, 1996. To appear in *Operations Research Letters*.

[225] M. Kojima, N. Megiddo, and S. Mizuno. A general framework of continuation methods for complementarity problems. *Mathematics of Operations Research*, 18:945–963, 1993.

[226] M. Kojima, N. Megiddo, and S. Mizuno. A primal–dual infeasible-interior–point algorithm for linear programming. *Math. Programming*, 61:263–280, 1993.

[227] M. Kojima, N. Megiddo, T. Noma and A. Yoshise. *A unified approach to interior point algorithms for linear complementarity problems*. Lecture Notes in Computer Science 538, Springer Verlag, New York, 1991.

[228] M. Kojima, N. Megiddo, and Y. Ye. An interior point potential reduction algorithm for the linear complementarity problem. *Math. Programming*, 54:267–279, 1992.

[229] M. Kojima, S. Mizuno, and T. Noma. A new continuation method for complementarity problems with uniform P–functions. *Math. Programming*, 43:107–113, 1989.

[230] M. Kojima, S. Mizuno, and A. Yoshise. A polynomial–time algorithm for a class of linear complementarity problems. *Math. Programming*, 44:1–26, 1989.

[231] M. Kojima, S. Mizuno, and A. Yoshise. A primal–dual interior point algorithm for linear programming. In N. Megiddo, editor, *Progress in Mathematical Programming : Interior Point and Related Methods*, Springer Verlag, New York, 1989, pages 29–47.

[232] M. Kojima, S. Mizuno, and A. Yoshise. An $O(\sqrt{n}L)$ iteration potential reduction algorithm for linear complementarity problems. *Math. Programming*, 50:331–342, 1991.

[233] M. Kojima, S. Shindoh, and S. Hara Interior-point methods for the monotone semidefinite linear complementarity problem in symmetric matrices. *SIAM J. Optimization*, 7:86–125, 1997.

[234] K. O. Kortanek, F. Potra, and Y. Ye. On some efficient interior point methods for nonlinear convex programming. *Linear Algebra and Its Applications*, 152:169–189, 1991.

[235] K. O. Kortanek and M. Shi. Convergence results and numerical experiments on a linear programming hybrid algorithm. *European Journal of Operational Research*, 32:47–61, 1987.

[236] K. O. Kortanek and J. Zhu. New purification algorithms for linear programming. *Naval Research Logistics Quarterly*, 35:571–583, 1988.

[237] K. O. Kortanek and J. Zhu. A polynomial barrier algorithm for linearly constrained convex programming problems. *Mathematics of Operations Research*, 18(1):116–127, 1993.

[238] V. V. Kovacevic–Vujcic. Improving the rate of convergence of interior point methods for linear programming. *Math. Programming*, 52:467–479, 1991.

[239] E. Kranich. Interior-point methods bibliography. *SIAG/OPT Views-and-News, A Forum for the SIAM Activity Group on Optimization*, 1:11, 1992.

[240] H. W. Kuhn and A. W. Tucker. Nonlinear programming. In J. Neyman, editor, *Proceedings of the Second Berkeley Symposium on Mathematical Statistics and Probability*, pages 481–492, University of California Press, Berkeley and Los Angeles, CA, 1961.

[241] J. C. Lagarias. The nonlinear geometry of linear programming, Part III : Projective Legendre transform coordinates and Hilbert geometry. *Transactions of the American Mathematical Society*, 320:193–225, 1990.

[242] J. C. Lagarias. A collinear scaling interpretation of Karmarkar's linear programming algorithm. *SIAM J. Optimization*, 3:630–636, 1993.

[243] K. Levenberg. A method for the solution of certain non-linear problems in least squares. *Quarterly Appl. Math.*, 2:164-168, 1963.

[244] A. Levin. On an algorithm for the minimization of convex functions. *Soviet Math. Doklady*, 6:286-290, 1965.

[245] J. W. -H. Liu. A generalized envelope method for sparse factorization by rows. *ACM Trans. Math. Software*, 17(1):112–129, 1991.

[246] M. S. Lobo, L. Vandenberghe, and S. Boyd. Second-order cone programming. Manuscript, Information Systems Laboratory, Dept. of Electrical Engineering, Stanford University, Stanford, CA, 1997.

[247] L. Lovász and A. Shrijver. Cones of matrices and setfunctions, and $0-1$ optimization. *SIAM J. Optimization*, 1:166-190, 1990.

BIBLIOGRAPHY

[248] D. G. Luenberger. *Linear and Nonlinear Programming*, 2nd Edition. Addison–Wesley, Menlo Park, CA, 1984.

[249] Z. . Luo and J. Sun. Cutting surfaces and analytic center: a polynomial algorithm for the convex feasibility problem defined by self-concordant inequalities. Working Paper, National Univ. of Singapore, 1996.

[250] Z. Q. Luo, J. Sturm and S. Zhang. Superlinear convergence of a symmetric primal-dual path following algorithm for semidefinite programming. Report 9607/A, Econometric Institute, Erasmus University, Rotterdam, 1995. To appear in *SIAM J. Optimization*.

[251] Z. Q. Luo, J. Sturm and S. Zhang. Duality and self-duality for conic convex programming. Report 9620/A, Econometric Institute, Erasmus University Rotterdam, 1996.

[252] Z. Q. Luo and P. Tseng. Error bounds and convergence analysis of matrix splitting algorithms for the affine variational inequality problem. *SIAM J. Optimization*, 2:43–54, 1992.

[253] Z. Q. Luo and Y. Ye. A genuine quadratically convergent polynomial interior-point algorithm for linear programming. In Ding-Zhu Du and Jie Sun, editors, *Advances in Optimization and Approximation*, pages 235–246, Kluwer Academic, Dordrecht, 1994

[254] I. J. Lustig. Feasibility issues in a primal–dual interior point method for linear programming. *Math. Programming*, 49:145–162, 1990/91.

[255] I. J. Lustig, R. E. Marsten, and D. F. Shanno. Computational experience with a primal–dual interior point method for linear programming. *Linear Algebra and Its Applications*, 152:191–222, 1991.

[256] I. J. Lustig, R. E. Marsten, and D. F. Shanno. The interaction of algorithms and architectures for interior point methods. In P. M. Pardalos, editor, *Advances in optimization and parallel computing*, pages 190–205. Elsevier Sciences Publishers B.V., 1992.

[257] I. J. Lustig, R. E. Marsten, and D. F. Shanno. Computational experience with a globally convergent primal–dual predictor–corrector algorithm for linear programming. *Math. Programming*, 66:123–135, 1994.

[258] I. J. Lustig, R. E. Marsten, and D. F. Shanno. Interior point methods: Computational state of the art. *ORSA Journal on Computing*, 6:1–14, 1994.

[259] I. J. Lustig, R. E. Marsten, and D. F. Shanno. On implementing Mehrotra's predictor–corrector interior point method for linear programming. *SIAM J. Optimization*, 2:435–449, 1992.

[260] O. L. Mangasarian. Global error bounds for monotone affine variational inequality problems. *Linear Algebra and Applications*, 174:153-164, 1992.

[261] O. L. Mangasarian and T.-H. Shiau. Error bounds for monotone linear complementarity problems. *Math. Programming*, 36:81–89, 1986.

[262] H. M. Markowitz. The elimination form of the inverse and its application to linear programming. *Management Sci.*, 3:255–269, 1957.

[263] D. W. Marquardt. An algorithm for least-squares estimation of nonlinear parameters. *J. SIAM*, 11:431-441, 1963.

[264] I. Maros and Cs. Mészáros. The role of the augmented system in interior point methods. Technical Report TR/06/95, Brunel University, Department of Mathematics and Statistics, London, 1995.

[265] W. F. Mascarenhas. The affine scaling algorithm fails for $\lambda = 0.999$. *SIAM J. Optimization*, 7:34–46, 1997.

[266] C. McDiarmid. On the improvement per iteration in Karmarkar's method for linear programming. *Math. Programming*, 46:299–320, 1990.

[267] L. McLinden. The analogue of Moreau's proximation theorem, with applications to the nonlinear complementarity problem. *Pacific Journal of Mathematics*, 88:101–161, 1980.

[268] K. A. McShane. Superlinearly convergent $O(\sqrt{n}L)$-iteration interior-point algorithms for linear programming and the monotone linear complementarity problem. *SIAM J. Optimization*, 4(2):247–261, 1994.

[269] K. A. McShane, C. L. Monma, and D. F. Shanno. An implementation of a primal–dual interior point method for linear programming. *ORSA Journal on Computing*, 1:70–83, 1989.

[270] N. Megiddo. Linear programming (1986). *Annual Review of Computer Science*, 2:119–145, 1987.

BIBLIOGRAPHY

[271] N. Megiddo. Pathways to the optimal set in linear programming. In N. Megiddo, editor, *Progress in Mathematical Programming : Interior Point and Related Methods*, pages 131–158. Springer Verlag, New York, 1989. Identical version in : *Proceedings of the 6th Mathematical Programming Symposium of Japan, Nagoya, Japan*, 1986, pages 1–35.

[272] N. Megiddo. On finding primal– and dual–optimal bases. *ORSA Journal on Computing*, 3:63–65, 1991.

[273] N. Megiddo, S. Mizuno and T. Tsuchiya. A modified layered-step interior-point algorithm for linear programming. IBM Technical Report, San Jose, 1996.

[274] N. Megiddo and M. Shub. Boundary behavior of interior point algorithms in linear programming. *Mathematics of Operations Research*, 14:97–146, 1989.

[275] S. Mehrotra. Quadratic convergence in a primal–dual method. *Mathematics of Operations Research*, 18:741-751, 1993.

[276] S. Mehrotra. On the implementation of a primal–dual interior point method. *SIAM J. Optimization*, 2(4):575–601, 1992.

[277] S. Mehrotra. Handling free variables in interior methods. Technical Report 91-06, Department of Industrial Engineering and Management Sciences, Northwestern University, Evanston, USA., March 1991.

[278] S. Mehrotra. High order methods and their performance. Technical Report 90-16R1, Department of Industrial Engineering and Management Sciences, Northwestern University, Evanston, USA., 1991.

[279] S. Mehrotra and J. Sun. An algorithm for convex quadratic programming that requires $O(n^{3.5}L)$ arithmetic operations. *Mathematics of Operations Research*, 15:342–363, 1990.

[280] S. Mehrotra and J. Sun. A method of analytic centers for quadratically constrained convex quadratic programs. *SIAM J. Numerical Analysis*, 28(2):529–544, 1991.

[281] S. Mehrotra and Y. Ye. Finding an interior point in the optimal face of linear programs. *Math. Programming*, 62:497–515, 1993.

[282] Cs. Mészáros. Fast Choleski factorization for interior point methods of linear programming. Technical report, Computer and Automation Institute, Hungarian Academy of Sciences, Budapest, 1994. To appear in *Computers & Mathematics with Applications*.

[283] Cs. Mészáros. The "inexact" minimum local fill–in ordering algorithm. Working paper WP 95-7, Computer and Automation Institute, Hungarian Academy of Sciences, Budapest, 1995.

[284] J. E. Mitchell. An interior point column generation method for linear programming using shifted barriers. *SIAM J. Optimization*, 4(2):423–440, 1994.

[285] J. E. Mitchell and M. J. Todd. Solving combinatorial optimization problems using Karmarkar's algorithm. *Math. Programming*, 56:245–284, 1992.

[286] B. S. Mityagin. Two inequalities for volumes of convex bodies. *Matematicheskie Zametki*, 1:99-106, 1969.

[287] S. Mizuno. $O(n^\rho L)$ iteration $O(n^3 L)$ potential reduction algorithms for linear programming. *Linear Algebra and Its Applications*, 152:155–168, 1991.

[288] S. Mizuno. A new polynomial time method for a linear complementarity problem. *Math. Programming*, 56:31–43, 1992.

[289] S. Mizuno. Polynomiality of infeasible–interior–point algorithms for linear programming. *Math. Programming*, 67:109–119, 1994.

[290] S. Mizuno. A superlinearly convergent infeasible-interior-point algorithm for geometrical LCPs without a strictly complementary condition. *Mathematics of Operations Research*, 21(2):382–400, 1996.

[291] S. Mizuno, M. Kojima, and M. J. Todd. Infeasible–interior–point primal–dual potential–reduction algorithms for linear programming. *SIAM J. Optimization*, 5:52–67, 1995.

[292] S. Mizuno, M. J. Todd, and Y. Ye. On adaptive step primal–dual interior–point algorithms for linear programming. *Mathematics of Operations Research*, 18:964–981, 1993.

[293] S. Mizuno, M. J. Todd, and Y. Ye. A surface of analytic centers and primal–dual infeasible–interior–point algorithms for linear programming. *Mathematics of Operations Research*, 20:135–162, 1995.

[294] C. L. Monma. Successful implementations of interior algorithms. *SIAM News*, 22(2):14–16, 1989.

[295] R. D. C. Monteiro. On the continuous trajectories for a potential reduction algorithm for linear programming. *Mathematics of Operations Research*, 17:225–253, 1992.

[296] R. D. C. Monteiro. Convergence and boundary behavior of the projective scaling trajectories for linear programming. *Mathematics of Operations Research*, 16:842–858, 1991.

[297] R. D. C. Monteiro. A globally convergent primal-dual interior point algorithm for convex programming. *Math. Programming*, 64:123–147, 1994.

[298] R. D. C. Monteiro and I. Adler. Interior path following primal–dual algorithms : Part I : Linear programming. *Math. Programming*, 44:27–41, 1989.

[299] R. D. C. Monteiro and I. Adler. Interior path following primal-dual algorithms : Part II : Convex quadratic programming. *Math. Programming*, 44:43–66, 1989.

[300] R. D. C. Monteiro and I. Adler. An extension of Karmarkar–type algorithm to a class of convex separable programming problems with global linear rate of convergence. *Mathematics of Operations Research*, 15:408–422, 1990.

[301] R. D. C. Monteiro, I. Adler, and M. G. C. Resende. A polynomial-time primal–dual affine scaling algorithm for linear and convex quadratic programming and its power series extension. *Mathematics of Operations Research*, 15:191–214, 1990.

[302] R. D. C. Monteiro and S. Mehrotra. A general parametric analysis approach and its implication to sensitivity analysis in interior point methods. *Math. Programming*, 47:65–82, 1996.

[303] R. D. C. Monteiro and J-S. Pang. On two interior-point mappings for nonlinear semidefinite complementarity problems. Working Paper, John Hopkins University, 1996.

[304] R. D. C. Monteiro and T. Tsuchiya. Limiting behavior of the derivatives of certain trajectories associated with a monotone horizontal linear complementarity problem. *Mathematics of Operations Research*, 21(4):793–814, 1996.

[305] R. D. C. Monteiro, T. Tsuchiya, and Y. Wang. A simplified global convergence proof of the affine scaling algorithm. *Ann. Oper. Res.*, 47:443–482, 1993.

[306] R. D. C. Monteiro and S. J. Wright. Local convergence of interior–point algorithms for degenerate monotone LCP. *Computational Optimization and Applications*, 3:131–155, 1994.

[307] R. D. C. Monteiro and Y. Zhang. A unified analysis for a class of path-following primal-dual interior-point algorithms for semidefinite programming. School of ISyE, Georgia Tech, GA 30332, 1995.

[308] J. J. Moré. The Levenberg-Marquardt algorithm: implementation and theory. In G. A. Watson, editor: *Numerical Analysis* Springer-Verlag, New York, 1977.

[309] K. G. Murty. *Linear Complementarity, Linear and Nonlinear Programming*. Heldermann, Verlag, Berlin, 1988.

[310] K. G. Murty. *Linear Complementarity, Linear and Nonlinear Programming*, volume 3 of *Sigma Series in Applied Mathematics*, chapter 11.4.1 : The Karmarkar's algorithm for linear programming, pages 469–494. Heldermann Verlag, Nassauische Str. 26, D–1000 Berlin 31, Germany, 1988.

[311] K. G. Murty and S. N. Kabadi. Some NP-complete problems in quadratic and nonlinear programming. *Math. Programming*, 39:117-129, 1987.

[312] M. Muramatsu. Affine scaling algorithm fails for semidefinite programming. Research report No.16, Department of Mechanical Engineering, Sophia University, Japan, July 1996.

[313] M. Muramatsu and T. Tsuchiya. An affine scaling method with an infeasible starting point: convergence analysis under nondegeneracy assumption. *Annals of Operations Research*, 62:325–356, 1996.

[314] S. G. Nash and A. Sofer. A barrier method for large–scale constrained optimization. *ORSA Journal on Computing*, 5:40–53, 1993.

[315] J. L. Nazareth. Homotopy techniques in linear programming. *Algorithmica*, 1(4):529–535, 1986.

[316] J. L. Nazareth. The homotopy principle and algorithms for linear programming. *SIAM J. Optimization*, 1:316–332, 1991.

[317] J. L. Nazareth. *Computer Solution of Linear Programs.* Oxford University Press, New York, 1987.

[318] D. J. Newman. Location of the maximum on unimodal surfaces. *J. Assoc. Comput. Math.*, 12:395-398, 1965.

[319] A. S. Nemirovskii. A new polynomial algorithm for linear programming. *Doklady Akademii Nauk SSSR*, 298(6):1321-1325, 1988. Translated in : *Soviet Mathematics Doklady*, 37(1):264-269, 1988.

[320] A. S. Nemirovskii. On polynomiality of the method of analytic centers for fractional problems. *Math. Programming*, 73(2):175-198, 1996.

[321] A. S. Nemirovsky and D. B. Yudin. *Problem Complexity and Method Efficiency in Optimization.* John Wiley and Sons, Chichester, 1983. Translated by E. R. Dawson from *Slozhnost' Zadach i Effektivnost' Metodov Optimizatsii*, 1979, Glavnaya redaktsiya fiziko-matematicheskoi literatury, Izdatelstva "Nauka".

[322] J. von Neumann. On a maximization problem. Manuscript, Institute for Advanced Studies, Princeton University, Princeton, NJ 08544, USA, 1947.

[323] Yu. E. Nesterov. Quality of semidefinite relaxation for nonconvex quadratic optimization. CORE Discussion Paper, #9719, Belgium, March 1997.

[324] Yu. E. Nesterov. Polynomial-time dual algorithms for linear programming. *Kibernetika*, 25(1):34-40, 1989. Translated in: *Cybernetics*, 25(1):40-49, 1989.

[325] Yu. E. Nesterov. Complexity estimates of some cutting plane methods based on the analytic barrier. *Math. Programming*, 69:149-176, 1995.

[326] Yu. E. Nesterov and A. S. Nemirovskii. Acceleration and parallelization of the path-following interior point method for a linearly constrained convex quadratic problem. *SIAM J. Optimization*, 1(4):548-564, 1991.

[327] Yu. E. Nesterov and A. S. Nemirovskii. *Interior Point Polynomial Methods in Convex Programming: Theory and Algorithms.* SIAM Publications. SIAM, Philadelphia, 1993.

[328] Yu. E. Nesterov and A. S. Nemirovskii. An interior-point method for generalized linear-fractional programming. *Math. Programming*, 69:177–204, 1995.

[329] Yu. E. Nesterov and M. J. Todd. Self-scaled barriers and interior-point methods for convex programming. *Mathematics of Operations Research*, 22(1):1–42, 1997.

[330] Yu. E. Nesterov and M. J. Todd. Primal-dual interior-point methods for self-scaled cones. Technical Report No. 1125, School of Operations Research and Industrial Engineering, Cornell University, Ithaca, NY, 1995, to appear in *SIAM J. Optimization*.

[331] Yu. Nesterov, M. J. Todd and Y. Ye. Infeasible-start primal-dual methods and infeasibility detectors for nonlinear programming problems. Technical Report No. 1156, School of Operations Research and Industrial Engineering, Cornell University, Ithaca, NY, 1996.

[332] E. Ng and B. W. Peyton. A supernodal Choleski factorization algorithm for shared–memory multiprocessors. *SIAM J. Sci. Statist. Comput.*, 14(4):761–769, 1993.

[333] T. Noma. *A globally convergent iterative algorithm for complementarity problems : A modification of interior point algorithms for linear complementarity problems*. Ph.D. Thesis, Dept. of Information Science, Tokyo Institute of Technology, 2–12–1 Oh–Okayama, Meguro–ku, Tokyo 152, Japan, 1991.

[334] J. M. Ortega and W. C. Rheinboldt. *Iterative Solution of Nonlinear Equations in Several Variables*. Academic Press, New York, New York, 1970.

[335] M. W. Padberg. A different convergence proof of the projective method for linear programming. *Operations Research Letters*, 4:253–257, 1986.

[336] V. Pan. The modified barrier function method for linear programming and its extension. *Computers and Mathematics with Applications*, 20(3):1–14, 1990.

[337] C. H. Papadimitriou and K. Steiglitz. *Combinatorial Optimization: Algorithms and Complexity*. Prentice–Hall, Englewood Cliffs, NJ, 1982.

[338] P. M. Pardalos and J. B. Rosen. *Constrained Global Optimization: Algorithms and Applications.* Springer-Verlag, Lecture Notes in Computer Sciences 268, 1987.

[339] P. M. Pardalos and S. Jha. Complexity of uniqueness and local search in quadratic 0-1 programming. *Operations Research Letters*, 11:119-123, 1992.

[340] P. M. Pardalos, Y. Ye, and C. G. Han. Algorithms for the solution of quadratic knapsack problems. *Linear Algebra and Its Applications*, 152:69–91, 1991.

[341] E. Polak, J. E. Higgins, and D. Q. Mayne. A barrier function method for minimax problems. *Math. Programming*, 54:155–176, 1992.

[342] S. Polijak, F. Rendl and H. Wolkowicz. A recipe for semidefinite relaxation for 0-1 quadratic programming. *Journal of Global Optimization*, 7:51-73, 1995.

[343] R. Polyak. Modified barrier functions (theory and methods). *Math. Programming*, 54:177–222, 1992.

[344] D. B. Ponceleon. *Barrier methods for large–scale quadratic programming.* Ph.D. Thesis, Computer Science Department, Stanford University, Stanford, CA, 1990.

[345] K. Ponnambalam, A. Vannelli, and S. Woo. An interior point method implementation for solving large planning problems in the oil refinery industry. *Canadian Journal of Chemical Engineering*, 70:368–374, 1992.

[346] L. Portugal, F. Bastos, J. Judice, J. Paixõ, and T. Terlaky. An investigation of interior point algorithms for the linear transportation problems. *SIAM J. Scientific Computing*, 17(5):1202–1223, 1996.

[347] F. A. Potra. A quadratically convergent predictor–corrector method for solving linear programs from infeasible starting points. *Math. Programming*, 67:383–406, 1994.

[348] F. Potra and R. Sheng. Homogeneous interior–point algorithms for semidefinite programming. Reports On Computational Mathematics, No. 82/1995, Department of Mathematics, The University of Iowa, 1995.

[349] F. Potra and R. Sheng. Superlinear convergence of a predictor-corrector method for semidefinite programming without shrinking central path neighborhood. Reports on Computational Mathematics, No. 91, Department of Mathematics, The University of Iowa, 1996.

[350] F. A. Potra and Y. Ye. A quadratically convergent polynomial algorithm for solving entropy optimization problems. *SIAM J. Optimization*, 3:843–860, 1993.

[351] F. A. Potra and Y. Ye. Interior point methods for nonlinear complementarity problems. *Journal of Optimization Theory and Application*, 88:617–642, 1996.

[352] M. J. D. Powell. On the number of iterations of Karmarkar's algorithm for linear programming. *Math. Programming*, 62:153–197, 1993.

[353] L. Qi and X. Chen. A globally convergent successive approximation method for severely nonsmooth equations. *SIAM J. Control Optim.*, 33:402–418, 1995.

[354] M. Ramana. An exact duality theory for semidefinite programming and its complexity implications. *Math. Programming*, 77:129–162, 1997.

[355] M. Ramana, L. Tuncel and H. Wolkowicz. Strong duality for semidefinite programming. CORR 95-12, Department of Combinatorics and Optimization, University of Waterloo, Waterloo, Ontario, Canada, 1995. To appear in *SIAM J. Optimization*.

[356] S. Ramaswamy and J. E. Mitchell. A long step cutting plane algorithm that uses the volumetric barrier. Department of Mathematical Science, RPI, Troy, NY, 1995.

[357] F. Rendl and H. Wolkowicz. A semidefinite framework for trust region subproblems with applications to large scale minimization. *Math. Programming*, 77:273–300, 1997.

[358] J. Renegar. A polynomial–time algorithm, based on Newton's method, for linear programming. *Math. Programming*, 40:59–93, 1988.

[359] J. Renegar. Some perturbation theory for linear programming. *Math. Programming*, 65:73–91, 1994.

BIBLIOGRAPHY

[360] J. Renegar and M. Shub. Unified complexity analysis for Newton LP methods. *Math. Programming*, 53:1–16, 1992.

[361] M. G. C. Resende and G. Veiga. An implementation of the dual affine scaling algorithm for minimum cost flow on bipartite uncapacitated networks. *SIAM J. Optimization*, 3:516–537, 1993.

[362] M. G. C. Resende and G. Veiga. An efficient implementation of a network interior point method. Technical report, AT&T Bell Laboratories, Murray Hill, NJ, USA, February 1992.

[363] S. Robinson. A linearization technique for solving the irreducible von Neumann economic model. In Loś and Loś, eds., *Mathematical Models in Economics*, page 139–150, PWN-Polish Scientific Publishers, Warszawa, 1974.

[364] R. T. Rockafellar. *Convex Analysis*. Princeton University Press, Princeton, New Jersey, 1970.

[365] E. Rothberg and A. Gupta. Efficient Sparse Matrix Factorization on High-Performance Workstations-Exploiting the Memory Hierarchy. *ACM Trans. Math. Software*, 17(3):313–334, 1991.

[366] C. Roos, T. Terlaky and J.-Ph. Vial. *Theory and Algorithms for Linear Optimization: An Interior Point Approach*. John Wiley & Sons, Chichester, 1997.

[367] C. Roos and J.-Ph. Vial. A polynomial method of approximate centers for linear programming. *Math. Programming*, 54:295–305, 1992.

[368] S. Sahni. Computationally related problems. *SIAM J. Compu.*, 3:262-279, 1974.

[369] R. Saigal. A simple proof of a primal affine scaling method. *Annals of Operations Research*, 62:303–324, 1996.

[370] R. Saigal. *Linear Programming: Modern Integrated Analysis*. Kluwer Academic Publisher, Boston, 1995.

[371] R. Saigal and C. J. Lin. An infeasible start predictor corrector method for semi-definite linear programming. Tech. Report, Dept. of Industrial and Operations Engineering, University of Michigan, Ann Arbor, MI, Dec 1995.

[372] B. K. Schmidt and T. H. Mattheiss. The probability that a random polytope is bounded. *Mathematics of Operations Research*, 2:292–296, 1977.

[373] A. Schrijver. *Theory of Linear and Integer Programming*. John Wiley & Sons, New York, 1986.

[374] R. Seidel. Small-dimensional linear programming and convex hulls made easy. *Discrete Comput. Geom.*, 6:423–434, 1991.

[375] D. F. Shanno. Computing Karmarkar's projection quickly. *Math. Programming*, 41:61–71, 1988.

[376] D. F. Shanno and E. Simantiraki. An infeasible interior-point method for linear complementarity problems. Rutcor Research Report 7-95, March 1995 (revised, February, 1996).

[377] H. D. Sherali. Algorithmic insights and a convergence analysis for a Karmarkar–type of algorithm for linear programming problems. *Naval Research Logistics Quarterly*, 34:399–416, 1987.

[378] H. D. Sherali, B. O. Skarpness, and B. Kim. An assumption–free convergence analysis of the scaling algorithm for linear programs, with application to the L_1 estimation problem. *Naval Research Logistics Quarterly*, 35:473–492, 1988.

[379] Masayuki Shida, Susumu Shindoh and Masakazu Kojima. Existence of search directions in interior-point algorithms for the SDP and the monotone SDLCP. Research Report B-310, Dept. of Mathematical and Computing Sciences, Tokyo Institute of Technology, Oh-Okayama, Meguro, Tokyo 152 Japan, 1996. To appear in *SIAM J. Optimization*.

[380] N. Z. Shor. Cut-off method with space extension in convex programming problems. *Kibernetika*, 6:94-95, 1977.

[381] M. Shub. On the asymptotic behavior of the projective rescaling algorithm for linear programming. *Journal of Complexity*, 3:258–269, 1987.

[382] S. Smale. On the average number of steps of the simplex method of linear programming. *Math. Programming*, 27:241–262, 1983.

[383] G. Sonnevend. An "analytic center" for polyhedrons and new classes of global algorithms for linear (smooth, convex) programming. In

A. Prekopa, J. Szelezsan, and B. Strazicky, editors, *System Modelling and Optimization : Proceedings of the 12th IFIP-Conference held in Budapest, Hungary, September 1985*, volume 84 of *Lecture Notes in Control and Information Sciences*, pages 866–876. Springer Verlag, Berlin, Germany, 1986.

[384] G. Sonnevend, J. Stoer, and G. Zhao. On the complexity of following the central path by linear extrapolation in linear programming. *Methods of Operations Research*, 62:19–31, 1990.

[385] G. Sonnevend, J. Stoer, and G. Zhao. On the complexity of following the central path of linear programs by linear extrapolation II. *Math. Programming*, 52:527–553, 1991.

[386] G. Sonnevend, J. Stoer, and G. Zhao. Subspace methods for solving linear programming problems. Technical report, Institut fur Angewandte Mathematik und Statistic, Universitat Wurzburg, Wurzburg, Germany, January 1994.

[387] D. C. Sorenson. Newton's method with a model trust region modification. *SIAM J. Numer. Anal.*, 19:409-426, 1982.

[388] G. W. Stewart. Modifying pivot elements in Gaussian elimination. *Math. Comp.*, 28:537–542, 1974.

[389] J. F. Sturm and S. Zhang. New complexity results for the Iri-Imai method. *Annals of Operations Research*, 62:539–564, 1996.

[390] J. F. Sturm and S. Zhang. An $O(\sqrt{n}L)$ iteration bound primal-dual cone affine scaling algorithm for linear programming. *Math. Programming*, 72:177–194, 1996.

[391] J. F. Sturm and S. Zhang. On a wide region of centers and primal-dual interior point algorithms for linear programming. Tinbergen Institute Rotterdam, Erasmus University Rotterdam, The Netherlands, 1995.

[392] J. F. Sturm and S. Zhang. Symmetric primal-dual path following algorithms for semidefinite programming. Report 9554/A, Econometric Institute, Erasmus University, Rotterdam, 1995

[393] R. Subramanian, R. P. S. Scheff Jr., J. D. Qillinan, D. S. Wiper, and R. E. Marsten. Coldstart: Fleet assignment at Delta Air Lines. *Interfaces*, 24(1), 1994.

[394] U. H. Suhl. MPOS - Mathematical optimization system. *European J. Oper. Res.*, 72(2):312–322, 1994.

[395] U. H. Suhl and L. M. Suhl. Computing sparse LU factorizations for large-scale linear programming bases. *ORSA J. on Comput.*, 2(4):325–335, 1990.

[396] J. Sun. A convergence proof for an affine–scaling algorithm for convex quadratic programming without nondegeneracy assumptions. *Math. Programming*, 60:69–79, 1993.

[397] J. Sun and L. Qi. An interior point algorithm of $O(\sqrt{n}|\ln \varepsilon|)$ iterations for C^1-convex programming. *Math. Programming*, 57:239–257, 1992.

[398] J. Sun and G. Zhao. A quadratically convergent polynomial long-step algorithm for a class of nonlinear monotone complementarity problems. Working Paper, National Univ. of Singapore, 1996.

[399] K. C. Tan and R. M. Freund. Newton's method for the general parametric center problem with applications. Sloan School of Management, Massachusetts Institute of Technology, Cambridge, MA, 1991.

[400] K. Tanabe. Complementarity–enforced centered Newton method for mathematical programming. In K. Tone, editor, *New Methods for Linear Programming*, pages 118–144, The Institute of Statistical Mathematics, 4–6–7 Minami Azabu, Minatoku, Tokyo 106, Japan, 1987.

[401] K. Tanabe. Centered Newton method for mathematical programming. In M. Iri and K. Yajima, editors, *System Modeling and Optimization*, Proceedings of the 13th IFIP Conference, pages 197–206, Springer-Verlag, Tokyo, Japan, 1988.

[402] R. A. Tapia. Current research in numerical optimization. *SIAM News*, 20:10–11, March 1987.

[403] S. P. Tarasov, L. G. Khachiyan, and I. I. Érlikh. The method of inscribed ellipsoids. *Soviet Mathematics Doklady*, 37:226–230, 1988.

[404] T. Terlaky, editor. *Interior-Point Methods in Mathematical Programming* Kluwer Academic Publisher, Boston, 1996.

[405] M. J. Todd, Improved bounds and containing ellipsoids in Karmarkar's linear programming algorithm. *Mathematics of Operations Research*, 13:650-659, 1988.

[406] M. J. Todd. Recent developments and new directions in linear programming. In M. Iri and K. Tanabe, editors, *Mathematical Programming : Recent Developments and Applications*, pages 109–157. Kluwer Academic Press, Dordrecht, The Netherlands, 1989.

[407] M. J. Todd. Interior–point algorithms for semi–infinite programming. *Math. Programming*, 65:217–245, 1994.

[408] M. J. Todd. A low complexity interior point algorithm for linear programming. *SIAM J. Optimization*, 2:198–209, 1992.

[409] M. J. Todd. Polynomial expected behavior of a pivoting algorithm for linear complementarity and linear programming problems. *Math. Programming*, 35:173–192, 1986.

[410] M. J. Todd. Probabilistic models for linear programming. *Mathematics of Operations Research*, 16:671-693, 1991.

[411] M. J. Todd. A lower bound on the number of iterations of primal–dual interior–point methods for linear programming. In G.A. Watson and D.F. Griffiths, editors, *Numerical Analysis 1993*, Volume 303 of *Pitman Research Notes in Mathematics*, pages 237–259, Longman Press, Burnt Hill, UK, 1994.

[412] M. J. Todd. Potential-reduction methods in mathematical programming. *Math. Programming*, 76:3–45, 1996.

[413] M. J. Todd and B. P. Burrell. An extension of Karmarkar's algorithm for linear programming using dual variables. *Algorithmica*, 1(4):409–424, 1986.

[414] M. J. Todd, K. Toh and R. Tutuncu. On the Nesterov-Todd direction in semidefinite programming. Technical Report No. 1154, School of Operations Research and Industrial Engineering, Cornell University, Ithaca, N.Y., March, 1996.

[415] M. J. Todd and Y. Ye. A centered projective algorithm for linear programming. *Mathematics of Operations Research*, 15:508–529, 1990.

[416] M. J. Todd and Y. Ye. A lower bound on the number of iterations of long-step and polynomial interior-point linear programming algorithms. *Annals of Operations Research*, 6:233–252, 1996.

[417] M. J. Todd and Y. Ye. Approximate Farkas lemmas and stopping rules for iterative infeasible interior-point linear programming. Technical Report, School of Operations Research and Industrial Engineering, Cornell University, Ithaca, NY, 1994. To appear in *Math. Programming*.

[418] K. Tone. An active–set strategy in interior point method for linear programming. *Math. Programming*, 59:345–360, 1993.

[419] J. F. Traub. Computational complexity of iterative processes. *SIAM J. Comput.*, 1:167–179, 1972.

[420] J. F. Traub, G. W. Wasilkowski and H. Wozniakowski. *Information-Based Complexity*. Academic Press, San Diego, CA, 1988.

[421] P. Tseng. A simple complexity proof for a polynomial–time linear programming algorithm. *Operations Research Letters*, 8:155–159, 1989.

[422] P. Tseng. Complexity analysis of a linear complementarity algorithm based on a Lyapunov function. *Math. Programming*, 53:297–306, 1992.

[423] P. Tseng. Search directions and convergence analysis of some infeasible path-following methods for the monotone semi-definite LCP. Report, Department of Mathematics, University of Washington, Seattle, Washington 98195, 1996.

[424] P. Tseng and Z. Q. Luo. On the convergence of the affine–scaling algorithm. *Math. Programming*, 52:301–319, 1992.

[425] T. Tsuchiya. Global convergence of the affine scaling methods for degenerate linear programming problems. *Math. Programming*, 52:377–404, 1991.

[426] T. Tsuchiya. Global convergence property of the affine scaling methods for primal degenerate linear programming problems. *Mathematics of Operations Research*, 17(3):527–557, 1992.

[427] T. Tsuchiya. Quadratic convergence of the Iri-Imai algorithm for degenerate linear programming problems. *Journal of Optimization Theory and Applications*, 87(3):703–726, 1995.

[428] T. Tsuchiya and R. D. C. Monteiro. Superlinear convergence of the affine scaling algorithm. *Math. Programming*, 75:77–110, 1996.

[429] T. Tsuchiya and M. Muramatsu. Global convergence of a long–step affine scaling algorithm for degenerate linear programming problems. *SIAM J. Optimization*, 5:525–551, 1995.

[430] T. Tsuchiya and K. Tanabe. Local convergence properties of new methods in linear programming. *Journal of the Operations Research Society of Japan*, 33:22–45, 1990.

[431] A. W. Tucker. Dual systems of homogeneous linear relations, in H. W. Kuhn and A. W. Tucker, Editors, *Linear Inequalities and Related Systems*, page 3–, Princeton University Press, Princeton, NJ, 1956.

[432] L. Tunçel. Constant potential primal–dual algorithms: A framework. *Math. Programming*, 66:145–159, 1994.

[433] L. Tunçel and M. J. Todd. Asymptotic behavior of interior-point methods: a view from semi-infinite programming. *Mathematics of Operations Research*, 21:354–381, 1996.

[434] K. Turner. Computing projections for Karmarkar algorithm. *Linear Algebra Appl.*, 152:141–154, 1991.

[435] R. Tutuncu. An infeasible-interior-point potential-reduction algorithm for linear programming. Ph.D. Thesis, School of Operations Research and Industrial Engineering, Cornell University, Ithaca, NY, 1995.

[436] P. M. Vaidya. A locally well–behaved potential function and a simple Newton–type method for finding the center of a polytope. In N. Megiddo, editor, *Progress in Mathematical Programming : Interior Point and Related Methods*, pages 79–90. Springer Verlag, New York, 1989.

[437] P. M. Vaidya. An algorithm for linear programming which requires $O((m+n)n^2 + (m+n)^{1.5}nL)$ arithmetic operations. *Math. Programming*, 47:175–201, 1990. Condensed version in : *Proceedings of the 19th Annual ACM Symposium on Theory of Computing*, 1987, pages 29–38.

[438] P. M. Vaidya. A new algorithm for minimizing a convex function over convex sets. *Math. Programming*, 73:291–341, 1996.

[439] P. M. Vaidya and D. S. Atkinson. A technique for bounding the number of iterations in path following algorithms. In P. Pardalos,

editor, *Complexity in Numerical Optimization* page 462–489, World Scientific, New Jersey, 1993.

[440] L. Vandenberghe and S. Boyd. A primal–dual potential reduction method for problems involving matrix inequalities. *Math. Programming*, 69:205–236, 1995.

[441] L. Vandenberghe and S. Boyd. Semidefinite programming. *SIAM Review*, 38(1):49–95, 1996.

[442] L. Vandenberghe, S. Boyd, and S.P. Wu. Determinant maximization with linear matrix inequality constraints. Technical Report, Information Systems Laboratory, Stanford University, CA, 1996.

[443] R. J. Vanderbei. *Linear Programming: Foundations and Extensions* Kluwer Academic Publishers, Boston, 1997.

[444] R. J. Vanderbei. LOQO: An Interior Point Code for Quadratic Programming. Program in Statistics & Operations Research, Princeton University, NJ, 1995.

[445] R. J. Vanderbei. Symmetric quasidefinite matrices. *SIAM J. Optimization*, 5:100–113, 1995.

[446] R. J. Vanderbei. Splitting dense columns in sparse linear systems. *Linear Algebra Appl.*, 152:107–117, 1991.

[447] R. J. Vanderbei and T. J. Carpenter. Symmetric indefinite systems for interior point methods. *Math. Programming*, 58:1–32, 1993.

[448] R. J. Vanderbei and J. C. Lagarias. I. I. Dikin's convergence result for the affine–scaling algorithm. In J. C. Lagarias and M. J. Todd, editors, *Mathematical Developments Arising from Linear Programming*, volume 114 of *Contemporary Mathematics*, pages 109–119. American Mathematical Society, Providence, RI, 1990.

[449] R. J. Vanderbei, M. S. Meketon, and B. A. Freedman. A modification of Karmarkar's linear programming algorithm. *Algorithmica*, 1(4):395–407, 1986.

[450] S. A. Vavasis. *Nonlinear Optimization: Complexity Issues*. Oxford Science, New York, NY, 1991.

[451] S. A. Vavasis. Polynomial time weak approximation algorithms for quadratic programming. In C. A. Floudas and P. M. Pardalos, editors, *Complexity in Numerical Optimization*. World Scientific, New Jersey, 1993.

BIBLIOGRAPHY

[452] S. A. Vavasis and Y. Ye. A primal-dual interior-point method whose running time depends only on the constraint matrix. *Math. Programming*, 74:79–120, 1996.

[453] S. A. Vavasis and R. Zippel. Proving polynomial time for sphere-constrained quadratic programming. Technical Report 90-1182, Department of Computer Science, Cornell University, Ithaca, NY, 1990.

[454] C. Wallacher and U. Zimmermann. A combinatorial interior point method for network flow problems. *Math. Programming*, 52:321–335, 1992.

[455] T. Wang, R. D. C. Monteiro, and J.-S. Pang. An interior point potential reduction method for constrained equations. *Math. Programming*, 74(2):159–196, 1996.

[456] Z. L. Wei. An interior point method for linear programming. *Journal of Computational Mathematics*, 5:342–351, 1987.

[457] M. H. Wright. A brief history of linear programming. *SIAM News*, 18(3):4, November 1985.

[458] S. J. Wright. An infeasible–interior–point algorithm for linear complementarity problems. *Math. Programming*, 67:29–51, 1994.

[459] S. J. Wright. *Primal-Dual Interior-Point Methods* SIAM, Philadelphia, 1996.

[460] S. Q. Wu and F. Wu. New algorithms for linear programming. *Acta Mathematica Applicate Sinica (English Series)*, 8(1):18–26, 1992.

[461] X. Xu, P-F. Hung and Y. Ye. A simplification of the homogeneous and self-dual linear programming algorithm and its implementation. *Annals of Operations Research*, 62:151–172, 1996.

[462] G. Xue and Y. Ye. Efficient algorithms for minimizing a sum of Euclidean norms with applications. Working Paper, Department of Management Sciences, The University of Iowa, 1995. To appear in *SIAM J. Optimization*.

[463] H. Yamashita. A polynomially and quadratically convergent method for linear programming. Mathematical Systems Institute, Tokyo, Japan, 1986.

[464] M. Yannakakis. Computing the minimum fill–in is NP–complete. *SIAM J. Algebraic Discrete Methods*, pages 77–79, 1981.

[465] Y. Ye. Karmarkar's algorithm and the ellipsoid method. *Operations Research Letters*, 6:177–182, 1987.

[466] Y. Ye. An $O(n^3 L)$ potential reduction algorithm for linear programming. *Math. Programming*, 50:239–258, 1991.

[467] Y. Ye. Further development on the interior algorithm for convex quadratic programming. Working Paper, Dept. of Engineering Economic Systems, Stanford University, Stanford, CA 94305, USA, 1987.

[468] Y. Ye. A class of projective transformations for linear programming. *SIAM J. Computing*, 19:457–466, 1990.

[469] Y. Ye. On an affine scaling algorithm for nonconvex quadratic programming. *Math. Programming*, 52:285–300, 1992.

[470] Y. Ye. A potential reduction algorithm allowing column generation. *SIAM J. Optimization*, 2:7–20, 1992.

[471] Y. Ye. On the Q-order of convergence of interior-point algorithms for linear programming. In F. Wu, editor, *Proc. of the 1992 Symp. on Applied Mathematics*. Institute of Applied Mathematics, Chinese Academy of Sciences, 1992, 534-543.

[472] Y. Ye. Toward probabilistic analysis of interior–point algorithms for linear programming. *Mathematics of Operations Research*, 19:38–52, 1994.

[473] Y. Ye. Combining binary search and Newton's method to compute real roots for a class of real functions. *Journal of Complexity*, 10:271–280, 1994.

[474] Y. Ye. On the von Neumann economic growth problem. *Mathematics of Operations Research*, 20:617–633, 1995.

[475] Y. Ye and K. M. Anstreicher. On quadratic and $O(\sqrt{n}L)$ convergence of a predictor–corrector algorithm for LCP. *Math. Programming*, 62:537–551, 1993.

[476] Y. Ye, O. Güler, R. A. Tapia, and Y. Zhang. A quadratically convergent $O(\sqrt{n}L)$–iteration algorithm for linear programming. *Math. Programming*, 59:151–162, 1993.

[477] Y. Ye and M. Kojima. Recovering optimal dual solutions in Karmarkar's polynomial algorithm for linear programming. *Math. Programming*, 39:305–317, 1987.

BIBLIOGRAPHY

[478] Y. Ye and P. M. Pardalos. A class of linear complementarity problems solvable in polynomial time. *Linear Algebra and Its Applications*, 152:3–17, 1991.

[479] Y. Ye, M. J. Todd and S. Mizuno. An $O(\sqrt{n}L)$–iteration homogeneous and self–dual linear programming algorithm. *Mathematics of Operations Research*, 19:53–67, 1994.

[480] Y. Ye and E. Tse. An extension of Karmarkar's projective algorithm for convex quadratic programming. *Math. Programming*, 44:157–179, 1989.

[481] S. Zhang. Convergence property of the Iri-Imai algorithm for some smooth convex programming problems. *Journal of Optimization Theory and Applications*, 82:121–138, 1994.

[482] Y. Zhang. On the convergence of a class of infeasible interior-point methods for the horizontal linear complementarity problem. *SIAM J. Optimization*, 4:208–227, 1994.

[483] Y. Zhang and R. A. Tapia. A superlinearly convergent polynomial primal–dual interior–point algorithm for linear programming. *SIAM J. Optimization*, 3:118–133, 1993.

[484] Y. Zhang, R. A. Tapia, and J. E. Dennis. On the superlinear and quadratic convergence of primal–dual interior point linear programming algorithms. *SIAM J. Optimization*, 2(2):304–324, 1992.

[485] Y. Zhang, R. A. Tapia, and F. Potra. On the superlinear convergence of interior point algorithms for a general class of problems. *SIAM J. Optimization*, 3(2):413–422, 1993.

[486] Y. Zhang and D. Zhang. On polynomiality of the Mehrotra-type predictor-corrector interior-point algorithms. *Math. Programming*, 68:303–317, 1995.

[487] G. Zhao. Interior point algorithms for linear complementarity problems based on large neighborhoods of the central path. Research Report No. 650, Dept of Mathematics, National University of Singapore, Singapore, 1996. To appear in *SIAM J. Optimization*.

[488] G. Zhao and J. Stoer. Estimating the complexity of path–following methods for solving linear programs by curvature integrals. *Applied Mathematics & Optimization*, 27(1):85–103, 1993.

[489] U. Zimmermann. Search directions for a class of projective methods. *Zeitschrift für Operations Research—Methods and Models of Operations Research*, 34:353–379, 1990.

Index

active constraints, **27**
adaptive path-following algorithm, 128
addition of hyperplanes, 56
affine potential algorithm, 98
affine scaling algorithm, 100, 136
affine scaling direction, 101
affine scaling transformation, 98
affine set $\mathcal{A}_\Omega$, 51
algorithm, 28
$\mathcal{A}_p$, 28
all-zero column, 338
all-zero row, 338
analytic center, 44, **49**, 76, 81
analytic central-section theorem, 53, 56
approximate analytic center, 81, 232, 235, 287
approximate Farkas' lemma, 177
approximate solution, 28
arithmetic sequence, 226
arithmetic-geometric mean, 14, 63, 292, 305
asymptotic behavior, 167
asymptotic complexity, 33
asymptotically feasible, 256
asymptotically solvable, 256
average complexity, 32, 196
average convergence order, **211**, 224
average-case analysis, 179

back-substitution, **35**

backward analysis, 30
ball, 8
ball-constrained linear problem, **37**, 89, 97, 99, 111, 121, 292, 305
ball-constrained quadratic problem, **37**, 313, 314
barrier function, **49**, 76
barrier function method, 38, 70
barrier parameter, 351, 355
basic feasible partition, 195
basis, 349
basis identification, 152
beta distribution, 183
beta random variable, 184
big M method, 148, 158
big O, 213
bisection method, **30**, 38, **43**, 279, 319, 354
bit-size L, 29
block Choleski factorization, 348
block diagonal matrix, 343

cache memory, 348
Carathéodory's theorem, 10, 194
Cauchy distribution, 190, 204
Cauchy sequence, 139
Cauchy-Schwarz inequality, 14
center, 44
$\mathcal{C}$, **70**, 128, 165
central path, **70**, 114, 176, 266
central path theorem, 72
central-section, 43

409

central-section algorithm, 90, 117, 286
central-section steps, 45, 56
central-section theorem, 44, 46, 47
Chebychev's inequality, 185
chi-square distribution, 205
chi-square variable, 185
Choleski decomposition, 36
Choleski factor, 36
column generation, 241
column generation method, 238
column-sufficient matrix, 301
combined Phase I and II, 158
combined primal and dual, 158
compact set, 12
complementarity, 270
complementarity condition, 356
complementarity gap, **20**, 170, 212, 311, 351
complementarity partition, **21**, 139, 153, 356, 359
complementary matrices, 26
complexity, 1, **29**, 38
    arithmetic operation, 147
    arithmetic operation $c_o(A, Z)$, 29
    condition-based, 31
    error-based, 30
    iteration $c_i(A, Z)$, 29
    size-based, 29
complexity lower bound, 174
concave function, 49
concavity of log, 293
condition number, 32, 167, 175
    $\gamma(M, q, \epsilon)$, 294
    $\lambda(M)$, **299**
    $\theta(M)$, **299**
    $\xi(A, b, c)$, 157
    $\zeta(A, b, c)$, 157
    $c(A)$, 175
    $c(A, b, c)$, 167

condition-based complexity, 31, 157, 175, 291
cone, 8
    convex cone, 8
    dual cone, 9
    finitely generated cone, 10
    polyhedral cone, 9
    positive orthant, 9
    positive semi-definite matrix cone, 9
    second-order cone, 9
constraints, 15
containing coordinate-aligned ellipsoid, 53, 60
containing set, 43
continuous function, **12**
convergence criteria, 169
convergence order, **33**, 210
convergence rate, 33, 209
convergence ratio, **34**, 211
convex body, 44
    center of gravity, 44
    volume, 44
convex function, **13**
    quasi-convex function, 13
convex inequalities, 237
convex optimization, 231
convex polyhedron, 11
convex set, 259
coordinate-aligned ellipsoid, 51
copositive, 297
corrector step, **131**, 166, 219, 225
C^p, 13
cube, 45, 236
cut, 238
cutting plane method, 238, 242

Dantzig, 2
data set, 167
data set $\mathcal{Z}_p$, 15
decision problem $\mathcal{P}$, 15
decision variables, 15

INDEX

deeper cut, 113
degeneracy, 358
dense matrix, 342
dense mode, 346
dense window, 346
dependent constraints, 338
determinant, 6
deterministic algorithm, 179
diagonal and PSD matrix, 295
diagonal matrix, 6
Dikin ellipsoid, 136
direct method, 340
    QR decomposition, 340
    augmented system, 340, 343
    Choleski decomposition, 340
    Choleski factor, 341
    column-Choleski factorization, 345
    minimum degree, 341
    minimum local fill-in, 341, 342
    normal equation, 340
    numerical factorization, 345
    pivoting order, 340
    row-Choleski factorization, 345
    stability and sparsity, 344
    submatrix-Choleski factorization, 346
    triangular decomposition, 340
$\mathcal{B}_{n+rho}(s,z)$, 66
$\mathcal{B}_{n+rho}(y,s,z)$, 66
dual NEG problem, 278, 290
dual Newton procedure, **88**, 89, 241
dual norm, **6**, 177
$\|\cdot\|^*$, 6
dual of PSP, 25
dual potential algorithm, 89
dual potential function, **49**
dual problem, 195
dual slack set, **50**
duality gap, **20**, 181
duality overlap, 278

duality theorem, 20, 112
duality theorem in PSP, 26
duplicate columns, 338
duplicate constraints, 338
$D_x d_s$, 129

e, **5**
eigenvalue, 7, 31, 311, 314
ellipsoid, 8, 45
    coordinate-aligned ellipsoid, 8
ellipsoid method, 2, 38, **47**, 61
$\mathcal{E}$, **12**
ϵ-minimizer, 311, 325
ϵ-KKT point, 304, 305, 311
error bound, 229
error function, 212
expected time of the simplex method, 206
exponential time, 2
extreme value theory, 186

$\mathcal{F}_d$, 19, 109
$\mathcal{F}_h$, 160
$\mathcal{F}_p$, 19, 109
$f[\cdot]$, 328
face, 11
Farkas' lemma, 17, 188
Farkas' lemma in PSP, 25
feasibility error, 170
feasible region, 2
feasible solution, 17
    strictly or interior, 17, 25
fill-in, 341, 345
fixed variable, 338
floating-point operation, 341
fractional program, 278
fractional programming, 277
free variable, **18**, 19, 338, 342
fully polynomial-time approximation scheme, **238**, 246, 303

$\mathcal{G}$, 297

γ-level set, **278**, 279
Gaussian elimination, 35, 341
Gaussian matrix, 204
general QP, 277
generalized linear complementarity problem (GLCP), 24, **303**
geometric convergence, **34**, 211
geometric mean, 212
geometric sequence, 226
global minimizer, 325
gradient method, 37
gradient projection, 83
gradient vector, **13**

Haar condition, 199
Haar matrix, 199
Hadamard inequality, 14
half ellipsoid, 47
half spaces H_+ and H_-, 11
Harmonic inequality, 14
Hessian, 82
Hessian matrix, **13**
high probability, 179, **180**, 189
high-order method, 350, 352
high-probability analysis, 32
homeomorphical mapping, 257
homogeneous, 159
homogeneous and self-dual, **158**, 196, 355
homogeneous linear system, 187, 193
homogeneous MCP model, 260
homogenizing variable, 160
HSDP, **160**
hyperplane H, 10

I_j, 326
ill-condition, 342
implementation, 355
infeasibility, 270, 339

infeasibility certificate, 15, **17**, 159, 162, 261
 dual, 18
 primal, 17
infeasible central path, 169
infeasible-starting algorithm, 169, 174
initial point, 109
initialization, 148, 157
inner product, **5**, 7
 $\cdot \bullet \cdot$, 7
 $\langle \cdot, \cdot \rangle$, **5**, 7
interior, 5, 7
interior-point algorithm, 3
inverse transformation, 111
iterate convergence, 229
iterative algorithm, 33
iterative method, 349
 conjugate gradient method, 349
 preconditioner, 349

Jacobian matrix, **13**, 28

κ, 160
Karmarkar's algorithm, 60, **96**, 109
Karmarkar's canonical form, **96**, 110, 159
Karmarkar's canonical set, 50
Karmarkar's potential function, 64, 76
Karmarkar's projective transformation, 110
Karush-Kuhn-Tucker (KKT), 22
KKT condition, 22, 216
KKT point, 23, **27**, 28, 303, 309, 311
KKT system, 340, 345, 355
KKT theorem, 27

Lagrange multiplier, 27
layered interior-point algorithm, 175
layered least-squares, 175

INDEX 413

least-squares, 155, 187, 215, 222
least-squares problem, 85, 215
level set, 13, 71
limit point, **7**
linear complementarity problem (LCP), **23**, 291
linear convergence, **34**, 211
linear equations, 15
linear inequalities, 1, 17
linear least-squares, 16
linear programming, 1, 109
    basic feasible solution, 21
    basic solution, 21
    basic variables, 21
    constraints, 19
    decision variables, 18
    degeneracy, 21, 152
    dual problem, 19
    feasible set, 19
    infeasible, 157
    nondegeneracy, 21, 152
    objective function, 18
    optimal basic solution, 21, 157
    optimal basis, 21, 152
    optimal face, 21, 153
    optimal solution, 19
    optimal vertex, 21
    unbounded, 19, 157
    vertex, 21
linear programming (LP), **18**
linear subspace, 15
local complexity, 33
local or asymptotic convergence, 209
loop rolling, 349
loop unrolling, 348
lower bound, 324
lower-triangular matrix L, 35
LP fundamental theorem, 21
LU-decomposition, 35

matrix

$\mathcal{M}^n$, 7
A_I, **6**
A_{IJ}, **6**
A_J, **6**
$a_{.j}$, **6**
$a_{i.}$, **6**
a_{ij}, **6**
$\det(A)$, **6**
I, **6**
$\mathcal{N}()$, **6**, 221
$\mathcal{R}()$, **6**, 221
$\operatorname{tr}(A)$, **6**
$\mathcal{R}^{m \times n}$, **6**
matrix factorization, 326
matrix inner product, 7, 24
matrix norm, 6, 248
    Frobenius matrix norm, 7, 248
max-potential, **52**, 119, 238, 279–284
$\mathcal{B}(\Omega)$, 52
max-potential reduction, 238, 288
max-volume coordinate-aligned ellipsoid, 53
max-volume ellipsoid, 46, 75
max-volume sphere, 75
maximal complementary solution, **24**, 28, 74, 266
maximal-rank complementarity solution, 75
maximum-spanning tree, 350
measure of proximity, 81
memory references, 349
merit or descent function, 43
min-volume ellipsoid, 46
minimizer, 12
$\overset{\circ}{\mathcal{M}}{}^n_+$, **7**
$\mathcal{M}^n$, **7**, 325
$\mathcal{M}^n_+$, **7**
modified Newton's method, 36
monotone complementarity problem (MCP), **28**, 256

monotone complementarity theorem, 28
monotone function, 28
monotone LCP, 24, 218
monotone linear complementarity theorem, 24
monotone mapping, 257
multiplicative barrier function method, 143

$N(0,1)$, 204
$\mathcal{N}_\infty^-$, **128**, 351, 352, 354
$\mathcal{N}_2$, **128**, 218
$\mathcal{N}_\infty$, **128**
necessary and sufficient condition, 314
NEG
    $\Delta\gamma$, 285
    all-zero column, 278
    all-zero row, 278
    duality theorem, 291
    good, 277
    growth factor, 277
    input matrix, 277
    intensity, 277
    irreducibility, 278
    irrelevant good, 283
    output matrix, 277
    process, 277
negative definite, **7**
    ND, 7
negative semi-definite, **7**, 308
    NSD, 7
$\mathcal{N}(\eta$, 169
$\mathcal{N}$, **73**, 128, 165, 181
neighborhood of the central path, 73, 128
nested polytopes, 231
network-flow problem, 349
Newton procedure, 92, 118
Newton step, 36
Newton's method, 36

node-arc incidence matrix, 349
non–symmetric PSD matrix, 220
non-homogeneous linear system, 196
non-monotone LCP, 277
nonconvex problem, 311
nondegeneracy, 153, 228
nonlinear complementarity problem (NCP), 27
nonlinear equations, 16
nonlinear programming (NP), 27
nonnegative matrix, 277, 279
norm, **6**
    $\|\cdot\|$, 6
    $\|z\|_\infty^+$, **128**
    $\|z\|_\infty^-$, **128**
    l_2, **5**
    l_∞, **6**, 248
    l_p, **6**, 177
    l_q, **6**
normal matrix, **17**, 148, 150, 250
normalizing constraint, 159, 160
NP, 304
NP theory, 38
NP-complete theory, 304
NP-complete problem, 311

$O^*()$, 246
$O(x)$, **7**
$\Omega(x)$, **7**
one-sided norm, 73
one-step analysis, 180, 206
open set, 257
optimal basic solution, 356
optimal basis, 356, 359
optimal face, 140, 153, 189, 356
optimal solution, 2, 162
optimal value z^*, 109
optimality conditions, 15, 38, 43, 49
oracle, 16
orthogonal $n \times n$ matrix, 183

INDEX

orthogonal matrix, 185
orthogonal transformation, 182
$o(x)$, **7**
$\theta(x)$, **7**

P matrix, 299
P, 304
P matrix number θ, 299
$p(1/n)$, 189
$p(n)$, 189
$\mathcal{P}(x)$, 82
P-matrix LCP, 299
P-matrix number θ, 299
parallel, 153
partial derivatives, 13
partial potential function, 139
path-following algorithm, 117
permanent, 45
permutation matrix, 341
perturbed LP problem, 358
#P-Hard, 45
Phase I, 158, 309
Phase I linear program, 187
Phase II, 158
pivot, 35, 341, 343, 345
pivot operation, 157
pivot order, 344
pivot step, 35
pivoting method, 356
$\mathcal{P}_{n+rho}(x,z)$, 64
$\mathcal{P}_{n+1}(x,z)$, 64
$\mathcal{P}(x,\Omega)$, **60**
polyhedron, 1
polynomial algorithm, **30**
polynomial approximation scheme, **30**
polynomial time, 2
polynomially solvable, 30
polytope, 11
positive definite, **7**, 25
    PD, 7
positive normal distribution, 186

positive semi-definite, **7**, 14, 25, 244
    PSD, 7
positive semi-definite matrix, 326
positive semi-definite programming, 325
positive semi-definite programming (PSP), **24**, 247
postsolver, 340
potential function, **49**, 81, 114, 304
    $\mathcal{B}(s,\Omega)$, 49
    $\mathcal{B}(y)$, 81
    $\mathcal{B}(y,\Omega)$, 49
potential reduction algorithm, 120, 126, 141, 150, 232, 247, 292, 312
potential reduction theorem, 123, 251
$\mathcal{B}_{n+\rho}(y,S,z)$, 69
$\mathcal{P}_{n+\rho}(X,z)$, 69
Pq, 129
$\|Pq\|$, 181
$\|Pq\|_\infty$, 181
$\|Pq\|_\infty^-$, 181
$\prec$, **7**
$\preceq$, **7**, 25, 331
predictor direction, 350, 359
predictor step, **131**, 165, 181, 186, 214, 219, 224
predictor-corrector, 350
predictor-corrector algorithm, 131, 165, 181, 213, 218
presolver, 337
primal Newton procedure, **95**
primal potential function, **60**, 63, 95, 233
primal-dual affine scaling algorithm, 225
primal-dual algorithm, 126, 141
primal-dual Newton procedure, **102**,

103, 286
primal-dual Newton's method, 268
primal-dual potential algorithm, 103
primal-dual potential function, **62**, 66, 69, 77, 121, 125, 292
probabilistic bound, 179
projection, 17, 151, 155, 166, 250, 293
projection matrix, 17
projection scheme, 199
projection theorem, 16
projective transformation, **97**
    inverse transformation, 97
proper subset, 278
proximity to analytic center, 81
    $\eta(x,s)$, 83
    $\eta_d(s)$, 82
    $\eta_p(x)$, 83
    $p(s)$, 82
    $p(x)$, 83
    $s(x)$, 83
    $x(s)$, 82
    $y(x)$, 83
PSD matrix, 295
pseudo-inverse, 17
pseudo-polynomial algorithm, **30**
$\psi(x,s)$, 67, 83
$\psi_n(x,s)$, **62**
$\psi_{n+\rho}(X,S)$, **69**
$\psi_{n+\rho}(x,s)$, **66**, 68
PSP
    $\overset{\circ}{\mathcal{M}}^n_+$, 246
    $\mathcal{M}^n$, 246
    $\mathcal{M}^n_+$, 246
    potential reduction algorithm, 252
    primal potential function, 247
    primal-dual algorithm, 254
    primal-dual potential function, 247
    scaling matrix D, 254
purification, 152

$\mathcal{Q}$, 298
Q-norm, **7**
$\|\cdot\|_Q$, 7, 31
quadratic convergence, **33**, 89, 95, 103, 211, 217, 359
quadratic equation, 135
quadratic programming, 310, 325, 344
    dual problem, 22
quadratic programming (QP), **22**
quartic equation, 219
quasi Newton method, 37

$\mathcal{R}$, **5**
$\mathcal{R}_+$, **5**
R_{++}, 257
r-order method, 352
random algorithm, 179, 206
random perturbation, 153
random problem, 179
random subspace U, 182
random subspace $U^\perp$, 182
rank-one update, 148
ray, 159
real number model, 30, 39, 147
real root, 86
recursive algorithm, 232
redundant constraints, 337
redundant inequality, 49
relaxation, 325
residual, 43
residual error, 351
$\rightarrow$, **7**
$\overset{\circ}{\mathcal{R}}_+$, **5**
$\overset{\circ}{\mathcal{R}}^n_+$, **5**
$\mathcal{R}^n$, **5**
$\mathcal{R}^n_+$, **5**
rotationally-symmetric distribution, 193

INDEX

row-sufficient matrix, 301, 310

S_{++}, 257
scaled gradient, 292, 305
scaled Lipschitz condition, 256, 270
scaled projection, 83
Schur complement, 35, 341, 342
second order condition, 324
second-order method, 350
self-central path, 165
self-complementary partition, 197, 199
self-complementary solution, 161, 202
self-dual, 159
self-dual cone, 9
self-duality, 160
semi-infinite problem, 238
sensitivity analysis, 153
separating hyperplane, 43, 242
separating hyperplane theorem, 11
separating oracle, 237
sequential, 152
set, 8
    boundary of Ω, 8
    bounded set, 8
    $\cap$, 8
    closed set, 8
    $\hat{\Omega}$, 8
    compact set, 8
    convex hull, 8
    convex set, 8
    $\cup$, 8
    $\in$, 8
    $\overset{\circ}{\Omega}$, 8
    $\notin$, 8
    open set, 8
Sherman-Morrison-Woodbury formula, 39, 342
$\sigma(\cdot)$, 327
$\Sigma^+(M, q)$, 296
sign$(\cdot)$, 327

simplex, 45, **50**
simplex method, 2, 38, 147, 356
simultaneously feasible and optimal, 159
singleton column, 338
singleton row, 338
size-based complexity, 29
skew-symmetric matrix, 160, 164
slack variable, **19**, 21, 49
solution set $\mathcal{S}_p$, 15
$\mathcal{S}_\Omega$, 51
sparse matrix, 341
sparse mode, 346
sparsity, 339
standard Gauss distribution, 189
standard Gaussian matrix, 193
standard normal distribution, 182
standard normal variable $N(0,1)$, 184
stationary point, **303**
step-size, 36, 120, 133, 165, 306
strict complementarity, 218
strict complementarity partition, **21**, 72, 153, 157, 214, 220
strict complementarity theorem, 20
strictly complementary solution, 153
strictly self-complementary solution, 197
strong duality theorem, 19
strong duality theorem in PSP, 26
strongly infeasible, 263
strongly polynomial algorithm, **30**
strongly polynomial time, 356
subgradient, 237
subsequence, 110, 264
$\succ$, **7**, 25, 328
$\succeq$, **7**, 244, 326
super-basic solution, 357
super-non-basic variable, 357
superfluous variables, 337

superlinear convergence, **34**, **211**, 218
supernode, 347
support, 20, 153
symbolic phase, 341
symmetric indefinite matrix, 343

τ, 160
Taylor expansion, 351, 352
Taylor expansion theorem, 14
termination, 147, 152, 166, 187, 197, 319
the conjugate gradient, **34**
the dual simplex method, 358
the primal simplex method, 357
the steepest descent, **31**
Todd's degenerate model, 196, **202**
topological interior, 50
trace, **6**, **25**, 248
translation of hyperplanes, 53, 110
transpose operation, 5
triangle inequality, 128
triangular matrix, 350
trust region method, 38
trust region procedure, 38
Turing machine model, 29

uniformly distributed, 183, 327
unit sphere, 183, 327
upper-triangular matrix U, 35

vector function, 13
vertex, 2
volume of a convex body, 76
volumetric center, 75
volumetric potential function, 175
von Neumann economic growth problem (NEG), **277**

weak duality theorem, 19
weak duality theorem in PSP, 25
Weierstrass theorem, 12

wide-neighborhood algorithm, 134, 181
wider neighborhood, 129
worst-case complexity, **29**

X, **6**
ξ, 188
$\xi(A, b, c)$, 153
xi, 203
$\xi_d(A, b, c)$, 153
$\xi_p(A, b, c)$, 153
$\{x^k\}_0^\infty$, **7**

z^*, 188

WILEY-INTERSCIENCE
SERIES IN DISCRETE MATHEMATICS AND OPTIMIZATION

ADVISORY EDITORS

RONALD L. GRAHAM
AT & T Bell Laboratories, Murray Hill, New Jersey, U.S.A.

JAN KAREL LENSTRA
Department of Mathematics and Computer Science,
Eindhoven University of Technology, Eindhoven, The Netherlands

ROBERT E. TARJAN
Princeton University, New Jersey, and
NEC Research Institute, Princeton, New Jersey, U.S.A.

AARTS AND KORST • Simulated Annealing and Boltzmann Machines: A Stochastic Approach to Combinatorial Optimization and Neural Computing

AARTS AND LENSTRA • Local Search in Combinatorial Optimization

ALON, SPENCER, AND ERDŐS • The Probabilistic Method

ANDERSON AND NASH • Linear Programming in Infinite-Dimensional Spaces: Theory and Application

ASENCOTT • Simulated Annealing: Parallelization Techniques

BARTHÉLEMY AND GUÉNOCHE • Trees and Proximity Representations

BAZARRA, JARVIS, AND SHERALI • Linear Programming and Network Flows

CHONG AND ZAK • An Introduction to Optimization

COFFMAN AND LUEKER • Probabilistic Analysis of Packing and Partitioning Algorithms

DASKIN • Network and Discrete Location: Modes, Algorithms and Applications

DINITZ AND STINSON • Contemporary Design Theory: A Collection of Surveys

ERICKSON • Introduction to Combinatorics

GLOVER, KLINGHAM, AND PHILLIPS • Network Models in Optimization and Their Practical Problems

GOLSHTEIN AND TRETYAKOV • Modified Lagrangians and Monotone Maps in Optimization

GONDRAN AND MINOUX • Graphs and Algorithms *(Translated by S. Vajdā)*

GRAHAM, ROTHSCHILD, AND SPENCER • Ramsey Theory, Second Edition

GROSS AND TUCKER • Topological Graph Theory

HALL • Combinatorial Theory, Second Edition

JENSEN AND TOFT • Graph Coloring Problems

LAWLER, LENSTRA, RINNOOY KAN, AND SHMOYS, Editors • The Traveling Salesman Problem: A Guided Tour of Combinatorial Optimization

LEVITIN • Perturbation Theory in Mathematical Programming Applications

MAHMOUD • Evolution of Random Search Trees

MARTELLO AND TOTH • Knapsack Problems: Algorithms and Computer Implementations

McALOON AND TRETKOFF • Optimization and Computational Logic

MINC • Nonnegative Matrices

MINOUX • Mathematical Programming: Theory and Algorithms *(Translated by S. Vajdā)*

MIRCHANDANI AND FRANCIS, Editors • Discrete Location Theory

NEMHAUSER AND WOLSEY • Integer and Combinatorial Optimization

NEMIROVSKY AND YUDIN • Problem Complexity and Method Efficiency in Optimization *(Translated by E. R. Dawson)*

PACH AND AGARWAL • Combinatorial Geometry

PLESS • Introduction to the Theory of Error-Correcting Codes, Second Edition

ROOS AND VIAL • Ph. Theory and Algorithms for Linear Optimization: An Interior Point Approach

SCHEINERMAN • Fractional Graph Theory: A Rational Approach to the Theory of Graphs

SCHRIJVER • Theory of Linear and Integer Programming

TOMESCU • Problems in Combinatorics and Graph Theory *(Translated by R. A. Melter)*

TUCKER • Applied Combinatorics, Second Edition

YE • Interior Point Algorithms: Theory and Analysis